Lecture Notes in Computer Science 11993

More information about this series at http://www.springer.com/series/7412

Alessandro Crimi · Spyridon Bakas (Eds.)

Brainlesion: Glioma, Multiple Sclerosis, Stroke and Traumatic Brain Injuries

5th International Workshop, BrainLes 2019
Held in Conjunction with MICCAI 2019
Shenzhen, China, October 17, 2019
Revised Selected Papers, Part II

 Springer

Editors
Alessandro Crimi ⓘ
University Hospital of Zurich
Zurich, Switzerland

Spyridon Bakas ⓘ
University of Pennsylvania
Philadelphia, PA, USA

ISSN 0302-9743   ISSN 1611-3349   (electronic)
Lecture Notes in Computer Science
ISBN 978-3-030-46642-8   ISBN 978-3-030-46643-5   (eBook)
https://doi.org/10.1007/978-3-030-46643-5

LNCS Sublibrary: SL6 – Image Processing, Computer Vision, Pattern Recognition, and Graphics

This Springer imprint is published by the registered company Springer Nature Switzerland AG
The registered company address is: Gewerbestrasse 11, 6330 Cham, Switzerland

Preface

This volume contains articles from the Brain-Lesion workshop (BrainLes 2019), as well as the (a) International Multimodal Brain Tumor Segmentation (BraTS 2019) challenge, (b) Computational Precision Medicine: Radiology-Pathology Challenge on Brain Tumor Classification (CPM-RadPath 2019) challenge, and (c) the tutorial session on Tools Allowing Clinical Translation of Image Computing Algorithms (TACTICAL 2019). All these events were held in conjunction with the Medical Image Computing for Computer Assisted Intervention (MICCAI 2019) conference during October 13–17, 2019, in Shenzhen, China.

The papers presented describe research of computational scientists and clinical researchers working on glioma, multiple sclerosis, cerebral stroke, trauma brain injuries, and white matter hyper-intensities of presumed vascular origin. This compilation does not claim to provide a comprehensive understanding from all points of view; however the authors present their latest advances in segmentation, disease prognosis, and other applications to the clinical context.

The volume is divided into four parts: The first part comprises the paper submissions to BrainLes 2019, the second contains a selection of papers regarding methods presented at BraTS 2019, the third includes a selection of papers regarding methods presented at CPM-RadPath 2019, and lastly papers from TACTICAL 2019.

The aim of the first chapter, focusing on BrainLes 2019 submissions, is to provide an overview of new advances of medical image analysis in all of the aforementioned brain pathologies. Bringing together researchers from the medical image analysis domain, neurologists, and radiologists working on at least one of these diseases. The aim is to consider neuroimaging biomarkers used for one disease applied to the other diseases. This session did not make use of a specific dataset.

The second chapter focuses on a selection of papers from BraTS 2019 participants. BraTS 2019 made publicly available a large (n = 626) manually annotated dataset of pre-operative brain tumor scans from 19 international institutions, in order to gauge the current state of the art in automated brain tumor segmentation using multi-parametric MRI modalities, and comparing different methods. To pinpoint and evaluate the clinical relevance of tumor segmentation, BraTS 2019 also included the prediction of patient overall survival, via integrative analyses of radiomic features and machine learning algorithms, as well as experimentally attempted to evaluate the quantification of the uncertainty in the predicted segmentations, as noted in: www.med.upenn.edu/cbica/brats2019.html.

The third chapter contains descriptions of a selection of the leading algorithms showcased during CPM-RadPath 2019 (www.med.upenn.edu/cbica/cpm2019-data.html). CRM-RadPath 2019 used corresponding imaging and pathology data in order to classify a cohort of diffuse glioma tumors into two sub-types of oligodendroglioma and astrocytoma. This challenge presented a new paradigm in algorithmic challenges, where data and analytical tasks related to the management of brain tumors were combined to arrive at a more accurate tumor classification. Data from both challenges were obtained from The Cancer Genome Atlas/The Cancer Imaging Archive (TCGA/TCGA) repository and the Hospital of the University of Pennslvania.

The final chapter comprises two TACTICAL 2019 papers. The motivation for the tutorial on TACTICAL is driven by the continuously increasing number of newly developed algorithms and software tools for quantitative medical image computing and analysis towards covering emerging topics in medical imaging and aiming towards the clinical translation of complex computational algorithms (www.med.upenn.edu/cbica/miccai-tactical-2019.html).

We heartily hope that this volume will promote further exiting research about brain lesions.

March 2020

Alessandro Crimi
Spyridon Bakas

Organization

BrainLes 2019 Organizing Committee

Spyridon Bakas — University of Pennsylvania, USA
Alessandro Crimi — African Institutes for Mathematical Sciences, Ghana
Keyvan Farahani — National Institutes of Health, USA

BraTS 2019 Organizing Committee

Spyridon Bakas — University of Pennsylvania, USA
Christos Davatzikos — University of Pennsylvania, USA
Keyvan Farahani — National Institutes of Health, USA
Jayashree Kalpathy-Cramer — Harvard University, USA
Bjoern Menze — Technical University of Munich, Germany

CPM-RadPath 2019 Organizing Committee

Spyridon Bakas — University of Pennsylvania, USA
Benjamin Bearce — Harvard University, USA
Keyvan Farahani — National Institutes of Health, USA
Jayashree Kalpathy-Cramer — Harvard University, USA
Tahsin Kurc — Stony Brook University, USA
MacLean Nasrallah — University of Pennsylvania, USA

TACTICAL 2019 Organizing Committee

Spyridon Bakas — University of Pennsylvania, USA
Christos Davatzikos — University of Pennsylvania, USA

Program Committee

Ujjwal Baid — Shri Guru Gobind Singhji Institute of Engineering and Technology, India
Jacopo Cavazza — Instituto Italiano di Tecnologia (IIT), Italy
Guray Erus — University of Pennsylvania, USA
Anahita Fathi Kazerooni — University of Pennsylvania, USA
Hugo Kuijf — Utrecht University, The Netherlands
Jana Lipkova — Technical University of Munich, Germany
Yusuf Osmanlioglu — University of Pennsylvania, USA
Sadhana Ravikumar — University of Pennsylvania, USA
Zahra Riahi Samani — University of Pennsylvania, USA

Aristeidis Sotiras   Washington University in St. Louis, USA
Anupa Vijayakumari   University of Pennsylvania, USA
Stefan Winzeck   University of Cambridge, UK

Sponsoring Institutions

Center for Biomedical Image Computing and Analytics, University of Pennsylvania, USA

Contents – Part II

Contents – Part I

Brain Tumor Image Segmentation

Brain Tumor Image Segmentation

Brain Tumor Segmentation Using Attention-Based Network in 3D MRI Images

Xiaowei Xu, Wangyuan Zhao, and Jun Zhao[✉]

School of Biomedical Engineering, Shanghai Jiao Tong University,
Shanghai, China
junzhao@sjtu.edu.cn

Abstract. Gliomas are the most common primary brain malignancies. Identifying the sub-regions of gliomas before surgery is meaningful, which may extend the survival of patients. However, due to the heterogeneous appearance and shape of gliomas, it is a challenge to accurately segment the enhancing tumor, the necrotic, the non-enhancing tumor core and the peritumoral edema. In this study, an attention-based network was used to segment the glioma sub-regions in multi-modality MRI scans. Attention U-Net was employed as the basic architecture of the proposed network. The attention gates help the network focus on the task-relevant regions in the image. Besides the spatial-wise attention gates, the channel-wise attention gates proposed in SE Net were also embedded into the segmentation network. This attention mechanism in the feature dimension prompts the network to focus on the useful feature maps. Furthermore, in order to reduce false positives, a training strategy combined with a sampling strategy was proposed in our study. The segmentation performance of the proposed network was evaluated on the BraTS 2019 validation dataset and testing dataset. In the validation dataset, the dice similarity coefficients of enhancing tumor, tumor core and whole tumor were 0.759, 0.807 and 0.893 respectively. And in the testing dataset, the dice scores of enhancing tumor, tumor core and whole tumor were 0.794, 0.814 and 0.866 respectively.

Keywords: Glioma segmentation · Multimodal MRI · Convolutional neural network · Attention mechanism

1 Introduction

Gliomas are the most common intracranial tumors, which comprise about 30% of all intracranial tumors and four fifths of them are malignant. They can lead to kinds of nonspecific symptoms, such as headaches, vomiting, seizures and so on. According to the grade of malignancy, gliomas can be divided into low-grade glioma (LGG), WHO grade I and II with low risk, and high-grade glioma (HGG), WHO grade III and IV with high risk. Surgery with the assistance of radiation therapy and chemotherapy is the main treatment for gliomas. However, there are heterogeneous histological sub-regions in gliomas, such as the peritumoral edema, necrotic core, enhancing and non-enhancing tumor core, and boundaries between gliomas and normal tissue are not clear. Therefore, it's hard for surgeons to determine the invasion area of lesion and surgery boundary,

© Springer Nature Switzerland AG 2020
A. Crimi and S. Bakas (Eds.): BrainLes 2019, LNCS 11993, pp. 3–13, 2020.
https://doi.org/10.1007/978-3-030-46643-5_1

which may affect the quality of surgery and prognosis of patients. For instance, resecting normal regions around tumors may have a serious effect on the patients' life quality, while residual lesions can lead to the recurrence of tumors. Therefore, identifying whole tumors before surgery is of great help and guiding significance for the clinical surgery planning. In that case, doctors will have a better chance to cut tumors off entirely with protecting brain functions of surrounding areas.

Magnetic Resonance Imaging (MRI), a medical imaging technique, is used widely in clinical diagnosis of central nervous system. It has many advantages, such as high resolution, high contrast among soft tissues, no radiation and the ability to image from any cross sections. Additionally, specific image appearance can be obtained by changing the MRI sequence. Thus, different contrasts can be generated among tissues by different MRI modalities, which is helpful for tumor detection.

Brain Tumor Segmentation (BraTS) [1–4] is a world-wide medical image segmentation challenge, which focuses on the evaluation of state-of-the-art segmentation methods of brain tumors. In this challenge, multimodal MRI scans are provided to segment the subregions of gliomas, including T1-weighted (T1), post-contrast T1-weighted (T1_CE), T2-weighted (T2) and T2 Fluid Attenuated Inversion Recovery (FLAIR). The enhancing tumor shows hyper-intensity in T1_CE when compared to T1, while the necrotic and the non-enhancing tumor core are hypo-intense in T1_CE when compared to T1. The peritumoral edema is usually detected in FLAIR due to its hyper-intensity in this modality.

Fully convolutional neural networks (FCNs) [5–7] have obtained remarkable achievement in medical image segmentation. For instance, Andriy Myronenko [8] proposed an encoder-decoder architecture with a variational auto-encoder branch to regularize the encoder part and won first place in BraTS 2018 challenge. Fabian Isensee [9] employed U-Net in the glioma segmentation task. Combined with suitable training process, the results obtained by this well-trained U-Net ranked in the second place in BraTS 2018 challenge.

Recently, attention mechanism has shown brilliant results in natural language processing. Bidirectional Encoder Representations from Transformers (BERT) [10], a language representation model based on attention mechanism, has become the new state-of the-art method for many language tasks. Inspired by this, attention mechanism has been transferred into the field of computer vision and achieved noticeable results in image analysis. For example, Ozan Oktay [11] proposed the attention U-Net by introducing self-attention mechanism into classical U-Net and significantly improved the performance of pancreas segmentation.

Inspired by the success of FCNs and attention mechanism, an attention-based fully convolutional neural network was proposed in this study. Combined with the sampling and training strategy, the network can achieve relatively good results in the segmentation task of gliomas. In Sect. 2, these strategies and network architectures used in our experiment will be depicted in detail. Then Sect. 3 will describe the quantitative and qualitive results obtained by the proposed method. The discussion and conclusion will be shown in Sects. 4 and 5 respectively.

2   Methods

Due to the limitation of the memory, 3D patches were used to train the segmentation network. In our experiment, suitable sampling, training strategies and network architecture were proposed to prompt the network to learn more features about the lesions and produce less false positives. In this section, these strategies and the network architecture will be elaborated.

2.1   Sampling Strategy

On account of the variance in intensity between different samples, Z-normalization was processed to normalize each sample to be centered around zero with a standard deviation of one before sampling. It should be noticed that each modality was normalized separately because of high variance of intensity between different modalities.

As mentioned above, patches would be employed to train the segmentation network. In our experiment, a patch whose center was in the region of the lesion was deemed as a positive patch, while a patch whose center is in the region of brain other than the lesion (background) was called a negative patch. Training the segmentation network only by positive patches helps the network focus on learning the features of glioma but tends to cause false positives during training. To address this issue, both positive patches and negative patches were sampled to train the network. During sampling, the centers for positive patches and negative patches were randomly selected in the regions of lesions and background respectively, and the patch size in this study was set to $128 \times 128 \times 128$. For each sample, 20 positive patches and negative patches were captured respectively, to some extent, which can be deemed as a method of data augmentation.

2.2   Network Architecture

Attention mechanism can help the network focus on the task-relevant regions and features, which will be beneficial for the segmentation task. Therefore, both spatial-wise and channel-wise attention gates were introduced into our segmentation network.

Attention Mechanism in Spatial Dimension. Attention U-Net [11], an extension of standard U-Net [7], is an attention gate model which can automatically learn to focus on the target regions while suppress the irrelevant regions. In this experiment, we employed attention U-Net which set attention gates in the spatial dimension as the basic architecture to better complete the segmentation of the enhanced tumor (ET), necrosis (NCR), non-enhanced tumor core (NET) and peritumoral edema (ED). The framework of this network is shown in Fig. 1. The left side of the network works as an encoder to extract the features of different levels, and the right component of the network acts as a decoder to aggregate the features and generate the segmentation mask. In the stage of encoding, the encoder extracts features at multiple scales and generates fine-to-coarse feature maps. Fine feature maps contain low-level features but more spatial information, while coarse feature maps contain the opposite. Skip connection is used to combine coarse and fine feature maps so as for accurate segmentation. Different from

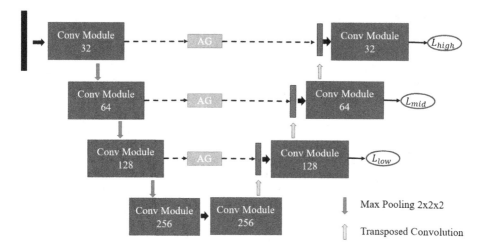

Fig. 1. The architecture of the segmentation network. AG means attention gate.

U-Net in which the fine feature maps are directly concatenated with the coarse feature maps, attention U-Net uses self-attention gating modules to generate an attention map for fine feature maps and thus select relevant regions in the fine feature maps to concatenate with the coarse feature maps. The structure of the attention gating module is shown in Fig. 2(a). It receives higher-level image representation x_g and lower-level image representation x_l as inputs. x_g which contains more contextual information generates the attention coefficients α, and then these attention coefficients are applied to disambiguate irrelevant and noisy responses in x_l.

Attention Mechanism in Feature Dimension. For the attention mechanism in the feature dimension, Squeeze-and-Excitation (SE) modules [12] were embedded into the convolution blocks. The SE module is a channel-wise attention gate in essence, which explicitly models channel-interdependencies and recalibrates features to help the network focus on the task-relevant feature maps. This channel-wise gating mechanism (Fig. 2(b)) is composed of three parts: Squeeze, Extraction and Scale. The Squeeze operation shrinks the feature maps through spatial dimension to get a global response in each channel. Then the Excitation operation uses convolutions (or fully connected layers) and activations to model the interdependencies among the feature maps and produce the channel-wise weights. Finally, the channel-wise weights are utilized by the Scale operation to reweight the feature maps. By embedding these SE modules into the convolution block, the network can automatically learn the importance of different feature maps, selectively enhance useful feature maps and suppress less useful ones.

Deep Supervision. Gradient vanishing often occurs when training a deep neural network. When gradient vanishing occurs, the gradient cannot be efficiently back propagated to the shallow layers and thus the weights and bias in these layers cannot be efficiently updated, which will affect the segmentation performance of the network. In order to alleviate the potential gradient vanishing problem during training and speed up

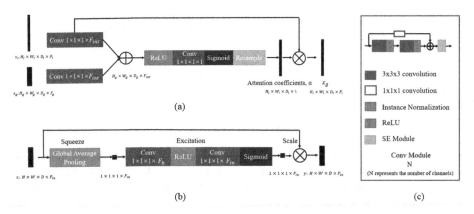

Fig. 2. The submodules of the segmentation network. (a) is the attention gate. (b) is the SE module. (c) is the convolution module used in both encoder and decoder.

the convergence, multi-level deep supervision was applied in the network training. Besides the main classifier in the upper final layer, two other classifiers in respective layers were involved in the decoder (Fig. 1).

2.3 Network Training

As mentioned in the section of sampling strategy, both positive patches and negative patches were extracted. Network trained by positive patches can learn the lesion features better, however, may produce some false positives. For instance, if a patch which contains no lesion is extracted in the testing set by sliding window, false positives may occur because all patches used for training are positive. To mitigate this problem, we proposed a training strategy by utilizing both positive patches and negative patches for network training.

In our approach, the training procedure can be partitioned into two stages. In the first stage, only positive patches were fed into the network to update the parameters. In this stage, the network can focus on learning the features of lesions and satisfyingly segment the lesion into enhanced tumor, necrosis, non-enhanced tumor core and peritumoral edema. Nonetheless, this trained network can produce false positives for some negative patches. Therefore, in the second stage, both positive and negative patches were used to train the network. This stage was aimed to finetune the parameters and reduce the false positives.

2.4 Loss Function

Owing to the class imbalance, weighted dice loss (Eq. (1)) and weighted cross-entropy loss (Eq. (2)) were used for network training. For the purpose of making the dice loss comparable with the cross-entropy loss, logarithmic operation was added in Eq. (1).

$$L_D = -\log\left(\frac{1}{\sum_{k=1}^{4} w_k} \sum_{k=1}^{4} w_k \frac{2\sum_{i=1}^{n} y_i^k \widehat{y}_i^k + \varepsilon}{\sum_{i=1}^{n} (y_i^k + \widehat{y}_i^k) + \varepsilon}\right) \tag{1}$$

$$L_{CE} = -\frac{1}{n\sum_{k=1}^{4} w_k} \sum_{k=1}^{4} w_k \sum_{i=1}^{n} y_i^k \log\left(\widehat{y}_i^k + \varepsilon\right) \tag{2}$$

$$\mathrm{L} = L_D + L_{CE} \tag{3}$$

where w_k is the weight for different class (ET, NCR and NET, ED, background). y is the ground truth and $\widehat{y}$ is the probability map. ε was set to 1e−5 in this experiment.

As mentioned above, deep supervision was adopted in our method, consequently, the final objective loss is shown in Eq. (4).

$$L_{total} = W_{high} \times L_{high} + W_{mid} \times L_{mid} + W_{low} \times L_{low} + L_{l2\_regularization} \tag{4}$$

where W_{high}, W_{mid} and W_{low} are the weights for different classifiers, which are set to 0.7, 0.2 and 0.1 respectively in our experiment.

3   Results

3.1   Datasets

The dataset used in this experiment was provided by BraTS 2019 Challenge. The training set contains 355 samples of which 259 are high-grade gliomas and the rest are low-grade gliomas. Each sample includes four modalities of brain MRI scans: T1, T1_CE, T2 and FLAIR. Furthermore, each sample in this dataset is equipped with the ground truth. The validation dataset provided by BraTS 2019 contains 125 subjects. Each subject contains the same four modalities of brain MRI scans, but no ground truth. In our experiment, the training set was applied to optimize the trainable parameters in the network, and the validation set was utilized to evaluate the performance of the trained network.

3.2   Segmentation Results

Implementation Details. The network was implemented by Tensorflow [13] and trained on two GeForce GTX 1080 TI with a batch size of 2. Adam optimizer with an initial learning rate 5e−4 was employed to optimize the parameters. The learning rate was decreased by 0.5 per 2000 epochs. As mentioned in the section of network training, there were two stages during training. In this experiment, the first 8000 epochs were deemed as the first stage, which means only positive patches were used to train the network in the first 8000 epochs. Then the rest 8000 epochs were deemed as the second stage where both positive and negative patches were fed into the network to finetune the parameters. Besides sampling multiple patches in each subject, no other

Table 1. Results of the proposed method on BraTS 2019 validation dataset

		ET	WT	TC
DSC	Mean	0.759	0.893	0.807
	Median	0.865	0.917	0.888
Sensitivity	Mean	0.776	0.903	0.790
	Median	0.880	0.926	0.883
Hausdorff95	Mean	4.193	6.964	7.663
	Median	2	3	3.742

Table 2. Results of the proposed method on BraTS 2019 testing dataset

	ET	WT	TC
DSC (mean)	0.794	0.866	0.814
Hausdorff95 (mean)	2.979	8.316	6.814

data augmentation methods were used in our experiment. As for post-processing, the predicted enhancing tumor whose volume is less than 500 voxels was discarded.

Performance. The performance of the trained network was evaluated on the validation set provided by BraTS 2019. The dice similarity coefficient (DSC), sensitivity and the Hausdorff distance (95%) of enhancing tumor (ET), tumor core (TC) and whole tumor (WT) were used as metrics to measure the similarity between the predicted masks and the ground truth. The quantitative and qualitative results are shown in Table 1 and Fig. 3 respectively. According to the mean dice similarity coefficients of ET, WT and TC, we can see that the proposed method achieved a relatively good segmentation performance.

We also evaluated the performance of the proposed method on the testing dataset provided by BraTS 2019. The testing dataset involves 166 subjects. For each subject, four modalities of MRI scans are available, but the ground truth is unknown. Table 2 shows the results obtained by our proposed network on this testing dataset.

Table 3. Comparison of mean DSCs and Hausdorff distances (95%) on BraTS 2019 validation dataset in ablation studies

	ET		WT		TC	
	Dice	Hausdorff	Dice	Hausdorff	Dice	Hausdorff
w/o TS/SS, AM, DS	0.729	12.610	0.856	22.063	0.772	24.297
w/o AM, DS	**0.764**	7.488	0.885	11.98	0.792	14.32
w/o DS	0.748	4.547	0.892	7.128	**0.819**	8.608
Proposed	0.759	**4.193**	**0.893**	**6.964**	0.807	**7.663**

Note: w/o is an abbreviation for without. TS, SS, AM and DS mean training strategy, sampling strategy, attention mechanism and deep supervision respectively. In our experiment, network without training strategy and sampling strategy means that the network was only trained on positive patches.

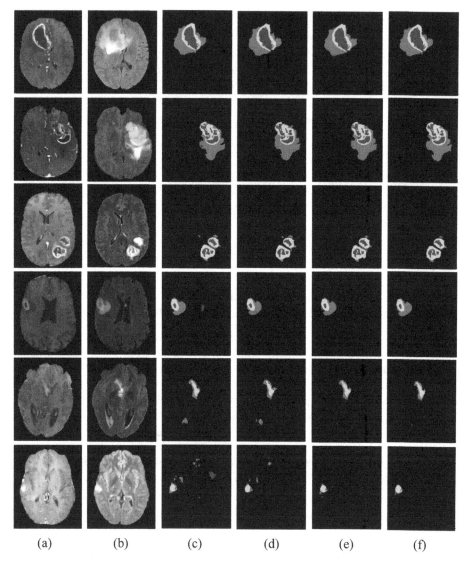

(a)   (b)   (c)   (d)   (e)   (f)

Fig. 3. Segmentation results of different methods. (a) T1_CE. (b) FLAIR. (c) the segmentation results obtained by the network without training strategy, attention mechanism and deep supervision. (d) the segmentation results obtained by the network without attention mechanism and deep supervision. (e) the segmentation results obtained by the network without deep supervision. (f) the segmentation results obtained by the proposed network. Yellow: the enhancing tumor. Red: the necrotic and non-enhancing tumor core. Green: the peritumoral edema. (Color figure online)

Ablation Studies. Ablation studies were conducted to assess the influence of the different modules on the segmentation performance. In our experiment, the training strategy combined with the sampling strategy was applied to reduce false positives. In addition, the attention mechanism was used in both spatial dimension and feature dimension to help the network focus on the task-relevant regions and features. Finally, deep supervision was employed to alleviate the potential gradient vanishing problem and speed up the convergence. The influence of these three modules on DSCs of enhancing tumor, whole tumor and tumor core were shown in Table 3.

According to Table 3 and Fig. 3, the training strategy combined with the sampling strategy can help the network reduce the false positives and significantly improve the DSCs of ET, WT and TC. Embedding spatial-wise and channel-wise attention gates into the segmentation network is also beneficial for the segmentation performance. Compared with the method without deep supervision, though the DSC of tumor core decreases in the proposed method, the DSCs of enhancing tumor and whole tumor increase and the Hausdorff distances of all sub-regions decline.

4 Discussion

In this study, we present a training strategy and an attention-based segmentation network. According to Table 1 and Table 2, the proposed method shows a potential in gliomas segmentation, which may be beneficial for the treatment and the prognosis of gliomas. As can be seen from the segmentation results on BraTS 2019 validation dataset and testing dataset, the median DSCs are significantly higher than the mean DSCs, which is caused by the poor segmentation results of some samples. Improving the segmentation performance on these challenging samples is a task to be accomplished in future.

The sampling and training strategy can reduce the false positive rate and help the network achieve better segmentation results (Table 3 and Fig. 3). Due to the limitation of the memory, 3D patches ($128 \times 128 \times 128$) were used to train the fully convolutional neural network. Training network only by positive patches produced some false positives when using sliding window to predict the mask of gliomas (Fig. 3(c)). In the proposed method, both negative patches and positive patches were involved in the training stage. This sampling strategy and two-stage training strategy effectively balance the sensitivity and the specificity of the network, and improve both volume-based and boundary-based metrics.

Self-attention mechanism was employed to help the segmentation network suppress the useless regions and features. As shown in Table 3, these attention gates in spatial and feature dimension can improve the similarity of the predicted mask and the ground truth, especially for whole tumor and tumor core. However, the dice scores of enhancing tumor decreased, which may be the side effect of post-processing. A few samples have no enhancing tumor but some bright spots or points exist in the tumor region of T1_CE scans. These bright spots or points may be mistaken for the enhancing tumor by the network. In order to remove these false positives, the post-processing which simply set a threshold to identify whether the bright spot is an enhancing tumor or not was designed in our experiment. Whereas this post-processing can lead to some false negatives for samples with small enhancing tumor, which may be the reason for the decrease of enhancing tumor's dice score. In future, a better network architecture or post-processing method is supposed to be proposed to solve this issue.

5 Conclusion

In this experiment, an attention-based network combined with a sampling and training strategy was applied to segment the sub-regions of gliomas. Attention U-Net was employed as the basic architecture which can prompt the network to focus on the task-relevant regions. And the Squeeze-and-Excitation module was embedded into the convolution blocks to recalibrate features and select the useful feature maps for glioma segmentation. Moreover, a training strategy combined with a sampling strategy was utilized to train the network, which reduced the false positive rate and enhanced the segmentation performance significantly. The proposed method was evaluated on the validation dataset and testing dataset provided by BraTS 2019. On the validation dataset, the DSCs of ET, TC and WT were 0.759, 0.807 and 0.893 respectively. And the dice scores of ET, TC and WT on the testing dataset were 0.794, 0.814 and 0.866 respectively.

References

1. Bakas, S., Akbari, H., Sotiras, A., Bilello, M., et al.: Segmentation labels and radiomic features for the pre-operative scans of the TCGA-LGG collection. Cancer Imaging Arch. **286** (2017)
2. Bakas, S., Akbari, H., Sotiras, A., Bilello, M., et al.: Advancing the cancer genome atlas glioma MRI collections with expert segmentation labels and radiomic features. Sci. Data **4** (1), 170117 (2017)
3. Bakas, S., Reyes, M., Jakab, A., Bauer, S., et al.: Identifying the Best Machine Learning Algorithms for Brain Tumor Segmentation, Progression Assessment, and Overall Survival Prediction in the BRATS Challenge, arXiv preprint arXiv:1811.02629 (2019)
4. Menze, B.H., Jakab, A., Bauer, S., Kalpathy-Cramer, J., et al.: The multimodal brain tumor image segmentation benchmark (BRATS). IEEE Trans. Med. Imaging **34**(10), 1993–2024 (2015)
5. Çiçek, Ö., Abdulkadir, A., Lienkamp, S.S., Brox, T., Ronneberger, O.: 3D U-Net: learning dense volumetric segmentation from sparse annotation. In: Ourselin, S., Joskowicz, L., Sabuncu, M., Unal, G., Wells, W. (eds.) MICCAI 2016. LNCS, vol. 9901, pp. 424–432. Springer, Cham (2016). https://doi.org/10.1007/978-3-319-46723-8_49
6. Milletari, F., Navab, N., Ahmadi, S.: V-Net: fully convolutional neural networks for volumetric medical image segmentation. In: 2016 Fourth International Conference on 3D Vision (3DV), pp. 565–571 (2016)
7. Ronneberger, O., Fischer, P., Brox, T.: U-Net: convolutional networks for biomedical image segmentation. In: Navab, N., Hornegger, J., Wells, W., Frangi, A. (eds.) MICCAI 2015. LNCS, vol. 9351, pp. 234–241. Springer, Cham (2015). https://doi.org/10.1007/978-3-319-24574-4_28
8. Myronenko, A.: 3D MRI brain tumor segmentation using autoencoder regularization, pp. 311–320 (2019)
9. Isensee, F., Kickingereder, P., Wick, W., Bendszus, M., Maier-Hein, K.H.: No New-Net, pp. 234–244 (2019)
10. Devlin, J., Chang, M.-W., Lee, K., Toutanova, K.: BERT: Pre-training of Deep Bidirectional Transformers for Language Understanding, arXiv preprint arXiv:1810.04805 (2018)

11. Oktay, O., Schlemper, J., Folgoc, L.L., Lee, M., et al.: Attention U-Net: learning where to look for the pancreas, arXiv preprint arXiv:1804.03999 (2018)
12. Hu, J., Shen, L., Sun, G.: Squeeze-and-excitation networks. In: Proceedings of the IEEE Conference on Computer Vision and Pattern Recognition, pp. 7132–7141 (2018)
13. Abadi, M., Barham, P., Chen, J., Chen, Z., et al.: Tensorflow: a system for large-scale machine learning. In: 12th {USENIX} Symposium on Operating Systems Design and Implementation ({OSDI} 16), pp. 265–283 (2016)

Multimodal Brain Image Segmentation and Analysis with Neuromorphic Attention-Based Learning

Woo-Sup Han and Il Song Han[✉]

Odiga, London SE1 4YL, UK
{phil.han,ishan}@odiga.co.uk

Abstract. Automated image analysis of brain tumors from 3D Magnetic Resonance Imaging (MRI) is necessary for the diagnosis and treatment planning of the disease, because manual practices of segmenting tumors are time consuming, expensive and can be subject to clinician diagnostic error. We propose a novel neuromorphic attention-based learner (NABL) model to train the deep neural network for tumor segmentation, which is with challenges of typically small datasets and the difficulty of exact segmentation class determination. The core idea is to introduce the neuromorphic attention to guide the learning process of deep neural network architecture, providing the highlighted region of interest for tumor segmentation. The neuromorphic convolution filters mimicking visual cortex neurons are adopted for the neuromorphic attention generation, transferred from the pre-trained neuromorphic convolutional neural networks(CNNs) for adversarial imagery environments. Our pre-trained neuromorphic CNN has the feature extraction ability applicable to brain MRI data, verified by the overall survival prediction without the tumor segmentation training at Brain Tumor Segmentation (BraTS) Challenge 2018. NABL provides us with an affordable solution of more accurate and faster image analysis of brain tumor segmentation, by incorporating the typical encoder-decoder U-net architecture of CNN. Experiment results illustrated the effectiveness and feasibility of our proposed method with flexible requirements of clinical diagnostic decision data, from segmentation to overall survival prediction. The overall survival prediction accuracy is 55% for predicting overall survival period in days, based on the BraTS 2019 validation dataset, while 48.6% based on the BraTS 2019 test dataset.

Keywords: Neuromorphic-attention · Image segmentation · Visual cortex

1 Introduction

Brain tumor severely harms the health of patients and causes high mortality, causing 85% death within five or more years of diagnosis. Early diagnosis of tumor is crucial for treatment planning, monitoring and analysis. Manual segmentation of brain tumors from Magnetic Resonance Imaging (MRI) in Fig. 1 is currently the key component in diagnosis process. Particularly, the accuracy of categorical estimates ranged from 23% up to 78% for the survival prediction among the expert clinicians [1, 2]. Automated

© Springer Nature Switzerland AG 2020
A. Crimi and S. Bakas (Eds.): BrainLes 2019, LNCS 11993, pp. 14–26, 2020.
https://doi.org/10.1007/978-3-030-46643-5_2

brain tumor segmentation would greatly aid the diagnosis, as the manual approach requires a high level of professional training and experience, in addition to reduction of significant time taken for analysis [3–7].

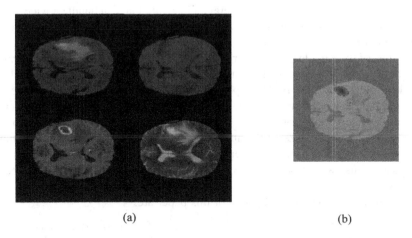

(a)   (b)

Fig. 1. Brain multi-modal MRI images in (a), and the segmentation of brain tumors

For brain tumor, segmentation task is particularly challenging due to the wide range of variables, such as size, position, shape, and other appearance properties. Tumors often overlapping with normal brain tissue in MRI, and growth of tumors can cause deformation in nearby structures, causing further complications. The approaches for automated brain tumor segmentation can be generally categorized as either generative or discriminative. Generative methods model the anatomy of brain tumors, and the appearance statistics. This requires more professional expertise and elaborate prior knowledge. Discriminate method is more reliant upon the training data, but can satisfy the task-relevant demands from labelled data directly. Machine Learning had been applied in this field previously, using various techniques including Support Vector Machines. Although those traditional classification methods reported high performances, new trend of fully automatic brain tumor segmentation techniques based on deep learning methods are also emerging with the state-of- the-art results [8].

Recent performances of deep learning methods, specifically Convolutional Neural Networks (CNNs), in several object recognition and biological image segmentation challenges increased their popularity among researches. In contrast to traditional classification methods, where hand crafted features are fed into, CNNs are able to extract relevant features for the task on their own by updating the networks gradually with gradient backpropagation, directly from the data itself [9, 10].

Yet the challenges faced in discerning between tumors and deformed or misshapen normal tissues remain. The issue of inter-class interference is due to the similar features shared among different tissues, leading to confusion in categorizing tumor and non-tumor tissues. We look to introduce concept of attention-based learning to solve this challenge of inter-class interferences in automated segmentation. One of the most curious facets of the human visual system is the presence of attention. Rather than compress an entire image into a static representation, attention allows for salient features to dynamically come to the forefront as needed. Attention-based learning has been a popular approach in recent computer vision research. In particular, attention-based learning was attributed as the general architecture that can handle a wide variety of modelling tasks, by taking into account the interdependence between modelling features [11–14].

The idea of focusing learning is particularly important when there is a lot of noises in the target data. We propose that attention-based learning can be introduced into biomedical image analysis. In this paper, we propose a novel network component of Neuromorphic Attention-Based Learner (NABL), that aims to address the challenges faced in automated segmentation tasks. Segmenting individual tumor tissue, each with their own organic shape, size and contrast, is highly challenging. But once the tumor had been determined, separating the entire tumor from the non-tumor region is relatively easy. Thus, we introduce NABL, which focus upon the region of attracting attentions.

In summary, the main contributions of this paper are:

(1) We propose NABL, which incorporates aspect of attention-based selective processing by neuromorphic neural networks, mimicking the brain. The neuromorphic attention has shown more effective processing for the visual object detection in adverse environments, as shown in Fig. 2.

(a)   (b)

Fig. 2. Attention-based selective processing by neuromorphic neural networks applied to detect vulnerable road users (a) in the blind spot of vehicle and (b) in the front on the road.

(2) We incorporate NABL into CNNs like U-net, and show that with a focus on particular features it is possible to guide the deep learning for the training with limited data set.

The remainder of this paper is organized as follows. The next section discusses the related works of deep learning in biomedical imaging, and attention-based learning approaches in applications. In Sect. 3, we introduce the NABL, and go over the method used for design of NABL. It will also show how NABL has been integrated into a CNN, gives an overview of the entire system. Section 4 provides details and results of experimentation on Brain Tumor Segmentation(BraTS) dataset, alongside analysis of the experimental results. Finally, Sect. 5 presents the conclusion on NABL.

2   Related Work

Recently, deep neural networks have been highly effective for image segmentation, and the semantic segmentation approaches generally rely on fully convolutional network architectures applied to image data [15, 16]. During the past two decades, many state-of-the-art methods have been proposed to design saliency models or attention modules for integrating multi-level convolutional features together. Some ambiguities of mimicking the human vision system have been addressed by adopting various network engineering such as U-net or addition of customized networks [17, 18].

Inter-class interference is a challenge commonly faced within medical image analysis. For brain tumors, tumor tissues can vary greatly in shape, size, location and contrast from patient to patient. Boundaries of tumor is also often unclear and irregular with discontinuities. Furthermore, because MRI devices and protocols used for acquisition can vary dramatically from scan to scan, automatic segmentation of brain tumors becomes a very challenging problem. Recently, several methods have been proposed for automated image segmentation in brain tumor analysis – in particular in glioma. Brain tumors can be categorised based on its point of origin. Primary brain tumors, like the gliomas, originate from brain glial cells, unlike the secondary types, which metastasize into the brain from the outer parts. Primary brain tumor, due to its aggressiveness and low survival rate, had gained the attention. These traditional brain tumor segmentation methods can be classified as either discriminative or generative methods. Discriminative methods try to learn the relationship between the input image and the ground truth. They focus on the choice of features, and its extraction from the MRIs [19].

Many discriminative methods implement a similar processing flow involving pre-processing, feature extraction, classification and post-processing steps. Pre-processing involves processes such as noise removal operations, skull-stripping, and intensity bias correction. Specific features that will split the different tissue types are then extracted. Choice of features is diverse, some using asymmetry-related features, Brownian motion features, first order statistical features, raw intensities, local image textures, intensity gradients, and edge-based features to name few examples. These features are then used as the basis of differentiating different classes. The type of classifiers is also diverse, ranging from neural networks. In contrast, generative methods generate probabilistic models by using prior knowledge like location and spatial extent of healthy tissues. However, converting prior knowledge into suitable probabilistic models is a complicated task, not least due to the small size of available dataset. While the traditional classification methods relied upon hand-crafted features to be fed into the classifiers, CNNs automatically learn representative complex features directly from the data itself. Efficient 3D CNN architectures were introduced for the multi-modal MRI glioma segmentation. Multimodality 3D patches extracted from the different brain MRI modalities are used as inputs to the CNN to predict the tissue label of the center voxel of the cube. New ways of selective processing have been introduced for focused analysis of medical images, using various approaches of attention-based algorithm. The U-net architecture remains as one of most popular deep neural networks for segmentation [20].

3   Convolutional Neural Networks with Neuromorphic Attention-Based Learner (NABL)

3.1   New Network Architecture: CNN with NABL

The new architecture of Fig. 3 is designed to control the training by introducing an additional guidance to the ordinary learning process, which is usually based on the error back propagation during the training process. There are two issues of deep learning for segmenting brain tumors, which are the lack of dataset and the limited clinical accuracy of training dataset. In this paper, the concept of NABL is introduced for the pre-determined mechanism of estimating the target object areas, where the neuromorphic neural networks act as fixed feature extractors from the visual object.

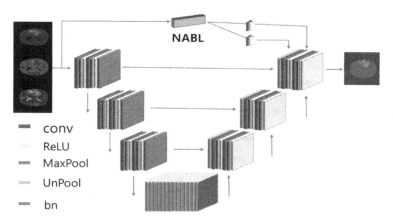

Fig. 3. Architecture of convolutional neural networks with neuromorphic attention-based learner.

The neuromorphic neural networks have been trained for the principal operation targeting various applications from vulnerable road user detection to dental 3D tooth segmentation [21]. The neuromorphic processing for generating the attention map has shown its effectiveness in the earlier brain tumor analysis, as the neuromorphic network performance shown in Fig. 2 was successfully applied to predict the patient's survival period in BraTS 2018 challenge [22]. In Fig. 4, the generated attention map are closely correspondent to the area of ground truth, which illustrates the principle of our neuromorphic neural network used in BraTS 2018 [22]. It can suggest a way of the feature extraction for estimating the tumor presence. The NABL in Fig. 3 aims to feed the deep neural network with the designated focusing areas, like attention areas in Fig. 4. The concept of Fig. 3 is to apply NABL to a popular deep neural network architecture of segmentation, that is, the U-net with 4-levels. The output signals of NABL are connected at the final stage of U-net, as the addition in two paths ($240 \times 240 \times 64$ and $240 \times 240 \times 4$ respectively). Here, the deep learning process never affects the module of NABL, as the neuromorphic neural networks of NABL is pre-trained and fixed one. The pre-trained NABL is transferred from the convolution kernels of earlier applications of neuromorphic deep neural network in [22]. The convolution kernels were either trained by image patches of dental X-ray CT or hand-crafted by mimicking the visual cortex neuron, where the combination of kernels was optimized for the application [22].

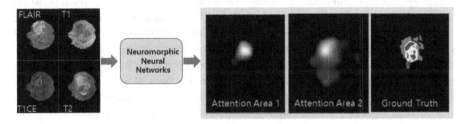

Fig. 4. Neuromorphic neural networks for generating attention maps of brain tumors [21].

It assists the training process for the segmentation, with less ambiguity as NABL maintains the feature map of $240 \times 240 \times 64$ and $240 \times 240 \times 4$. Two signal paths can become a more effective feedback to the ordinary deep learning process, as it likely introduces the attention processing similar to those shown in Fig. 4.

3.2 Neuromorphic Neural Networks and Neuromorphic Attention

The neuromorphic neural network is designed for the automatic attention/saliency map generation using the architecture of Fig. 5. The double V-shape of NABL provides the attention generation process based on the down-up resizing process, where the convolution neural network is based on neuromorphic neural network of orientation selective kernel filters. The kernels of 13×13 are transferred from the neuromorphic neural networks, which have been used in [22]. The same kernels were used for applications shown in Fig. 2 and Fig. 4. With the repeated V-shape of down-up resizing function, the nonlinear neuron function of ReLU works for the attention map generation in Fig. 3.

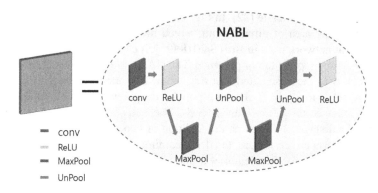

Fig. 5. NABL network structure.

3.3   Brain Tumor Segmentation Using CNN with NABL

The concept of CNN with NABL is to train the network with the partially fixed network, providing the embedded and pre-trained intelligent/feature extracting characteristics. The ratio of mixing the conventional deep learning process and the effect of fixed pre-trained network is optimized by the training of integrated two convolutional networks. The key components of NABL are the deconvolution network layer or UnPool layer similar to U-net. For an effective implementation, we apply the 3 images of FLAIR, T1CE and T2 instead of 4, as our existing deep neural networks training for images generally allow the single channel input or 3 channel inputs, without a serious degradation.

We configure the input as in Fig. 6, and maintains the compatibility of NABL to standard deep neural networks. The repeated V-shape NABL was based on the neuron type of threshold controlled Rectifier Linear Unit as in Eq. (1) and (2).

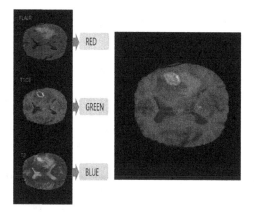

Fig. 6. Input design for the compatibility.

The pooling layer of NABL is designed with the ratio of 4, for the effective processing. The final module of U-net in Fig. 3, is connected to NABL via two signal paths, which are merged to the last convolution network layer and SoftMax layer as in Eq. (1) and (2). The NABL itself is applicable to general deep neural networks, which can be flexibly integrated or interfaced to U-net.

$$
\begin{aligned}
&\text{NABL: Conv (4 1x1x2)} > \text{ReLU} > \\
&\quad \text{(addition of NABL \& prior U-net)} > \text{U-net: SoftMax}
\end{aligned}
\tag{1}
$$

$$
\begin{aligned}
&\text{NABL: Conv (64 1x1x2)} > \text{ReLU} > \\
&\quad \text{(addition of NABL \& prior U-net)} > \text{U-net: Convolution (4 3x3x64)}
\end{aligned}
\tag{2}
$$

3.4 Overall Survival Period Prediction Based on Multiple Modes of Tumor Segmentation

The fully connected feedforward neural network in Fig. 7 is our 5-layer neural network based on our previous fully connected feedforward network of overall survival period prediction at BraTS challenge 2018 [21], with the change of number of input data of neural network. The new input layer framework has the 6 input data with 4 additional inputs compared to the neural network used in 2018. The new input layer framework represents the additional input data to the previous two input data of 'Neuromorphic Segmentation I' and 'Neuromorphic Segmentation II' of earlier neural network in [21]. The added input data are the provided patients 'age' and segmentation results (ET, WT, TC) generated by the U-net with NABL in Fig. 3. Input segmentation data other than age is applied as the segmentation volume data.

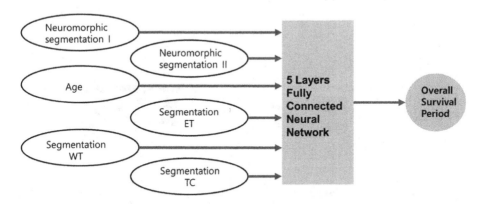

Fig. 7. Fully connected neural network for the diagnosis of overall survival period

4 Experiment

The U-net with NABL is evaluated by training the network using BraTS dataset. The test is to analyze the NABL for the segmentation performance and training effectiveness. The designs of NABL network is configured as a repeated V-shape, based on the learning performance under the limited number of dataset. The benefit of repeated V-shape is the consistent learning performance in a limited dataset.

The prediction accuracy is summarized in Table 1, which shows 59% based on BraTS challenge 2019 training dataset. The configuration of fully connected feedforward neural network in Fig. 7 is $200 \times 200 \times 150 \times 50 \times 3/8$, where the last layer of 3/8 neurons is introduced to generate the 3 or 8 binary outputs as the number of days of diagnosed overall survival period. There are two modes of training, which are for 3 outputs (Short-survivors, Med-survivors, Long-survivors) or 8 outputs (representing 255 levels of survival period).

Table 1. Prediction of overall survival (OS).

OS	Patient numbers	Prediction accuracy
Short-survivors (<10 months)	82	62 (76%)
Med-survivors (between 10 and 15 months)	54	24 (44%)
Long-survivors (>15 months)	76	40 (53%)
Total patients	212	126 (59%)

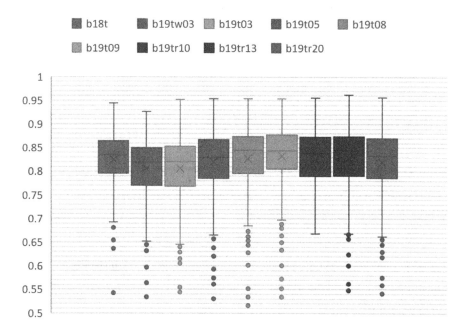

Fig. 8. Overall performance of tumor segmentation by deep neural network with Neuromorphic Attention-Based Learner (NABL), for different training dataset and training parameter of epoch number (b19t for BraTS 2019 dataset, b18t for BraTS 2018 dataset, last two digits for epoch numbers, and b19tw for whole training dataset of BraTS 2019).

The result in Fig. 8 shows the better performance with particular training procedure in respects of dataset selection and training parameters. The proposed U-net with Neuromorphic Attention-Based Learner in Fig. 3 has 72 layers in total.

The experimentation is mainly based on the training dataset BraTS challenge 2019. The different selection of training data or learning parameters were applied for evaluating the performance, where the last two digits represent the epoch number of training parameter. The other two digits represent the year of training dataset, that is, b18t means that the network trained by 2018 dataset was applied to segment the MRI images of 2019 dataset. The result (b18t in Fig. 8) based on BraTS 2018 training dataset (210 patients) is included, which demonstrated the comparable result to the other networks (trained by 2019 dataset) for evaluating the 2019 dataset. Higher accuracy (DICE coefficient) was observed in the case (b19t09 in Fig. 8) based on HGG dataset and 9 iterations for training. The 85% accuracy was simulated as the average DICE coefficient of ET, WT and TC. The computation time required was less than 20 s per person, based on the computer with a GPU (Nvidia's GTX1060).

5   Validation and Test Result

The validation process was used to assess the designs of Fig. 3 and Fig. 7 for their practical aspects. The segmentation analysis of validation data is omitted due to the presence of substantial difference between the validation result and the training result. The test result is in Table 2, where there exists the performance gap compared to the result in Fig. 8.

Table 2. Test result of segmentation.

Label	Dice_ET	Dice_WT	Dice_TC
Mean	0.59442	0.80277	0.6944
Median	0.68716	0.85589	0.81409
25 quantile	0.53623	0.77706	0.68333
75 quantile	0.75911	0.89753	0.88143

For the design of Fig. 3, the performance gap in the overall survival analysis is observed but much less than the segmentation case. Considering the limited segmentation performance, the different configurations of input data were evaluated. Two types of combination are evaluated as reasonable approaches (Table 3), which are the one with 4 inputs (Segmentation TC, Neuromorphic segmentation I, Neuromorphic segmentation II, Age) and the other with 3 inputs (Neuromorphic segmentation I, Neuromorphic segmentation II, Age). The neural network of 3 input data has the accuracy of 64% for the training data set, while the accuracy of 55.2% is observed at the validation as shown in Table 3. For different configurations of 4 inputs and 3 inputs, the neural network in Fig. 7 is trained by using the same training dataset, and the selected input data.

Table 3. Validation result of overall survival.

Label	Case evaluated	Accuracy
4 inputs	29	0.483
3 inputs	29	0.552

For the test, the 3 input neural network in Fig. 7 and Table 3 is applied and the result of Table 4 is attained. It is noted that the test dataset produced the different statistics of intermediate data of neuromorphic segmentation I and II during the evaluation, comparing to those during the training process or the validation stage. The distribution trend of intermediate data with test dataset is a rather sparse pattern compared to those with training data or validation data, which can cause the computational limitation on accuracy. Hence, for the test stage, the segmentation threshold value is controlled to improve the evaluation of interim data, which performs the conversion of attention maps in Fig. 3 into the neuromorphic segmentation I and II in Fig. 7.

Table 4. Test result of overall survival.

Label	Case evaluated	Accuracy
3 inputs	107	0.486

6 Conclusion

The NABL model is proposed to train the deep neural network for tumor segmentation, which is with challenges of typically small datasets and the difficulty of exact segmentation class determination. The neuromorphic neural network mimicking visual cortex neurons is adopted for the neuromorphic attention generation, inspired by the performance of pre-trained neuromorphic CNN applied to the overall survival analysis in BraTS 2018 [22]. The NABL algorithm assists the training of U-net CNN under the limited condition of deep learning, without relying on 3D voxel based CNN. The deep neural network of Fig. 3 has the faster and more reliable learning characteristics as shown in Fig. 9, which is influenced by the attention-generating training guide. The advantage is the relatively less possibility to be subject to unanimous overfitting, as far as the attention module is controlled from such symptom. Although the demonstrated efficiency of neuromorphic NABL, the 2D based processing can be improved substantially by introducing 3D based visual features in [21].

Another challenge is unexpected difference of intermediate processing results found between the validation data and test data, regardless of training data. There can be many causes of difference, however it would be ambiguous to eliminate the issue without the essential understanding of image data for the designated task such as the patient overall survival period. Deep learning or neural network training has the close relationship to the training data in quality and quantity, and the performance is

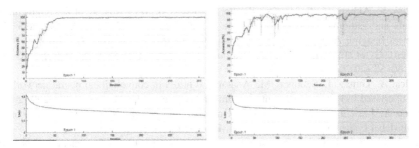

Fig. 9. Training progress by deep neural network with Neuromorphic Attention-Based Learner (NABL), illustrating faster speed of learning (left) than by deep neural network without NABL (right).

determined by learning. Hence it is essential to design the neuromorphic attention module based on the consultation with the end users, in this case, medical clinicians.

We illustrated the effective and practical approach of Deep Learning using BraTS dataset and introduced a new way of neuromorphic attention-based algorithm that can promote Deep Neural Networks applications to bio-medical image analysis for new clinical problems. The performance accuracy of neural network is mostly based on the network design with training data, and the design optimization of neural network engineering is under the further development for the improved performance and extended applications.

References

1. White, N., Reid, F., Harris, A., Harries, P., Stone, P.: A systematic review of predictions of survival in palliative care: how accurate are clinicians and who are the experts? PLoS ONE **11**(8), e0161407 (2016)
2. Cheon, S., et al.: The accuracy of clinicians' predictions of survival in advanced cancer: a review. Ann. Palliat. Med. **5**(1), 22–29 (2016). https://doi.org/10.3978/j.issn.2224-5920.2015.08.04
3. Menze, B., Jakab, A., Bauer, S., Kalpathy-Cramer, J., Farahani, K., Kirby, J., et al.: The multimodal brain tumor image segmentation benchmark (BRATS). IEEE Trans. Med. Imaging **34**(10), 1993–2024 (2015). https://doi.org/10.1109/TMI.2014.2377694
4. Bakas S., Akbari H., Sotrias A., Bilello M., Rozycki M., Kirby J., et al: Data descriptor: advancing the cancer genome atlas glioma MRI collections with expert segmentation labels and radiomic features. Nat. Sci. Data **117** (2017) https://doi.org/10.1038/sdata.2017.117
5. Bakas S., Reyes M., Jakab A., Bauer S., Rempfler M., Crimi A., et al: Identifying the Best Machine Learning Algorithms for Brain Tumor Segmentation, Progression Assessment, and Overall Survival Prediction in the BRATS Challenge, arXivpreprint arXiv:1811.02629.2018 (2018)
6. Bakas, S., Akbari, H., Sotiras, A., Bilello, M., Rozycki, M., Kirby, J., et al.: Segmentation labels and radiomic features for the pre-operative scans of the TCGA-GBM collection. Cancer Imaging Arch. (2017). https://doi.org/10.7937/K9/TCIA.2017.KLXWJJ1Q

7. Bakas, S., Akbari, H., Sotiras, A., Bilello, M., Rozycki, M., Kirby, J., et al.: Segmentation labels and radiomic features for the pre-operative scans of the TCGA-LGG collection. Cancer Imaging Arch. (2017). https://doi.org/10.7937/K9/TCIA.2017.GJQ7R0EF

8. Litjens, G., et al.: A survey on deep learning in medical image analysis. Med. Image Anal. **42**, 60–68 (2017)

9. Dalca, A., Guttag, J., Sabuncu, M.: Anatomical priors in convolutional neural networks for unsupervised biomedical segmentation. In: Proceedings of CVPR, pp. 9290–9299 (2018)

10. Chen, X., Liew, J., Xiong, W., Chui, C., Ong, S.: Segment and erase: an efficient network for multi-label brain tumor segmentation. In: Proceedings of ECCV (2018)

11. Gorji, S., Clark, J.: Going from image to video saliency: augmenting image salience with dynamic attentional push. In: Proceedings of CVPR (2018)

12. Woo, S., Park, J., Lee, J., Kweon, I.: CBAM: convolutional block attention module. In: Proceedings of ECCV (2018)

13. Lim, J., Yoo, Y., Heo, B., Choi, J.: Pose transforming network: learning to disentangle human posture in variational auto-encoded latent space. Pattern Recogn. Lett. **112**, 91–97 (2018)

14. Liu J., Gao C., Meng D., Hauptmann A.: DecideNet: counting varying density crowds through attention guided detection and density estimation. In: Proceedings of CVPR (2018)

15. Xu, X., et al.: Quantization of fully convolutional networks for accurate biomedical image segmentation. In: Proceedings of CVPR, pp. 8300–8308 (2018)

16. Long, J., Shelhamer, E., Darrell, T.: Fully convolutional networks for semantic segmentation. In: Proceedings of CVPR (2015)

17. Kohl, S., et al.: A probabilistic U-Net for segmentation of ambiguous images. In: Proceedings of Neural Information Processing Systems (NeurIPS) (2018)

18. Isensee, F., Kickingereder, P., Wick, W., Bendszus, M., Maier-Hein, K.: Brain Tumor Segmentation and Radiomics Survival Prediction: Contribution to the BRATS 2017 Challenge, arXiv:1802.10508v [cs.CV] (2018)

19. Sun, L., Zhang, S., Chen, H., Luo, L.: Brain tumor segmentation and survival prediction using multimodal MRI scans with deep learning. Front. Neurosci. **13**, Article 810 (2019)

20. Myronrnko, A.: 3D MRI Brain Tumor Segmentation Using Autoencoder Regularization, arXiv preprint arXiv:1810.11654v3 (2018)

21. Han, W.-S., Han, I.S.: Object segmentation for vehicle video and dental CBCT by neuromorphic convolutional recurrent neural network. In: Bi, Y., Kapoor, S., Bhatia, R. (eds.) IntelliSys 2016. SCI, vol. 751, pp. 264–284. Springer, Cham (2018). https://doi.org/10.1007/978-3-319-69266-1_13

22. Han, W.-S., Han, I.S.: Neuromorphic neural network for multimodal brain image segmentation and overall survival analysis. In: Crimi, A., Bakas, S., Kuijf, H., Keyvan, F., Reyes, M., van Walsum, T. (eds.) BrainLes 2018. LNCS, vol. 11384, pp. 178–188. Springer, Cham (2019). https://doi.org/10.1007/978-3-030-11726-9_16

Improving Brain Tumor Segmentation in Multi-sequence MR Images Using Cross-Sequence MR Image Generation

Guojing Zhao[1,2], Jianpeng Zhang[2], and Yong Xia[1,2(✉)]

[1] Research & Development Institute of Northwestern Polytechnical University
in Shenzhen, Shenzhen 518057, China
`yxia@nwpu.edu.cn`
[2] National Engineering Laboratory for Integrated Aero-Space-Ground-Ocean Big
Data Application Technology, School of Computer Science and Engineering,
Northwestern Polytechnical University, Xi'an 710072, China

Abstract. Accurate brain tumor segmentation using multi-sequence magnetic resonance (MR) imaging plays a pivotal role in clinical practice and research settings. Despite their prevalence, deep learning-based segmentation methods, which usually use multiple MR sequences as input, still have limited performance, partly due to their insufficient ability to image representation. In this paper, we propose a brain tumor segmentation (BraTSeg) model, which uses cross-sequence MR image generation as a self-supervision tool to improve the segmentation accuracy. This model is an ensemble of three image segmentation and generation (ImgSG) models, which are designed for simultaneous segmentation of brain tumors and generation of T1, T2, and Flair sequences, respectively. We evaluated the proposed BraTSeg model on the BraTS 2019 dataset and achieved an average Dice similarity coefficient (DSC) of 81.93%, 87.80%, and 83.44% in the segmentation of enhancing tumor, whole tumor, and tumor score on the testing set, respectively. Our results suggest that using cross-sequence MR image generation is an effective self-supervision method that can improve the accuracy of brain tumor segmentation and the proposed BraTSeg model can produce satisfactory segmentation of brain tumors and intra-tumor structures.

Keywords: Brain tumor segmentation · Deep learning · Multi-sequence MR imaging · MR image generation

1 Introduction

Brain tumors are cells that grow abnormally in the brain or skull, which are divided into primary and metastatic tumors according to different sources. Over 700,000 Americans are living with a brain tumor and nearly 80,0000 people were diagnosed with a primary brain tumor in 2019. As the most common

A. Crimi and S. Bakas (Eds.): BrainLes 2019, LNCS 11993, pp. 27–36, 2020.
https://doi.org/10.1007/978-3-030-46643-5_3

type of primary brain tumors, gliomas can affect the brain function and be life-threatening. Accurate segmentation of gliomas and its internal structures, including the enhancing tumor and tumor core, plays a critical role in diagnosis, treatment planning, and prognosis. Magnetic resonance (MR) imaging provides a non-invasive and non-radiation way to visualize the brain anatomy with a highly soft tissue contrast than other structure imaging techniques such as the computed tomography, and hence is the most widely used imaging modality for gliomas segmentation.

Automated segmentation of gliomas using multi-sequence MR data has been extensively investigated in recent years. A large number of methods have been published in the literature. Among them, the deep learning techniques, which usually take up to four MR sequences with equal importance as input, take the dominant position [5,6,10–12,15,17,18]. Kamnitsas et al. [12] proposed an ensemble of multiple deep learning models and architectures, in which the predictions made by different models are integrated to form the robust segmentation of brain tumors. Wang et al. [17] decomposed this four-class segmentation task into three binary segmentation sub-tasks, and thus cascaded the anisotropic convolution neural networks for those sub-tasks into a stronger segmentation model while considering the constraints on sub-regions. Fabian et al. [10] employed U-Net and several useful training strategies, such as the extensive data augmentation and Dice loss function, and achieved competitive performance in brain tumor segmentation. Zhou et al. [19] proposed a variety of deep architectures to learn the contextual and attentive information, and then used the ensemble of the predictions made by these models as a robust segmentation result. Andriy [15] developed a semantic segmentation network with an encoder-decoder architecture to perform this task and adopted a variational auto-encoder (VAE) to regularize the shared encoder. In our previous work [9], the cascaded U-Nets was constructed for the joint brain tumor detection and intra-tumor structure segmentation. We further proposed the multi-level upsampling network (MU-Net) to learn the image presentations of transverse, sagittal, and coronal views and fused them to segment brain tumors automatically [8]. Despite the improved performance, these deep learning-based brain tumor segmentation models still suffer from limited accuracy. The sub-optimal performance can be largely attributed to the less ability to represent multi-sequence MR data, which has been learned on a small dataset.

In this paper, we propose a self-supervised method to improve the accuracy of brain tumor segmentation using cross-sequence MR image generation. Specifically, we select three MR sequences as the observed data and the other sequence as the latent data, and feed the concatenation of observed data to our image segmentation-generation (ImgSG) model, which has an encoder and two decoders, for simultaneous brain tumor segmentation and latent MR sequence generation. In this way, the self-supervised task of MR sequence generation facilitates the encoder to extract rich semantic information. Next, we construct three ImgSG models to generate T1, T2, and Flair sequences, respectively, while performing segmentation. Finally, we use the ensemble of three ImgSG models to further improve the segmentation accuracy. We have evaluated the proposed brain tumor segmentation (BraTSeg) model on the BraTS 2019 validation and testing datasets and achieved satisfactory performance in the segmentation of the enhanced tumor core, whole tumor, and tumor core, respectively.

2   Method

The proposed BraTSeg model is an ensemble of three ImgSG models, each containing an encoder and two decoders. The encoder is used for image feature extraction, and two decoders are designed for image segmentation and generation, respectively. Compared to other MR sequences, the T1ce sequence contains more information about the brain tumor and is the sequence on which radiologists mainly focus in clinical practices. Therefore, we always treat the T1ce sequence as observed data and the other three sequences as latent data, which need to be generated by ImgSG models, respectively. A diagram that summarizes our BraTSeg model is shown in Fig. 1. We then delve into the details.

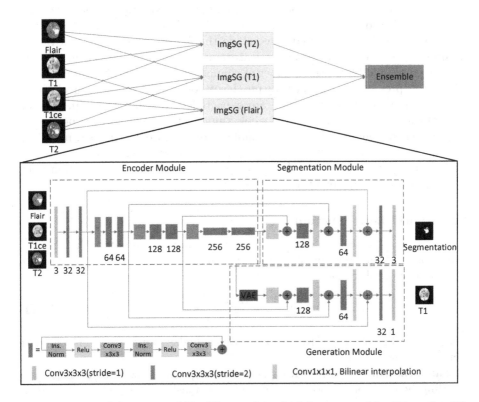

Fig. 1. Diagram of the proposed BraTSeg model, which is an ensemble of three ImgSG models. Each ImgSG model is designed for simultaneous brain tumor segmentation and MR sequence generation. The T1, T2, and Flair sequences are generated by three ImgSG models, respectively.

2.1   ImgSG Model

As shown in the bottom part of Fig. 1, the ImgSG model is composed of an encoder, a segmentation decoder, and a generation decoder.

The encoder takes three MR sequences as input and consists of a convolutional layer with kernel size of $3 \times 3 \times 3$ and stride of 1, eight convolutional blocks, and three downsampling layers, which are represented by yellow squares, blue squares, and orange squares, respectively, in Fig. 1. Each convolutional block contains sequentially an instance normalization layer [16], a ReLU layer, a convolutional layer, another instance normalization layer, another ReLU layer, another convolutional layer, and a residual connection [7]. After every two successive convolutional blocks, there is a downsampling layer, in which the kernel size is $3 \times 3 \times 3$ and the stride is 2. It should be noted that we replace the group normalization with the instance normalization due to the limited GPU memory.

The segmentation decoder has three successive pairs of upsampling layers (represented by green squares in Fig. 1) and convolutional blocks, followed by a output layer. An upsampling layer contains convolutional kernels of size $1 \times 1 \times 1$, followed by the bi-linear interpolation. After each upsampling layer, the skip connection is adopted to combine the feature maps produced by the upsampling layer with the same-size feature maps generated by the encoder. In each convolutional block, the number of channels is reduced to half of the channels in the previous block. The output layer contains $3 \times 3 \times 3$ convolutional kernels and three channels, which are expected to produce the segmentation mask of the enhancing tumor core, whole tumor, and tumor core, respectively.

The generation decoder has a similar architecture to the segmentation decoder, except that there is a variational autoencoder (VAE) module before the first upsampling layer and there is only one channel and no activation function in last $3 \times 3 \times 3$ convolutional layer.

2.2   Hybrid Loss Function

There are several performance metrics to measure the quality of image segmentation from different perspectives. DSC measures the similarity between a segmentation mask S and a ground truth G, defined as follows

$$L_{Dice} = \frac{2|S \cap G|}{|S| + |G|} \tag{1}$$

It takes a value from the range $[0, 1]$, and a higher value represents a more accurate segmentation result.

The mean square error (MSE) is the average of the squares of the corresponding point errors between a segmentation mask S and a ground truth G, shown as follows

$$L_{MSE} = \frac{1}{N} \sum_{i=1}^{n} (S_i - G_i)^2 \tag{2}$$

where N is the number of voxels. Obviously, a smaller MSE means a more accurate segmentation result.

The binary cross entropy (BCE) between a segmentation mask S and a ground truth G also measures the segmentation quality. It can be calculated as follows

$$L_{BCE} = \frac{1}{N} \sum_{i=1}^{n} -[G_i \log(S_i) + (1 - G_i) \log(1 - S_i))] \tag{3}$$

The value of BCE decreases if the a segmentation mask S gets closer to the ground truth G.

The Kullback-Leibler divergence (KL divergence) is a good measure of the distance between two probability distributions $p(i)$ and $q(i)$, shown as follows

$$L_{KL} = \sum_{i=1}^{n} P_i \ln \frac{P_i}{G_i} \tag{4}$$

The value of KL divergence is non-negative. If the distribution $p(i)$ approximates $q(i)$, the KL distance is approaching to 0.

For this study, we jointly used all these image segmentation metrics and proposed the following hybrid loss to train each ImgSG model.

$$L = L_{Dice} + w_1 \times L_{MSE} + w_2 \times L_{KL} + w_3 \times L_{BCE} \tag{5}$$

where the weighting parameters w_1, w_2, w_3 were empirically set to 0.1, 0.1, and 10, respectively.

2.3 Testing

In the testing stage, the image generation module in each ImgSG model is disabled. We first feed each combination of three MR sequences to the corresponding trained ImgSG model to generate a segmentation mask, and then fuse the obtained three segmentation masks using majority voting. It should be noted that there might be no enhanced tumor core in many LGG cases. Therefore, if the enhanced tumor core we detected is samller than 400 voxels, we believe that there might be no enhanced tumor core and hence remove those detected voxels from the segmentation result.

3 Dataset

The 2019 Brain Tumor Segmentation (BraTS 2019) challenge is focusing on the evaluation of state-of-art methods for the segmentation of brain tumors in multi-sequence MRI scans [1–4,14]. The BraTS 2019 dataset was collected from multiple institutions with the different clinical protocols and various scanners. This dataset has been partitioned into a training set, a validation set, and a testing set, which have 335, 125, and 166 cases, respectively. Each case has four MR sequences (i.e. Flair, T1, T1ce, and T2) and is annotated by professional radiologists. Each annotation contains the non-enhancing tumor core, tumor core, and enhancing tumor. All MR sequences have been interpolated to the voxel size of $1 \times 1 \times 1\,\text{mm}^3$ and the dimension of $155 \times 240 \times 240$, and have been made publicly available. However, the annotations of validation cases and testing cases are withheld for online evaluation, and the segmentation results obtained on the testing set can only be evaluated once online. The training set includes 259 HGG and 76 LGG cases, and the case distributions in the validation set and testing set are unknown.

4 Experiments and Results

4.1 Experiment Settings

We aim to segment each brain tumor into three sub-regions, including the tumor core, enhancing tumor, and whole tumor (i.e. the tumor core, enhancing tumor, and non-enhancing tumor core). Our BraTSeg Model was implemented in PyTorch and trained on a desktop with a NVIDIA TITAN Xp GPU. Every MR sequence was normalized by its mean and standard deviation. Simple data augmentation techniques, such as random scaling with a scaling factor from $[0.9, 1.1]$ and random axis mirror flipping, were employed to increase the number of training samples. Each ImgSG model takes three input volumes of size $80 \times 160 \times 160$, and was optimized by the Adam optimizer [13]. We set the batch size to 1, the maximum epochs to 200, and the initial learning rate to $lr = 1e - 4$, which is decayed as $lr = lr \times (1 - epoch/epochs)^{0.9}$.

4.2 Results on Validation Set

We first evaluated the proposed BraTSeg model on the BraTS 2019 validation set. The test time augmentation (TTA) was used in this experiment. Figure 2 shows four transverse slices from a T1ce sequence and the corresponding segmentation results produced by our model. In each result, the segmented the enhancing tumor, tumor core, and non-enhancing tumor core were highlighted in red, yellow, and green, respectively. Table 1 shows the average DSC and Hausdorff distance obtained by our model and each of its three ImgSG models, where ImgSG (#) refers to the ImgSG model trained to generate the # sequence. Our model achieved the averaged DSC of 0.781, 0.908, 0.843 for the enhanced tumor core, whole tumor, and tumor core, respectively. Compared with the best-performed single model (i.e. ImgSG (T2)), our model improves the DSC by 0.8% for the segmentation of enhance tumor and by 1.8% for the segmentation of tumor core. However, our model performs slightly worse than ImgSG (T2) in the segmentation of whole tumor (i.e. 90.81% vs. 90.83% in DSC and 5.2652 vs. 5.1694 in Hausdorff distance).

A unique feature of the proposed BrsTSeg model is to perform brain tumor segmentation and cross-sequence MR image generation simultaneously. To demonstrate the effectiveness of this strategy, we compared our model to its baseline, which is almost the same to our model except for not containing image

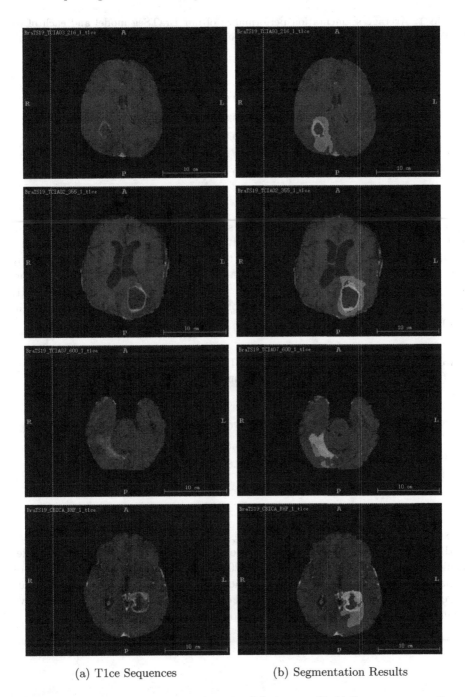

(a) T1ce Sequences   (b) Segmentation Results

Fig. 2. Segmentation results on BraTS 2019 validation set: (Left) Four transverse slices from a T1ce sequences and (Right) segmentation results produced by our BraTSeg model. (Color figure online)

Table 1. Average segmentation performance of our BraTSeg model and each of its three ImgSG models on BraTS 2019 validation set (ET: Enhancing Tumor, WT: Whole Tumor, and TC: Tumor Core).

	DSC			Hausdorff distance		
	ET	WT	TC	ET	WT	TC
ImgSG (T2)	0.7723	**0.9083**	0.8256	**2.7937**	**5.1694**	6.7467
ImgSG (T1)	0.7566	0.9052	0.8046	4.3852	5.5244	7.3070
ImgSG (Flair)	0.7709	0.8853	0.8238	3.25825	6.1143	7.0040
Our BraTSeg	**0.7809**	0.9081	**0.8432**	2.8802	5.2692	**5.7391**

generation decoders and taking four MR sequences as input. The performance of our model and its baseline was compared in Table 2. It shows that using the MR sequence generation as a self-supervised training strategy improves the DSC by 0.88%, 0.08%, and 3.10% in the segmentation of enhancing tumor, whole tumor, and tumor score, respectively. Therefore, we suggest that the cross-sequence MR image generation can be used as a kind of self-supervision to facilitate the brain tumor segmentation task.

Table 2. Performance of our BraTSeg model and its baseline on BraTS 2019 validation set (ET: Enhancing Tumor, WT: Whole Tumor, and TC: Tumor Core).

	DSC			Hausdorff distance		
	ET	WT	TC	ET	WT	TC
Our BraTSeg	**0.7809**	**0.9081**	**0.8432**	**2.8802**	5.2692	**5.7391**
Baseline	0.7721	0.9073	0.8122	3.2259	**5.0250**	6.7742

4.3   Results on Testing Set

Next, we verified the proposed BraTSeg model on the BraTS 2019 testing set and displayed the obtained performance in Table 3. It shows that our model achieves an average DSC of 81.93%, 87.80%, and 83.44% in the segmentation of enhancing tumor, whole tumor, and tumor score, respectively. The performance is generally consistent with that obtained on the validation set. Hence, we believe that the performance of the proposed model is robust.

Table 3. Performance of our BraTSeg model on BraTS 2019 testing set (ET: Enhancing Tumor, WT: Whole Tumor, and TC: Tumor Core).

	DSC			Hausdorff distance		
	ET	WT	TC	ET	WT	TC
Mean	0.8193	0.8780	0.8344	2.9201	6.1206	4.7604
StdDev	0.1851	0.1538	0.2503	6.8308	11.4028	9.6738
Median	0.8629	0.9244	0.9251	1.4142	2.9142	2
25quantile	0.7921	0.8796	0.8737	1	1.4142	1.4142
75quantile	0.9207	0.9508	0.9568	2.2361	5	3.7417

5 Conclusion

This paper proposes the BraTSeg model for brain tumor segmentation using multi-sequence MR scans. This model is an ensemble of three ImgSG models, each performing brain tumor segmentation and cross-sequence MR image generation simultaneously. Our results on the BraTS 2019 validation set and testing set suggest that using MR sequence generation as a self-supervision tool can improve the segmentation accuracy and the proposed model can produce satisfactory segmentation of brain tumors and intra-tumor structures. Our future work will focus on extending the proposed model to solving the missing data problem, i.e. segmenting brain tumors in a multi-sequence MR study where one or more sequences are missing.

Acknowledgement. This work was supported in part by the Science and Technology Innovation Committee of Shenzhen Municipality, China, under Grants JCYJ20180306171334997, in part by the National Natural Science Foundation of China under Grants 61771397, and in part by the Project for Graduate Innovation team of NPU. We appreciate the efforts devoted by BraTS 2019 Challenge organizers to collect and share the data for comparing brain tumor segmentation algorithms for multi-sequence MR sequences.

References

1. Bakas, S., et al.: Segmentation labels and radiomic features for the pre-operative scans of the TCGA-GBM collection. The cancer imaging archive (2017)
2. Bakas, S., et al.: Segmentation labels and radiomic features for the pre-operative scans of the TCGA-LGG collection. Cancer Imaging Archive **286** (2017)
3. Bakas, S., et al.: Advancing the cancer genome atlas Glioma MRI collections with expert segmentation labels and radiomic features. Sci. Data **4**, 170117 (2017)
4. Bakas, S., et al.: Identifying the best machine learning algorithms for brain tumor segmentation, progression assessment, and overall survival prediction in the brats challenge. arXiv preprint arXiv:1811.02629 (2018)
5. Dong, H., Yang, G., Liu, F., Mo, Y., Guo, Y.: Automatic brain tumor detection and segmentation using U-Net based fully convolutional networks. In: Valdés

Hernández, M., González-Castro, V. (eds.) MIUA 2017. CCIS, vol. 723, pp. 506–517. Springer, Cham (2017). https://doi.org/10.1007/978-3-319-60964-5_44

6. Havaei, M., et al.: Brain tumor segmentation with deep neural networks. Med. Image Anal. **35**, 18–31 (2017)

7. He, K., Zhang, X., Ren, S., Sun, J.: Deep residual learning for image recognition. In: Proceedings of the IEEE Conference on Computer Vision and Pattern Recognition, pp. 770–778 (2016)

8. Hu, Y., Liu, X., Wen, X., Niu, C., Xia, Y.: Brain tumor segmentation on multi-modal MR imaging using multi-level upsampling in decoder. In: Crimi, A., Bakas, S., Kuijf, H., Keyvan, F., Reyes, M., van Walsum, T. (eds.) BrainLes 2018. LNCS, vol. 11384, pp. 168–177. Springer, Cham (2019). https://doi.org/10.1007/978-3-030-11726-9_15

9. Hu, Y., Xia, Y.: 3D deep neural network-based brain tumor segmentation using multimodality magnetic resonance sequences. In: Crimi, A., Bakas, S., Kuijf, H., Menze, B., Reyes, M. (eds.) BrainLes 2017. LNCS, vol. 10670, pp. 423–434. Springer, Cham (2018). https://doi.org/10.1007/978-3-319-75238-9_36

10. Isensee, F., Kickingereder, P., Wick, W., Bendszus, M., Maier-Hein, K.H.: Brain tumor segmentation and radiomics survival prediction: contribution to the BRATS 2017 challenge. In: Crimi, A., Bakas, S., Kuijf, H., Menze, B., Reyes, M. (eds.) BrainLes 2017. LNCS, vol. 10670, pp. 287–297. Springer, Cham (2018). https://doi.org/10.1007/978-3-319-75238-9_25

11. Işın, A., Direkoğlu, C., Şah, M.: Review of MRI-based brain tumor image segmentation using deep learning methods. Procedia Comput. Sci. **102**, 317–324 (2016)

12. Kamnitsas, K., et al.: Ensembles of multiple models and architectures for robust brain tumour segmentation. In: Crimi, A., Bakas, S., Kuijf, H., Menze, B., Reyes, M. (eds.) BrainLes 2017. LNCS, vol. 10670, pp. 450–462. Springer, Cham (2018). https://doi.org/10.1007/978-3-319-75238-9_38

13. Kingma, D.P., Ba, J.: Adam: a method for stochastic optimization. arXiv preprint arXiv:1412.6980 (2014)

14. Menze, B.H., et al.: The multimodal brain tumor image segmentation benchmark (BRATS). IEEE Trans. Med. Imaging **34**(10), 1993–2024 (2014)

15. Myronenko, A.: 3D MRI brain tumor segmentation using autoencoder regularization. In: Crimi, A., Bakas, S., Kuijf, H., Keyvan, F., Reyes, M., van Walsum, T. (eds.) BrainLes 2018. LNCS, vol. 11384, pp. 311–320. Springer, Cham (2019). https://doi.org/10.1007/978-3-030-11726-9_28

16. Ulyanov, D., Vedaldi, A., Lempitsky, V.: Instance normalization: the missing ingredient for fast stylization. arXiv preprint arXiv:1607.08022 (2016)

17. Wang, G., Li, W., Ourselin, S., Vercauteren, T.: Automatic brain tumor segmentation using cascaded anisotropic convolutional neural networks. In: Crimi, A., Bakas, S., Kuijf, H., Menze, B., Reyes, M. (eds.) BrainLes 2017. LNCS, vol. 10670, pp. 178–190. Springer, Cham (2018). https://doi.org/10.1007/978-3-319-75238-9_16

18. Zhao, X., Wu, Y., Song, G., Li, Z., Zhang, Y., Fan, Y.: A deep learning model integrating FCNNs and CRFs for brain tumor segmentation. Med. Image Anal. **43**, 98–111 (2018)

19. Zhou, C., Chen, S., Ding, C., Tao, D.: Learning contextual and attentive information for brain tumor segmentation. In: Crimi, A., Bakas, S., Kuijf, H., Keyvan, F., Reyes, M., van Walsum, T. (eds.) BrainLes 2018. LNCS, vol. 11384, pp. 497–507. Springer, Cham (2019). https://doi.org/10.1007/978-3-030-11726-9_44

Ensemble of CNNs for Segmentation of Glioma Sub-regions with Survival Prediction

Subhashis Banerjee[1,2(✉)], Harkirat Singh Arora[3], and Sushmita Mitra[1]

[1] Machine Intelligence Unit, Indian Statistical Institute, Kolkata, India
mail.sb88@gmail.com,sushmita@isical.ac.in
[2] Department of CSE, University of Calcutta, Kolkata, India
[3] Department of Chemical Engineering, Indian Institute of Technology Roorkee, Roorkee, India
harora@ch.iitr.ac.in

Abstract. Gliomas are the most common malignant brain tumors, having varying level of aggressiveness, with Magnetic Resonance Imaging (MRI) being used for their diagnosis. As these tumors are highly heterogeneous in shape and appearance, their segmentation becomes a challenging task. In this paper we propose an ensemble of three Convolutional Neural Network (CNN) architectures *viz.* (i) P-Net, (ii) U-Net with spatial pooling, and (iii) ResInc-Net for glioma sub-regions segmentation. The segmented tumor Volume of Interest (VOI) is further used for extracting spatial habitat features for the prediction of Overall Survival (OS) of patients. A new aggregated loss function is used to help in effectively handling the data imbalance problem. The concept of modeling predictive distributions, test time augmentation and ensembling methods are used to reduce uncertainty and increase the confidence of the model prediction. The proposed integrated system (for Segmentation and OS prediction) is trained and validated on the Brain Tumor Segmentation (BraTS) Challenge 2019 dataset. We ranked among the top performing methods on Segmentation and Overall Survival prediction on the validation dataset, as observed from the leaderboard. We also ranked among the top four in the Uncertainty Quantification task on the testing dataset.

Keywords: Deep learning · Convolutional Neural Network · Brain Tumor Segmentation · Survival prediction · Spatial habitat · Class imbalance handling · Uncertainty quantification

1 Introduction

Gliomas are the most commonly occurring and aggressive malignant brain tumors originating in the glial cells of the Central Nervous System of the body. Based on their aggressiveness in infiltration, they are broadly classified into two categories, namely High-Grade Glioma or GlioBlastoma Multiforme

© Springer Nature Switzerland AG 2020
A. Crimi and S. Bakas (Eds.): BrainLes 2019, LNCS 11993, pp. 37–49, 2020.
https://doi.org/10.1007/978-3-030-46643-5_4

(HGG/GBM) and Low-Grade Glioma (LGG). Magnetic Resonance Imaging (MRI) has been extensively utilized over the years for diagnosing brain and nervous system abnormalities; largely because of its improved soft tissue contrast. Typically the MR sequences include $T1$-weighted, $T2$-weighted, $T1$-weighted Contrast enhanced ($T1C$), and $T2$-weighed with FLuid-Attenuated Inversion Recovery (FLAIR). The reason behind using all four sequences is that different tumor regions are more visible in different sequences; therefore helping in achieving more accurate demarcation of the tumor sub-region [5,6].

Accurate delineation of tumor regions in MRI sequences are of great significance since it allows: *i)* volumetric measurement of the tumor, *ii)* monitoring of tumor growth in patients between multiple MRI scans, over treatment span, and *iii)* treatment planning with follow-up evaluation, including the prediction of overall survival (OS). Manual segmentation of tumor regions from MRI sequences is highly tedious, expensive, time-consuming and error-prone task, mainly due to factors such as human fatigue, overabundance of MRI slices per patient, and an increasing number of patients. Such manual operations often lead to inaccurate segmentation. The need for an automated or semi-automated Computer Aided Diagnosis thus becomes apparent [7,8,14]. Presence of large structural and spatial variability among tumor regions in the brain makes automated segmentation a challenging problem. The distinctive segmentation of both HGG and LGG by the same model is in itself a difficult task to be carried out by automated methods.

A variety of Convolutional Neural Network (CNN) architectures have been proposed in recent literature for segmentation of different sub-regions of glioma *viz.* gadolinium enhancing tumor (ET), peritumoral edema (ED), and the necrotic and non-enhancing tumor (NCR/NET), using multi-modal MR images of the brain. Various architectures, presented over the last few years, commonly differ in their depth, number of filters and the processing of multi-scale context, among others. Model behavior is found to be biased towards architectural choices. For instance, models with large receptive fields may show improved localization capabilities but can be less sensitive to fine texture than models emphasizing local information.

In order to overcome the above problem, we propose an ensemble of three Convolutional Neural Networks (CNNs) *viz.* (i) P-Net [17], (ii) U-Net with spatial pooling [9], and (iii) ResInc-Net [11] for glioma segmentation. The segmented tumor region, commonly called Volume of Interest (VOI), is used to extract spatial habitat features [12,18], for predicting the Overall Survival of patients. A new loss function helps in handling the problem of class imbalance. Segmentation mask is used to quantify uncertainty.

The rest of the paper is organized as follows. Section 2 provides details about the dataset, CNN models for Glioma sub-region segmentation, extraction and preparation of patches for CNN training, as well as the proposed ensembling approach to segmentation with aggregated loss function for handling class imbalance. Habitat feature extraction from the segmented VOI for OS prediction and estimation of uncertainty in segmentation are also discussed. Section 3 describes the preliminary experimental results of the segmentation, OS prediction and uncertainty quantification tasks, demonstrating their effectiveness both qualitatively and quantitatively. The paper is concluded in Sect. 4.

2 Materials and Methods

In this section we discuss the BraTS 2019 data, and the procedure for tumor segmentation, survival rate prediction and uncertainty quantification. The proposed method comprises extraction of patches, following by the training and testing of various segmentation models, their post-processing and ensembling; thereafter, spatial habitat features are extracted for training and testing of classifiers towards prediction of overall survival, along with the quantification of the inherent uncertainty.

2.1 Dataset

Brain tumor MRI scans and Overall Survival (OS) data, used in this research, were obtained from BraTS 2019 Challenge [1–4,13]. It consists of 259 HGG/GBM and 76 LGG glioma cases as a part of the training dataset and 125 combined cases of HGG/GBM and LGG as the validation dataset. The OS data was available with correspondences to the pseudo-identifiers of the GBM/HGG imaging data, having 241 and 125 validation data points respectively. Each patient MRI scan set consist, of four MRI sequences or channels, encompassing native ($T1$) and post-contrast enhanced $T1$-weighted ($T1C$), $T2$-weighted ($T2$), and $T2$ FLuid-Attenuated Inversion Recovery ($FLAIR$) volumes, having 155 slices of 240×240 resolution images. The data is already aligned to the same anatomical template, skull-stripped, and interpolated to $1\,\mathrm{mm}^3$ voxel resolution. The manual segmentation of volume structures have been performed by experts following the same annotation protocol, and their annotations revised and approved by board-certified neuro-radiologists. Annotation labels included are for ET, ED, and NCR/NET. The predicted labels are evaluated by merging three regions, $viz.$ whole tumor (WT: all the three labels), tumor core (TC: ET and NCR/NET) and enhancing tumor (ET).

The OS data is defined in terms of days, and also includes the age of patients along with their resection status. Only those subjects with resection status GTR (Gross Total Resection) are considered for evaluating OS prediction. Based on the number of survival days, the subjects are grouped into three classes viz. long-survivors (>15 months), short-survivors (<10 months), and mid-survivors (between 10 to 15 months).

2.2 Segmentation of Glioma Sub-regions

An ensemble of three CNNs, $viz.$ (i) P-Net [17], (ii) U-Net with spatial pooling [9], and (iii) ResInc-Net [11], were used for segmentation of the sub-regions ET, ED, an NCR/NET from multi-modal MR images of the brain. These CNN architectures are discussed below.

P-Net. The architecture used was inspired from the P-Net [17] with dilated convolution filters, as illustrated in Fig. 1 for slice wise segmentation of the glioma and its sub-regions. The dropout layers were eliminated from the original architecture. A 3×3 convolution filter with a dilation rate of 2 covers a 5×5 area, leading to a larger receptive field and better delineation of regions in an image. Increasing the dilation rate enhances the receptive field and may lead to better results, especially in these type of cases when neighbouring pixels are related.

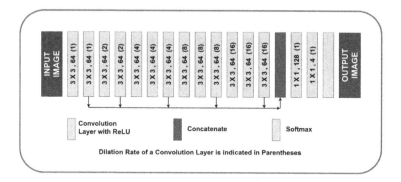

Fig. 1. P-Net with dilated convolution filters.

U-Net with Spatial Pooling. U-Net [9] with spatial pooling is an encoder-decoder type network architecture, which stores the max-pooling indices (i.e, the location where maximum feature value is placed in each pooling window) for each encoder feature map to be subsequently used by the corresponding

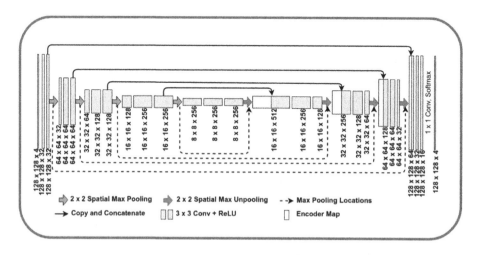

Fig. 2. U-Net with spatial pooling.

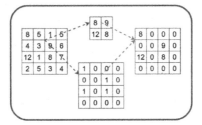

Fig. 3. Concept of spatial pooling.

decoder to up-sample its input feature map(s). The model is depicted in Fig. 2 with the concept of spatial pooling highlighted in Fig. 3. The encoder network uses spatial pooling layers for down sampling an image into a set of high-level features, followed by a decoder network which uses the feature information from the encoder network to construct a pixel-wise segmentation mask.

ResInc-Net. Residual Networks introduced a new way of training deeper CNN with Inception networks providing networks the freedom to learn representations. The merits of Residual Networks and Inception networks are leveraged together in the ResInc-Net [11] architecture, as shown in Fig. 4. The detailed ResNet and ResInc blocks are highlighted in Fig. 5.

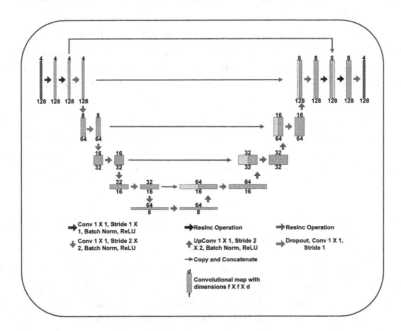

Fig. 4. ResNet with Inception module architecture.

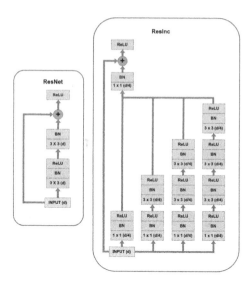

Fig. 5. ResNet and ResInc blocks.

The network architecture is of encoder-decoder type, involving an extensive use of 1×1 convolutions. During down-sampling 1×1 convolutions are used with stride 2, instead of the traditional MaxPooling layer. Analogously, in lieu of traditional up-sampling, the 1×1 convolutions are used with stride 2.

Patch Extraction. Each CNN is trained on patches of size $128 \times 128 \times 4$, extracted from all four MRI sequences corresponding to a particular plane, for efficient training. An algorithm for random patch extraction, is employed for the purpose. The patch selection is done using an entropy based criterion.

Ensembling. Average ensembling was employed to generate the final predictions based on these three segmentation models involving equal weightage. Ensembling assists on aggregating the differing learning capability of the constituents models, where each may get better tuned to certain set(s) of representations.

Quantitative Evaluation. "Dice score", and the "Hausdorff distance (95%)" were used as the metrics for evaluating the three segmented sub-regions WT, ET and TC. As the dataset was highly imbalanced, the standard loss functions available in literature were unsuitable for training and optimizing the CNN's. Majority of the classifiers, using standard loss functions, focus on learning representations of larger classes; thereby providing poor classification scores for the smaller classes. Hence a new loss function was proposed for efficient training. This entails an aggregation of two individual loss components; *viz.* – Generalized Dice loss [16] and Weighted Cross-entropy [15].

2.3   Prediction of Overall Survival with Spatial Habitat Features

It is often observed that (epi)genetic properties of individual cancer cells are highly variable, even within the same tumor, such that pre-existing resistant clones emerge and proliferate after therapeutic selection targeting the sensitive clones. Such heterogeneity within tumors is found to offer resistance to conventional therapies, by accelerating the unopposed proliferation of resistant subpopulations at the expense of the elimination of corresponding dose-susceptible subpopulations as well as their attendant competition for space and substrate.

Tumors can therefore be modeled as spatially heterogeneous complex adaptive systems [10], in which tumor growth and response to therapy are governed by eco-evolutionary interactions between the tumor micro-environment and phenotypic properties of local cellular subpopulations. Temporal and spatial cellular heterogeneities are due to clonal evolution from accumulating random mutations in cancer cell populations. Volumetric differential imaging holds promise in characterizing tumor heterogeneity at a holistic level, as compared to tissue biopsy which is constrained to sampling only a small fraction of the tumor.

Spatial heterogeneity within a tumor occurs mainly due to variations in cell density, necrosis, blood flow, etc. Such heterogeneous tumor regions are the habitats, which can be formulated in terms of their varying intensity profiles when using multimodal MRI scans (like $T1, T1C, T2, FLAIR$). Each tumor is thereby quantified as some combination of distinct habitats. Therefore habitat imaging, combining radiomics with multiparametric imaging, has the potential to provide early the noninvasive longitudinal biomarkers of intratumoral evolutionary and ecological dynamics for the informed application of adaptive therapy to manage tumors.

In this research the tumor voxels were partitioned into high and low intensity groups, based on the voxel intensity, by employing k-means clustering over each MR sequence. A total of 16 imaging habitats were generated from the four MR sequences, by considering all possible combinations of high and low intensity groups over each such. The 16 habitats are represented by a string of four binary numbers from 0000 to 1111, where "0" and "1" represent the low and high intensity groups in each of the four MRI sequences in the order "$T1, T1C, T2, FLAIR$". Next this binary representation was converted to decimal form, to be used as the label of the habitat. Thereby habitat H_0 represents the region where all four MR sequences have low intensity (0000), and H_{15} corresponds to the region where all the MR sequences have high intensity (1111). Therefore, 16 volumetric habitat features quantifying the volumes of the 16 habitats were extracted. These are provided as input to a Multilayer Perceptron (MLP), having two hidden layers, to predict the number of survival days; which is further used to determine the survival class (short, mid or long).

Quantification of Uncertainty in Segmentation. Deep learning models have been able to achieve state-of-the-art results on medical datasets. But apart from being robust and efficient in segmentation, the model needs to be certain about its prediction. Quantifying uncertainty enables us to estimate the confidence of a predictions. Although Monte Carlo Dropout is a way of estimating predictive uncertainty, it improves several has various constraints on the network architecture, activation function as well as addition of dropout layers. Therefore, instead of following the Bayesian and Monte Carlo approach for estimating predictive uncertainty, we combined the concepts of modeling predictive distributions, test time augmentation (i.e. complement to adversarial training) and ensembling methods for reducing uncertainty and increasing confidence of the prediction.

Although the BraTS challenge poses a segmentation problem, in a nutshell it boils down to a classification problem at the pixel-level. Although the BraTS challenge poses a segmentation problem, in a nutshell it boils down to a classification problem at the pixel-level. Let us consider the dataset to have m voxels with n dimension such that $\{x_i, y_i\}$ represents ith voxel, where x_i is four-dimensional due to the presence of data from four modalities for ith voxel, and y_i representing annotated label of voxel which will have one value out of 0,1,2 or 4.

The CNN approach is used to model the predictive distribution $p_w(y|x)$, where w is the weight vector of the network. The three different CNN networks were employed for each of the WT, TC and ET predictions. Developing a metric for measuring the quality of predictive uncertainty is a necessity. A particular score needs to be assigned to differentiate between good and bad predictions. It is experimentally observed that many loss functions are good metrics by themselves. For classification problems (as mentioned above), log-loss i.e. $-\log(p_w(y|x))$ is itself a great parameter and metric to quantify uncertainty.

Adversarial training is known to be an efficient method to improve model predictions. Adversarial training provides providing fake examples to fool a model it to make bad predictions. Here, instead of generating fake examples before training, we employ test time augmentation to generate more examples involving random change in rotation, brightness, contrast in the images. These steps serve the same purpose as that of adversarial training.

Ensembling has always been a boon for various complex tasks involving machine learning and deep learning, particularly for better predictions. It is done in two ways, $viz.$ randomization and boosting. We followed the randomization approach, training models on the whole dataset and then combining their predictions.

We trained the different models for WT, TC and ET segmentation. Three model ensemble was used, i.e. a total of nine models were trained for the task. Averaging various probabilities is one of the best and efficient ways to get a prediction of the ensemble model in classification. The uncertainty in predictions were quantified by the concept of entropy and variance of the distribution of prediction labels, in lieu of the averaged probability map resulting from N Monte Carlo samples that typically do not reflect the diversity information.

For estimating uncertainty, we use the concept of entropy, to represent voxel-wise variance and diversity information.

$$H(U_i|X) = -p^i * \log_2(p^i) + (1 - p^i) * \log_2(1 - p^i)$$

where U_i is resulting uncertainty in ith voxel of image X. As per uncertainty quantification task, we need to estimate uncertainty in WT, TC and ET segmentation. We calculated uncertainty for each segmentation using above equation. p^i represent probability of having WT, TC, ET depending upon segmentation for we which we are calculating uncertainty. The resulting uncertainty values were scaled such that they lie between 0 to 100 as per requirements of the Uncertainty quantification challenge.

We ranked among the top four methods on the testing dataset of this subtask.

3   Preliminary Experimental Results

The CNN models were developed and trained using Keras with Tensorflow backend in Python. All experiments were performed on the Intel AI DevCloud platform having a cluster of Intel Xeon Scalable processors. Codes of our experiment will soon be uploaded over Github. The proposed ensemble segmentation model was trained and validated on the training and validation datasets as respectively provided, by the BraTS 2019 [1, 3, 4] organizers. We used 80% of the training data (193 patients) for training, and the remaining 20% (48 patients) for validation of the proposed ensemble network.

The preliminary quantitative evaluation results, obtained by our segmentation model on the BraTS 2019 validation dataset, are reported in Table 1. Quantitative Performance metrics used for evaluating the segmentation output are (i) Dice score and (ii) Hausdorff distance, as computed for WT, TC and ET. Mean and standard deviation of Dice score and Hausdorff distance are given. Segmentation uncertainty quantification scores of the proposed ensembled CNNs, on the BraTS 2019 validation dataset, are reported in Table 2. Preliminary results of the proposed OS prediction method on the BraTS 2019 validation dataset is presented in Table 3. Qualitative segmentation results, obtained by our method for sample HGG and LGG patients from the BraTS 2019 training dataset, are shown in Fig. 6(b). The amount of uncertainity in the different sub-regions is qualitatively depicted in Fig. 6(c)-(e) in terms of the degree of the inherent whiteness. Performance evaluation of the proposed model on the BraTS 2019 testing dataset for the segmentation and OS prediction tasks are reported in Tables 4 and 5.

Table 1. Performance evaluation of the ensembled CNNs on the BraTS 2019 validation dataset (rounded off to two decimals).

Evaluation metrics	Dice			Hausdorff95		
	ET (mean ± sd)	WT (mean ± sd)	TC (mean ± sd)	ET (mean ± sd)	WT (mean ± sd)	TC (mean ± sd)
P-Net	0.73 ± 0.30	0.89 ± 0.09	0.77 ± 0.24	3.84 ± 4.71	5.73 ± 9.89	8.96 ± 11.44
ResInc-Net	0.76 ± 0.27	0.89 ± 0.08	0.84 ± 0.18	4.08 ± 7.25	7.10 ± 13.15	6.91 ± 12.75
U-Net	0.82 ± 0.21	0.90 ± 0.08	0.84 ± 0.19	3.66 ± 7.37	9.95 ± 20.69	7.52 ± 12.62
Ensemble	**0.93 ± 0.10**	**0.95 ± 0.08**	**0.94 ± 0.10**	**1.31 ± 2.61**	**5.97 ± 15.97**	**4.28 ± 11.59**

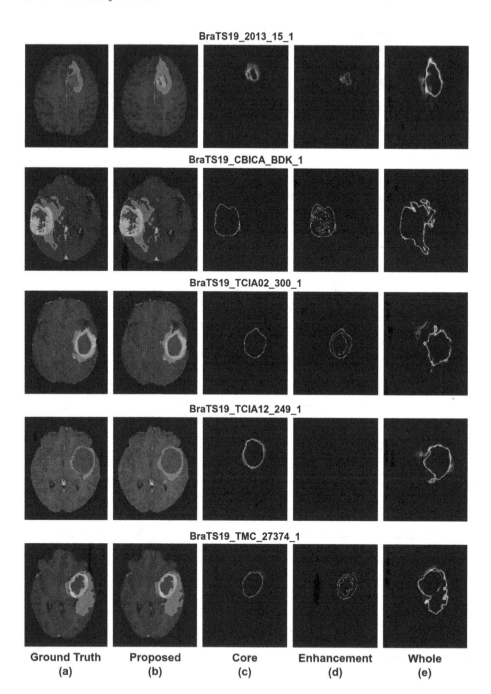

Fig. 6. Segmentation and uncertainty results of five sample patients from Training Set. (a) Ground truth, (b) output by proposed ensembling, with uncertainity in detecting (c) TC, (d) ET, and (e) WT.

Table 2. Performance evaluation of the ensemble model on the BraTS 2019 validation dataset for Uncertainty Estimation (rounded off to two decimals).

Evaluation metrics		Dice_AUC			FTP_AUC		
		WT	TC	ET	WT	TC	ET
Ensemble	Mean	0.87	0.82	0.71	0.03	0.03	0.04
	StdDev	0.06	0.12	0.27	0.02	0.02	0.03

Table 3. OS prediction result on the BraTS 2019 validation dataset.

Accuracy	MSE	medianSE	stdSE	SpearmanR
0.655	72705.127	17497.403	116969.56	0.369

Table 4. Performance evaluation of the ensembled CNNs on the BraTS 2019 testing dataset (rounded off to two decimals).

Evaluation metrics		Dice			Hausdorff95		
		ET	WT	TC	ET	WT	TC
Ensemble	Mean	0.79	0.89	0.85	2.73	4.64	5.46
	StdDev	0.22	0.10	0.23	4.54	6.01	9.18

Table 5. OS prediction result on the BraTS 2019 testing dataset.

Accuracy	MSE	medianSE	stdSE	SpearmanR
0.467	421208.874	59099.555	1212057.852	0.366

4   Conclusion

A new deep ensemble–based framework was developed for automated segmentation of brain tumor regions from multi-sequence MR images. The concepts of Dilated convolutions, Spatial pooling, Residual networks and Inception networks were incorporated via ensembling of three models. Dilation helped enhance the receptive field, spatial pooling enabled more accurate segmentation, Residual network allowed in training of deeper networks, and while Inception network let the model decide what representations it wanted to learn by providing various pathways. We have currently implemented all three models along one plane and believe that their applications across different planes can surely improve the performance of the ensembled model.

Acknowledgment. We gratefully acknowledge the support of Intel Corporation for providing access to the Intel AI DevCloud platform used in this work.

S. Banerjee acknowledges the support provided to him by the Intel Corporation, through the Intel AI Student Ambassador Program.

This publication is an outcome of the R&D work undertaken project under the Visvesvaraya PhD Scheme of Ministry of Electronics & Information Technology, Government of India, being implemented by Digital India Corporation.

References

1. Bakas, S., et al.: Segmentation labels and radiomic features for the pre-operative scans of the TCGA-GBM collection. Cancer Imaging Arch. (2017). https://doi.org/10.7937/K9/TCIA.2017.KLXWJJ1Q
2. Bakas, S., et al.: Identifying the best machine learning algorithms for brain tumor segmentation, progression assessment, and overall survival prediction in the BraTS challenge. arXiv preprint arXiv:1811.02629 (2018)
3. Bakas, S., Akbari, H., et al.: Advancing the cancer genome atlas glioma MRI collections with expert segmentation labels and radiomic features. Sci. Data **4**, 170117 (2017)
4. Bakas, S., Bakas, S., et al.: Segmentation labels and radiomic features for the pre-operative scans of the TCGA-LGG collection. Cancer Imaging Arch. (2017). https://doi.org/10.7937/K9/TCIA.2017.GJQ7R0EF
5. Banerjee, S., Mitra, S., Uma Shankar, B.: Single seed delineation of brain tumor using multi-thresholding. Inf. Sci. **330**, 88–103 (2016)
6. Banerjee, S., Mitra, S., Uma Shankar, B.: Synergetic neuro-fuzzy feature selection and classification of brain tumors. In: Proceedings of IEEE International Conference on Fuzzy Systems (FUZZ-IEEE), pp. 1–6 (2017)
7. Banerjee, S., Mitra, S., Uma Shankar, B.: Automated 3D segmentation of brain tumor using visual saliency. Inf. Sci. **424**, 337–353 (2018)
8. Banerjee, S., Mitra, S., Uma Shankar, B., Hayashi, Y.: A novel GBM saliency detection model using multi-channel MRI. PLoS ONE **11**(1), e0146388 (2016)
9. Banerjee, S., Mitra, S., Shankar, B.U.: Multi-planar spatial-ConvNet for segmentation and survival prediction in brain cancer. In: Crimi, A., Bakas, S., Kuijf, H., Keyvan, F., Reyes, M., van Walsum, T. (eds.) BrainLes 2018. LNCS, vol. 11384, pp. 94–104. Springer, Cham (2019). https://doi.org/10.1007/978-3-030-11726-9_9
10. Dextraze, K., et al.: Spatial habitats from multiparametric MR imaging are associated with signaling pathway activities and survival in glioblastoma. Oncotarget **8**(68), 112992 (2017)
11. Doshi, J., Erus, G., Habes, M., Davatzikos, C.: DeepMRSeg: a convolutional deep neural network for anatomy and abnormality segmentation on MR images. arXiv preprint arXiv:1907.02110 (2019)
12. Gillies, R.J., Kinahan, P.E., Hricak, H.: Radiomics: images are more than pictures, they are data. Radiology **278**, 563–577 (2015)
13. Menze, B.H., Menze, B.H., et al.: The multimodal Brain Tumor image Segmentation benchmark (BraTS). IEEE Trans. Med. Imaging **34**, 1993–2024 (2015)
14. Mitra, S., Banerjee, S., Hayashi, Y.: Volumetric brain tumour detection from MRI using visual saliency. PLoS ONE **12**, 1–14 (2017)
15. Ronneberger, O., Fischer, P., Brox, T.: U-Net: convolutional networks for biomedical image segmentation. In: Navab, N., Hornegger, J., Wells, W.M., Frangi, A.F. (eds.) MICCAI 2015. LNCS, vol. 9351, pp. 234–241. Springer, Cham (2015). https://doi.org/10.1007/978-3-319-24574-4_28

16. Sudre, C.H., Li, W., Vercauteren, T., Ourselin, S., Jorge Cardoso, M.: Generalised dice overlap as a deep learning loss function for highly unbalanced segmentations. In: Cardoso, M.J., et al. (eds.) DLMIA/ML-CDS -2017. LNCS, vol. 10553, pp. 240–248. Springer, Cham (2017). https://doi.org/10.1007/978-3-319-67558-9_28
17. Wang, G., Wang, G., et al.: Interactive medical image segmentation using deep learning with image-specific fine tuning. IEEE Trans. Med. Imaging **37**(7), 1562–1573 (2018)
18. Zhou, M., Scott, J., et al.: Radiomics in brain tumor: image assessment, quantitative feature descriptors, and machine-learning approaches. Am. J. Neuroradiol. **39**, 208–216 (2017)

Brain Tumor Segmentation Based on Attention Mechanism and Multi-model Fusion

Xutao Guo[1], Chushu Yang[1], Ting Ma[1,2,3,4(✉)], Pengzheng Zhou[1], Shangfeng Lu[1], Nan Ji[5], Deling Li[5], Tong Wang[1], and Haiyan Lv[6]

[1] Department of Electronic and Information Engineering, Harbin Institute of Technology at Shenzhen, Shenzhen, Guangdong, China
tmahit@outlook.com
[2] Advanced Innovation Center for Human Brain Protection, Capital Medical University, Beijing, China
[3] National Clinical Research Center for Geriatric Disorders, Xuanwu Hospital, Capital Medical University, Beijing, China
[4] Peng Cheng Laboratory, Shenzhen, Guangdong, China
[5] Department of Neurosurgery China National Clinical Research Center for Neurological Diseases, Beijing Tiantan Hospital, Capital Medical University, Beijing, China
[6] Mindsgo Life Science Shenzhen Ltd Member, IEEE, Haiyan Lv6, Shenzhen, China

Abstract. Brain tumor are uncontrollable and abnormal cells in the brain. The incidence and mortality of brain tumors are very high. Among them, gliomas are the most common primary malignant tumors with different degrees of invasion. The segmentation of brain tumors is a prerequisite for disease diagnosis, surgical planning and prognosis. According to the characteristics of brain tumor data, we designed a multi-model fusion brain tumor automatic segmentation algorithm based on attention mechanism [1]. Our network architecture is slightly modified based on 3D U-Net [2]. At the same time, the attention mechanism was added to the 3D U-Net model. According to the patch size and attention mechanism in the training process, four independent networks are designed. Here, we use $64 \times 64 \times 64$ and $128 \times 128 \times 128$ patch sizes to train different sub-networks. Finally, the results of the four models in the label layer are combined to get the final segmentation results. This multi model fusion method can effectively improve the robustness of the algorithm. At the same time, the attention method can improve the feature extraction ability of the network and improve the segmentation accuracy. Our experimental study on the newly released brats data set (brats 2019) shows that our method accurately describes brain tumors.

Keywords: Brain tumor · Attention mechanism · Segmentation · U-Net · CNN

1 Introduction

Brain tumors are one of the deadliest cancers. Among them, glioma is the most common tumor. According to its invasiveness, it can be divided into high-grade glioma (HGG) and low-grade glioma (LGG). High grade glioma (HGG) is a kind of malignant

A. Crimi and S. Bakas (Eds.): BrainLes 2019, LNCS 11993, pp. 50–60, 2020.
https://doi.org/10.1007/978-3-030-46643-5_5

and aggressive tumor. It often leads to necrotic nucleus with edema and swelling around the tumor [3]. Even with the best treatment available, the average survival rate is less than two years. Low grade glioma (LGG) grows slowly. However, it is also possible to develop into HGG [4]. Therefore, it is necessary to pay attention to LGG.

The automatic segmentation algorithm provides fast, accurate and objective segmentation results. Traditional methods have made some important contributions, such as outlier detection method based on atlas [5] and fuzzy region growth method [6, 7]. But in recent years, they have not made a breakthrough. With the development of deep learning methods, more and more fields try to solve practical problems through deep learning methods. In recent years, CNN has made amazing achievements in detection, classification and segmentation tasks. It shows a broad application prospect. In the field of brain tumor segmentation, full convolution neural network has achieved remarkable results. The U-Net model based on encoding/decoding structure has been widely used. At present, most of the networks used for segmentation are based on the improvement of U-Net structure. At the same time, 3D convolution is widely used in multimodal brain tumor segmentation. For the 3D voxel data of multimodal brain tumors, 2D convolution can not effectively extract the information between slices. In order to meet the limit of storage space, the current processing methods are based on patch. Recent work has shown that large patch can help the network extract context information effectively. Thus, the accuracy of network segmentation is improved.

1.1 Contribution

In this paper, a multi-model fusion brain tumor automatic segmentation algorithm based on attention mechanism is designed. Our network architecture is slightly modified based on 3D U-Net. At the same time, attention mechanism is added to the model. The results of multiple models are fused to get the final results. This multi model fusion method can effectively improve the robustness of the algorithm.

The contribution of this paper can be summarized as follows:

- Attention mechanism is introduced into image semantic segmentation and applied to 3D u-nets model. This method can improve the feature extraction ability of the network and the segmentation accuracy of the network.
- We split the original data into $64 \times 64 \times 64$ and $128 \times 128 \times 128$ sizes. Then, multiple network models are trained based on the above two patch sizes. This method provides two different information scales for model fusion. This method can effectively use context information and improve the robustness of the algorithm.
- According to the patch size and attention mechanism in the training process, four independent networks are designed. The results of the four models in the label layer are combined to get the final segmentation results. This multi model fusion method can effectively improve the robustness of the algorithm.
- We validated our technology on the latest released brain tumor segmentation dataset (brats 2019) and showed that they provide high quality segmentation.

The organizational structure of this paper is as follows. In Sect. 2, we discuss techniques related to brain tumor segmentation. Our method is described in detail in Sect. 3. Our experimental results were analyzed in Sect. 4. The fifth part is the discussion of the paper. This part focuses on the analysis of the model based on the results of the experiment.

2   Related Literature

2.1   Full Convolutional Neural Network

The automatic segmentation algorithm of brain tumor can be divided into supervised learning, semi supervised learning, unsupervised learning and hybrid learning. At present, the mainstream method based on supervised learning has the highest accuracy. Deep neural network has established the most advanced technology in a large number of image processing and image recognition tasks and successfully segmented different types of brain tissue. Full convolution network (FCN) [8] is regarded as a milestone in the development of convolution neural network (CNN). FCN can classify the image pixel by pixel, which solves the problem of image semantic level segmentation. It can segment any size image. Based on FCN architecture, u-net extends the network to achieve higher segmentation accuracy with fewer data sets. U-net uses a encoder/decoder structure. In the encoding part, the traditional feature extraction network is used, such as VGG, ResNet and so on. In the decoding process, u-net combines shallow texture features and high-level semantic features to achieve high segmentation accuracy. Our network architecture is a slightly modified 3D u-net example. The overall model structure is shown in Fig. 1.

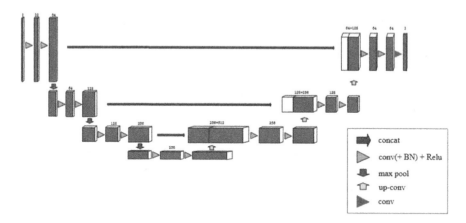

Fig. 1. The 3D U-Net architecture. Blue boxes represent feature maps. The number of channels is denoted above each feature map. (Color figure online)

2.2   Attention Mechanism

In the last two years, the Attention Model has been widely used in various types of deep learning tasks such as natural language processing, image recognition and speech recognition. It is the technology of deep learning technology that deserves the most attention and deep understanding. It mainly imitates human visual attention mechanism. This method focuses on regions of interest and weakens the focus on general regions.

This approach evolved into two different attention mechanisms. One is soft attention mechanism. The other is hard attention mechanism.

The key to soft attention is to pay more attention to spatial location or channel. Soft attention is a deterministic attention mechanism. After learning, it can be directly generated through the network. The key is learnability. Attention is derivative, which is a very important place. Attention mechanism can learn the weight of attention by calculating gradient and feedbacks through neural network.

The difference between strong attention and soft attention is that hard attention is more concerned. Every point in the image can be noticed. Strong attention is a random prediction process. Of course, most importantly, intense attention is a non different focus. The training process is often completed through reinforcement learning.

Three types of attention can be generated by changing the normalization of the activation function and adding different constraints to the Attention in the mask before the mask is output.

- channel attention: The value of the spatial position corresponding to each channel is directly normalized using the L2 normal form, and the spatial information is removed.
- spatial attention: Normalize the feature maps of all channels, then use the sigmoid function to obtain attention weights with only spatial information.
- mixed attention: Use the sigmoid function directly for all channel spatial locations without adding other constraints.

$$f_1(x_{i,c}) = \frac{1}{1 + \exp(-x_{i,c})} \tag{1}$$

$$f_2(x_{i,c}) = \frac{x_{i,c}}{\|x_i\|} \tag{2}$$

$$f_3(x_{i,c}) = \frac{1}{1 + \exp(-(x_{i,c} - mean_c)/std_c)} \tag{3}$$

Where i ranges over all spatial positions and c ranges overall channels. $mean_c$ and std_c denotes the mean value and standard deviation of feature map from c-th channel x_i denotes the feature vector at the i-th spatial position.

3   Method

3.1   Network Architecture

The 3D U-Net architecture is capable of combining different scale features to achieve accurate segmentation. Our network architecture is an example of 3D U-Net with minor modifications. At the same time, the attention mechanism was added to the 3D U-Net model. Four independent networks were designed based on the size of the patch during training and the mechanism of attention.

The backbone network used in this study is a five-layer VGG network. Due to the limitation of the size of the memory, we use the patch to train the network. The size of the patch is $64 \times 64 \times 64$ and $128 \times 128 \times 128$. The two patch sizes provide two different receptive fields. When the network is fused, the context information can be effectively used to improve the segmentation accuracy. At the same time, the attention mechanism was added to the 3D U-Net model. Four independent networks were designed based on the size of the patch during training and the mechanism of attention. When the patch size of the input data is $64 \times 64 \times 64$, the number of the first layer output feature channels is 32, the number of the second layer output feature channels is 64, and the number of the third layer output feature channels is 128. The fourth layer outputs characteristic channels of 256, and the fifth layer outputs characteristic channels of 512. Each layer of the network has a normalization operation, as well as a relu activation function, a $2 \times 2 \times 2$ pool layer. When the patch size of the input data is $128 \times 128 \times 128$, the number of the first layer output feature channel is 8, the second layer output feature channel number is 16, and the third layer output feature channel number is 32. The fourth layer outputs characteristic channels of 64, and the fifth layer outputs characteristic channels of 128. The size of the convolution kernel in the network is set to $3 \times 3 \times 3$ and the step size is set to 2. Each layer of the network has a normalization operation and a relu activation function, followed by a $2 \times 2 \times 2$ pool layer. The network decoding process is implemented by deconvolution.

3.2   Attention Model

The attention module is composed of a spatial attention sub-module and a channel attention sub-module, which respectively correspond to the two branches in Fig. 2. For the channel attention module, the input characteristic graph with the input size of $H \times W \times L \times C$ is first transformed into the characteristic graph with the size of $1 \times 1 \times 1 \times C$ through the global pooling layer. The module compresses the spatial direction of the feature map into one dimension. Then the feature map is passed through the sigmoid layer, so that the active value range of the feature map is $[0, 1]$. Finally, the output characteristic map is multiplied by the original input characteristic map to get the characteristic map with the size of $H \times W \times L \times C$. The channel attention module generates weights for each feature channel. The parameters are learned to explicitly model the correlation between feature channels. The multiplication of the weight output through the sigmoid layer and the original input graph can be regarded as the weight of different channels of the graph. It improves the target related channel weight and suppresses the uncorrelated channel weight. For the spatial attention module, the convolution layer with convolution kernel size of $3 \times 3 \times 3$ and step size of 1 is used to convolute with the characteristic graph with input size of $H \times W \times L \times C$. The output size is $H \times W \times L \times 1$. It compresses the channel direction of the graph into one dimension. Then, the output characteristic map is passed through the sigmoid layer, so that the activation value range of the characteristic map is $[0, 1]$. Finally, the output characteristic map is multiplied by the original input characteristic map to get the characteristic map with the size of $H \times W \times L \times C$. Spatial attention gives different weight values to different points of spatial position in the feature map, which makes the spatial position points related to the target get larger weight and reduces the weight of the irrelevant spatial position points.

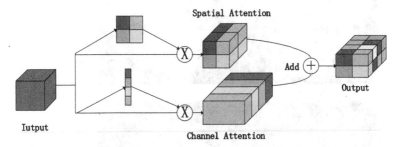

Fig. 2. Attention module. The upper branch in the figure is spatial attention, and the lower branch is channel attention.

4   Experimental Validation

4.1   Data

We use brain tumor data from the 2019 International MICCAI BraTS Challenge to validate our algorithm. The Brats public data set was established after two workshops at MICCAI 2012 and 2013 to evaluate the benchmarks for brain tumor image classification and segmentation algorithms. The data set is updated every year, with Brats2019 [9–13] including 259 high-grade gliomas (HGG) and 76 low-grade gliomas (LGG) and corresponding tissue segmentation images. The dataset was acquired from 19 institutions with different protocols and MRI scanners. Each case is provided with a T1-weighted (T1), a post-contrast T1-weighted (T1Gd), T2-weighted (T2) and a T2 Fluid Attenuated Inversion Recovery (FLAIR) volumes. All the sequences are skull-striped and re-sampled to an isotropic resolution of 1 mm * 1 mm * 1 mm. In addition, each case had been segmented manually by one to four raters and all annotations had been approved by experienced neuro-radiologists. The annotations consist of the four following labels: GD-enhancing tumor (ET–label 4), peritumoral edema (ED–label 2), necrotic and non-enhancing tumor (NCR/NET – label 1) and background (label 0). The ET is described by areas that show hyper-intensity in T1Gd when compared to T1, but also when compared to "healthy" white matter in T1Gd. The tumor core (TC) describes the bulk of the tumor, which is what is typically resected. The TC entails the ET, as well as the necrotic (fluid-filled) and the non enhancing (solid) parts of the tumor. The appearance of the necrotic (NCR) and the non-enhancing (NET) tumor core is typically hypo-intense in T1-Gd when compared to T1. The whole tumor (WT) describes the complete extent of the disease, as it entails the TC and the peritumoral edema (ED), which is typically depicted by hyper-intense signal in FLAIR. The labels in the provided data are: 1 for NCR & NET, 2 for ED, 4 for ET, and 0 for everything else [9, 10] (Fig. 3).

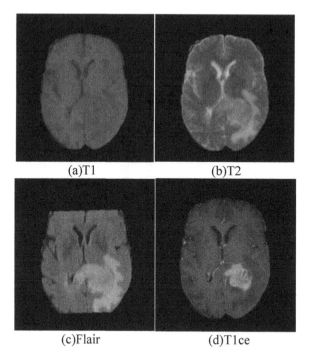

(a)T1   (b)T2

(c)Flair   (d)T1ce

Fig. 3. A brain tumor example ("BraTS19_2013_18_1") in BraTS 2019 dataset. (a–d) show four slices with the same position (77th slice) in different MRI scans.

4.2 Experimental Setup

Data normalization is critical given that the data comes from different organizations and MRI scanners with different protocols and is processed in the same network. Here, we use z-score to normalize, each value has an average of 0 and a standard deviation of 1. The original data is divided into $64 \times 64 \times 64$ and $128 \times 128 \times 128$ to train different sub-networks. At the same time, considering the imbalance of sample distribution, only the patch that the number of lesions is greater than a certain threshold is used here. The threshold is set to 30 when the size of the patch is $64 \times 64 \times 64$. The threshold is set to 60 when the size of the patch is $128 \times 128 \times 128$. This method alleviates the problem of uneven distribution of samples to some extent.

We built our network using PyTorch and trained it on 12G RAM on the Nvidia 1080Ti GPU. The batch size is set to 4. Compared with others, the initial learning rate is 1e−5, which is very small. Once the loss increases, the learning rate is set to 2–5 times the current. An adaptive moment estimate (Adam) with 1e−4 weight attenuation is added to our optimizer. Considering the imbalance of the labels, the loss function we use is the weighted cross entropy loss, and its weight is assigned to the background: necrosis: edema: the order of enhancement is 1:5:2:3.

4.3   Experimental Results

The segmentation performance was quantitatively evaluated using the dice coefficient (DC) of each person. In this paper, we calculated the dice coefficients of whole tumor (including edema, necrotic and non-enhancing tumor and enhancing tumor), tumor core (including necrotic and non-enhancing tumor and enhancing tumor) and enhancing tumor.

Table 1 presents quantitative evaluation of our segmentation in our testing set. The result of this table is that we randomly divide the BraTS_2019_Data_Training into a training set and a test set with a ratio of 8:2. And the test set does not participate in the training. Due to time, we only show the results of the Dice assessment in Table 1, but we have been able to demonstrate the advantages of our results. Among them, Crop_64 corresponds to the performance of a single network segmentation when the patch size is 64 × 64 × 64, and Crop_64_attention corresponds to the performance of a single network segmentation with attention mechanism when the patch size is 64 × 64 × 64. Crop_128 corresponds to the performance of a single network segment when the patch size is 128 × 128 × 128, and Crop_128_attention corresponds to the performance of a single network segmentation with attention mechanism when the patch size is 128 × 128 × 128. The last one is the result of the fusion of these four models.

Table 1. Mean values of dice coefficient on our testing set.

Dice	Whole tumor (WT)	Tumor core (TC)	Enhanced tumor (ET)
Crop_64	0.8672	0.8435	0.8330
Crop_64_attention	0.8596	0.8772	0.8165
Crop_128	0.8854	0.8461	0.8291
Crop_128_attention	0.8832	0.8524	0.8024
Ensemble	0.8846	0.8631	0.8301

As can be seen from Table 1, when the size of patch is 128 × 128 × 128, the segmentation accuracy of the network is higher than that of the size of patch is 64 × 64 × 64. The introduction of attention mechanism only improves the accuracy of some types of tumors, and the overall improvement will not be too great. When the size of patch is 64 × 64 × 64, attention mechanism is introduced to improve the segmentation accuracy of tumor core, but the segmentation accuracy of whole tumor and enhanced tumor decreases. When the size of patch is 128 × 128 × 128, attention mechanism is introduced to improve the segmentation accuracy of tumor core, but the segmentation accuracy of whole tumor and enhanced tumor decreases. Although the introduction of attention mechanism does not bring the overall performance improvement, it can expand the diversity of models and improve the overall segmentation accuracy in model fusion. The model trained based on multiple patch sizes can utilize the information of multiple scales and effectively utilize the context information of data. As shown in the table, our fusion model achieved averaged dice coefficients (DC) of 0.8846, 0.8631, 0.8301 for whole tumor (WT), tumor core (TC), and enhancing tumor (ET), respectively in the our testing set (Table 3).

Table 2. Evaluation result of our model for segmentation on the BraTS_2019_Data_Validation

	Dice ET	Dice WT	Dice TC	Sensitivity ET	Sensitivity WT	Sensitivity TC
Mean	0.677	0.872	0.728	0.755	0.872	0.697
StdDev	0.307	0.118	0.291	0.244	0.156	0.319
Median	0.816	0.904	0.859	0.823	0.925	0.867
25quantile	0.657	0.863	0.667	0.686	0.835	0.523
75quantile	0.870	0.933	0.926	0.917	0.967	0.934

Table 3. Evaluation result of our model for segmentation on the BraTS_2019_Data_Validation

	Specificity ET	Specifiity WT	Specifiity TC	Hausdorff95 ET	Hausdorff95 WT	Hausdorff95 TC
Mean	0.998	0.993	0.998	6.209	5.408	7.740
StdDev	0.002	0.005	0.001	10.991	5.251	8.643
Median	0.998	0.996	0.999	2.236	4	4.242
25quantile	0.997	0.991	0.997	1.485	2.236	2.236
75quantile	0.999	0.997	0.999	5.099	6	9.874

Some of the segmentation results are presented in Fig. 4. Here we randomly select slices from two different people. For simplification, we only display ground truth and our segmentation results.

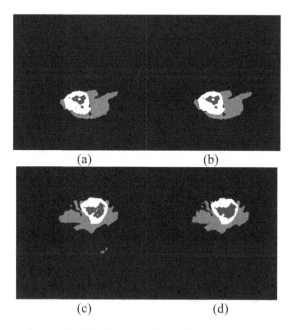

(a)   (b)

(c)   (d)

Fig. 4. Our segmentation results. The columns from left to right represent our segmentation results and ground truth

Our network is also evaluated online on the BraTS_2019_Data_Validation dataset. We prepared for this challenge later. Due to time constraints, only one online evaluation was conducted, and some cases where the submitted data was illegal were invalid, resulting in poor results on the verification set. We will modify our model in time, and get better results after the release of the test set. Table 2 shows the results of our model for val processing, online evaluation.

5   Conclusions

We designed a multi-model fusion brain tumor automatic segmentation algorithm based on attention mechanism. Our network architecture is an example of 3D U-Net with minor modifications. At the same time, the attention mechanism was added to the 3D U-Net model. The results of multiple models are fused to get the final result. This multi-model fusion method can effectively improve the robustness of the algorithm. First, we cut the raw data into $64 \times 64 \times 64$ and $128 \times 128 \times 128$ sizes. Then, we train multiple network models based on the above two patch sizes. Finally, the results of multiple models are fused to get the final result. This multi-model fusion method can effectively improve the robust-ness of the algorithm. At the same time, we add attention mechanism to the 3D U-Nets model. The introduction of attention mechanism only improves the accuracy of some types of tumors, and the overall improvement will not be too great. Although the introduction of attention mechanism does not bring the overall performance improvement, it can expand the diversity of models and improve the overall segmentation accuracy in model fusion. The model trained based on multiple patch sizes can utilize the information of multiple scales and effectively utilize the context information of data.

Acknowledgment. This study was supported by the National Key Research and Development Program of China (2018YFC1312000), The Basic Research Foundation Key Project Track of Shenzhen Science and Technology Program (JCYJ20160509162237418, JCYJ20170413110 656460).

References

1. Vaswani, A., et al.: Attention is all you need. In: Advances in Neural Information Processing Systems (2017)
2. Ronneberger, O., Fischer, P., Brox, T.: U-Net: convolutional networks for biomedical image segmentation (2015). CoRR, abs/1505.04597. http://arxiv.org/abs/1505.04597
3. Bauer, S., Wiest, R., Nolte, L.P., Reyes, M.: A survey of MRI-based medical image analysis for brain tumor studies. Phys. Med. Biol. **58**(13), R97 (2013)
4. Louis, D., et al.: The 2016 world health organization classification of tumors of the central nervous system: a summary. Acta Neuropathol. **131**(6), 803–820 (2016)
5. Prastawa, M., Bullitt, E., Ho, S., Gerig, G.: A brain tumor segmentation framework based on outlier detection. Med. Image Anal. **8**(3), 275–283 (2004)
6. Udupa, J.K., et al.: Multiple sclerosis lesion quantification using fuzzy-connectedness principles. IEEE Trans. Med. Imag. **16**(5), 598–609 (1997)

7. Udupa, J.K., et al.: Relative fuzzy connectedness and object definition: theory, algorithms, and application in image segmentation. IEEE Trans. Pattern Anal. Mach. Intell. **24**(11), 1485–1500 (2002)

8. Long, J., Shelhamer, E., Darrell, T.: Fully convolutional networks for semantic segmentation. In: 2015 IEEE Conference on Computer Vision and Pattern Recognition (CVPR), Boston, MA, pp. 3431–3440 (2015)

9. Menze, B.H., Jakab, A., Bauer, S., Kalpathy-Cramer, J., Farahani, K., Kirby, J., et al.: The multimodal brain tumor image segmentation benchmark (BRATS). IEEE Trans. Med. Imaging **34**(10), 1993–2024 (2015). https://doi.org/10.1109/TMI.2014.2377694

10. Bakas, S., Akbari, H., Sotiras, A., Bilello, M., Rozycki, M., Kirby, J.S., et al.: Advancing The Cancer Genome Atlas glioma MRI collections with expert segmentation labels and radiomic features. Nat. Sci. Data **4**, 170117 (2017). https://doi.org/10.1038/sdata.2017.117

11. Bakas, S., Reyes, M., Jakab, A., Bauer, S., Rempfler, M., Crimi, A., et al.: Identifying the best machine learning algorithms for brain tumor segmentation, progression assessment, and overall survival prediction in the BRATS challenge. arXiv preprint arXiv:1811.02629 (2018)

12. Bakas, S., Akbari, H., Sotiras, A., Bilello, M., Rozycki, M., Kirby, J., et al.: Segmentation labels and radiomic features for the pre-operative scans of the TCGA-GBM collection. Cancer Imaging Arch. (2017). https://doi.org/10.7937/K9/TCIA.2017.KLXWJJ1Q

13. Bakas, S., Akbari, H., Sotiras, A., Bilello, M., Rozycki, M., Kirby, J., et al.: Segmentation labels and radiomic features for the pre-operative scans of the TCGA-LGG collection. Cancer Imaging Arch. (2017). https://doi.org/10.7937/K9/TCIA.2017.GJQ7R0EF

Automatic Brain Tumour Segmentation and Biophysics-Guided Survival Prediction

Shuo Wang[1(✉)], Chengliang Dai[1], Yuanhan Mo[1], Elsa Angelini[2],
Yike Guo[1], and Wenjia Bai[1,3]

[1] Data Science Institute, Imperial College London, London, UK
{shuo.wang,c.dai}@imperial.ac.uk
[2] ITMAT Data Science Group, Imperial College London, London, UK
[3] Department of Brain Sciences, Imperial College London, London, UK

Abstract. Gliomas are the most common malignant brain tumours with intrinsic heterogeneity. Accurate segmentation of gliomas and their sub-regions on multi-parametric magnetic resonance images (mpMRI) is of great clinical importance, which defines tumour size, shape and appearance and provides abundant information for preoperative diagnosis, treatment planning and survival prediction. Recent developments on deep learning have significantly improved the performance of automated medical image segmentation. In this paper, we compare several state-of-the-art convolutional neural network models for brain tumour image segmentation. Based on the ensembled segmentation, we present a biophysics-guided prognostic model for patient overall survival prediction which outperforms a data-driven radiomics approach. Our method won the second place of the MICCAI 2019 BraTS Challenge for the overall survival prediction.

1 Introduction

Gliomas are the most common malignant brain tumours in adults, characterised by intrinsic heterogeneity and dismal prognostics [15]. Sub-regions with various biological properties coexist within the tumour and cause inconsistent treatment response. Multi-parametric magnetic resonance imaging (mpMRI) provides valuable information for characterising gliomas and their sub-regions, such as necrosis (NCR), non-enhancing tumour (NET), enhancing tumour (ET) and peritumoural edema (ED). Imaging phenotypes of these sub-regions show great potential in patient stratification [10]. However, due to the highly heterogeneous shape and appearance, accurate segmentation of the tumour sub-regions requires expertise from experienced radiologists.

S. Wang and C. Dai—The two authors contributed equally to this paper.

© Springer Nature Switzerland AG 2020
A. Crimi and S. Bakas (Eds.): BrainLes 2019, LNCS 11993, pp. 61–72, 2020.
https://doi.org/10.1007/978-3-030-46643-5_6

Automatic segmentation of brain tumour has drawn a lot of attention in the recent years due to the availability of open medical image datasets and the rapid development of convolutional neural networks (CNNs). A well-trained CNN model can finish the segmentation task in minutes with acceptable accuracy. However, there are still a few challenges including limited manually-annotated training data, variations in image acquisition protocols and MRI scanners etc. Apart from challenges for image segmentation, another challenge lies in building a robust prognostic model from the high-dimensional medical image phenotypes. Data-driven radiomics approach has demonstrated promising results while the reproducibility and explainability are still questionable [16].

To push the boundaries of automatic segmentation and survival prediction, Multimodal Brain Tumour Segmentation Challenge (BraTS) has been organised for the recent few years [1–4,13]. The BraTS datasets consist of mpMRI scans for glioblastoma (GBM/HGG) and low grade glioma (LGG). The modalities include T1-weighted scan (T1), post-contrast T1-weighted scan (T1Gd), T2-weighted scan (T2) and T2 Fluid Attenuated Inversion Recovery (T2-FLAIR) scan. These scans were acquired preoperatively with different clinical protocols from multiple institutions and annotated by experienced radiologists.

In this paper, we compare several different neural network models for brain tumour image segmentation on the BraTS 2019 dataset and investigate the influence of attention units, loss function and post-processing on segmentation performance. Based on the ensembled segmentation, we develop a robust biophysics-guided model for survival prediction.

2   Methods

In this section, we first present our segmentation models. Based on the tumour sub-region segmentation, we define a series of tumour features, perform feature selection and finally propose a prognostic model.

2.1   Tumour Sub-region Segmentation

Background: Winning Methods in BraTS 2017 and 2018. UNet and UNet-like models are adopted by most of the top participants in BraTS 2017 and 2018. The method described in [14] won the first place in BraTS 2018 with a UNet architecture plus an additional decoder branch derived from a generic variational autoencoder (VAE). Given this modified network structure, the loss function used by [14] consists of a soft Dice loss for the segmenter branch, the KL divergence loss and reconstruction loss for the reconstruction branch. A patch size of $160 \times 192 \times 128$ was chosen to make the most use of the Nvidia Tesla V100 graphic card with 32 GB GPU memory. The second place of BraTS 2018 used a vanilla UNet with minor modifications tailored for the BraTS dataset. The network was trained with both the BraTS dataset and an auxiliary dataset from their own institution to improve the Dice score of enhanced tumour region.

The winner of BraTS 2017 proposed a scheme that ensembles DeepMedic, FCN, and UNet models to minimize the bias introduced by using each single model and to improve the segmentation robustness [7]. For the second place in BraTS 2017, [19] adopted a cascaded training scheme with 3 similar CNNs, each segmenting one of three tumour sub-regions. Each subsequent network takes the cropped output from the previous network as the input. The subsequent network is trained to segment a different tumour sub-region from the input. Most well performing submissions also adopted the ensemble learning method to minimize the bias introduced by different network architectures, hyper-parameters and loss functions.

Proposed Network Architectures. Given the good performance of UNet in the previous BraTS challenges, we chose the vanilla UNet as one of the architecture used in this work, shown in Fig. 1. Unlike many previous submissions that used transposed convolution block in the decoder, we use linear upsampling block to reduce the number of parameters and save GPU memory for training. For activation function we empirically chose group normalisation.

The second architecture we used is a UNet with attention blocks (UNet-AT), shown in Fig. 2. The attention mechanism has been demonstrated to be effective in improving the network performance across different tasks [18,20]. Our attention UNet model leverages the ability of attention blocks to concentrate on components that are more informative to achieve a better segmentation performance. Instance normalisation was empirically chosen for the attention UNet. Both UNet and UNet-AT used leaky ReLU with a leakiness of 0.01 as the activation function.

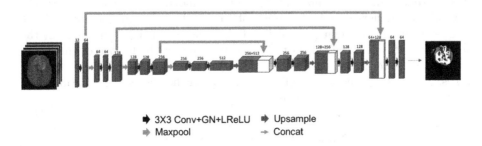

Fig. 1. 3D UNet for tumour segmentation.

Image Pre-processing and Augmentation. It was pointed out in [7] that different image normalisation methods have not shown significant impact on segmentation performance. Therefore we simply applied z-score normalisation onto the raw images. Random image rotation and horizontal flipping were performed for data augmentation during training.

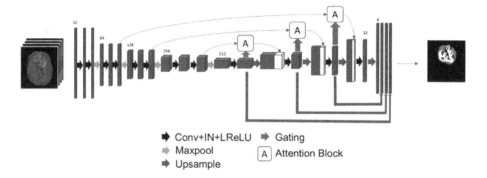

Fig. 2. 3D UNet with attention gates for tumour segmentation.

Ensemble of Network Models. We trained a number of different models (Table 1) for building the ensemble, using different loss functions (cross entropy or soft Dice) and post-processing steps (with or without conditional random field (CRF)). Due to the constraint of GPU memory, it was not possible for us to use the whole 3D volume as the input to any of our networks, so patch extraction was used for creating the training samples. The patches were randomly extracted with a 50% chance from background and the other 50% chance from any of the tumour sub-regions. Adam optimizer was used for training the models with the learning rate set to 10^{-4} and weight decay set to 10^{-5}. All the models were trained for 60,000 iterations and it took approximately 40 h to train a model on Nvidia Titan X.

Post-processing. We empirically adopted some automatic post-processing methods to further improve accuracy of the prediction from ensemble of the models, including removing small isolated whole/enhancing tumour from the prediction, adjusting the size of tumour core in line with the size of enhancing tumour and filtering based on the intensity distributions of different tumour labels. The post-processing significantly improved the Dice score and Hausdorff distance of enhancing tumour and tumour core.

Table 1. List of models for ensemble.

Model	Network structure	Patch size	Batch size	Loss function	CRF
1	UNet	96	4	Cross entropy	Yes
2	UNet	96	4	Cross entropy	No
3	UNet	96	4	Soft dice	Yes
4	UNet	96	4	Soft dice	No
5	UNet-AT	128	2	Soft dice	No
6	UNet-AT	128	2	Cross entropy	Yes

2.2   Image Feature Extraction

Based on the segmentation, we constructed a tumour structure map with four discrete values (1 for NCR and NET, 2 for ET, 3 for ED and 4 for normal tissue) for each patient. The spatial distribution of tumour sub-regions provides the information of tumour heterogeneity and tumour invasiveness [9]. Quantitative features were extracted from the tumour structure map for survival prediction (Fig. 3).

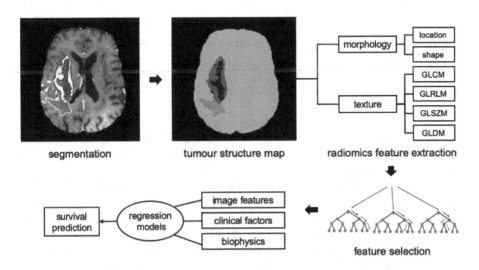

Fig. 3. Workflow of feature extraction and survival prediction.

81 radiomics features were extracted for each region-of-interest (ROI), consisting of 13 morphological features and 68 texture features. The morphological features describe the location and shape of the ROI, including X-, Y-, Z- coordinates of ROI centroid with respect to the brain centroid, volume, surface area, surface-area-to-volume ratio, sphericity, maximal 3D diameter, major axis length, minor axis length, least axis length, elongation and flatness. The textures features reveal the spatial distribution of tumour sub-regions within the ROI, which include 22 grey level occurrence matrix (GLCM) featuress, 16 gray level run length matrix (GLRLM) features, 16 grey level size zone matrix (GLSZM) features and 14 gray level dependence (GLDM) features. We extracted features for two ROIs, the whole tumour (WT) and the tumour core (TC), which amounted to 162 features in total. Feature extraction was implemented with Python package *PyRadiomics* [17].

Apart from radiomics features, we also considered the biophysics modelling of tumour growth [5]. The relative invasiveness coefficient (RIC) is of particular interest, which is defined as the extent ratio between the hypoxic tumour core and infiltration front according to the profile of tumour diffusion [12]. The characteristic extent of each ROI is calculated from the minimum volume ellipsoid.

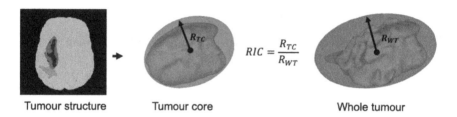

Tumour structure   **Tumour core**   **Whole tumour**

Fig. 4. Illustration of the calculation of relative invasiveness coefficient (RIC).

In this study, the ratio of the second semi-axis length between TC and WT was calculated as the RIC (Fig. 4).

2.3   Feature Selection

The large number of extracted radiomics features provide the capability for prognostic modelling, while it poses a risk of over-fitting in such a small training set. We apply feature selection techniques on the training set seeking a subset of radiomics features.

First, we detected highly correlated features (Pearson $r > 0.95$) and removed redundant features. The number of features was further reduced through a recursive feature elimination (RFE) scheme with a random forest regressor. The feature importance was evaluated and less important features were eliminated iteratively. The optimal number of features was determined by achieving the best performance on cross-validation results. The feature selection procedure was performed with the R package *caret* on the whole training data [8].

2.4   Prognostic Models

We compared three prognostic models, the baseline model which only uses age, the radiomics model and the biophysics-driven tumour invasiveness model.

Baseline Model. Age is the only available clinical factor and significantly correlated with the survival time (Pearson $r = -0.486$, $p < 1e-5$) on the training set. We constructed a linear regression model using age as the only predictor.

Radiomics Model. The selected radiomics features and age were integrated into a random forest (RF) model. This data-driven model included the largest number of image features.

Tumour Invasiveness Model. RIC derived from the tumour structure map was used to describe tumour invasiveness. An epsilon-support vector regression (ϵ-SVR) model was built using age and RIC as two predictors. SVR model was used as the number of features is small.

Classification and regression metrics were used to evaluate the prognostic performance. For classification, overall survival time was quantitised into three survival categories: short survival (<10 months), intermediate survival (10–15 months) and long survival (>15 months). The 3-class accuracy metric wass evaluated. Regression metrics include the mean squared error (MSE), median squared error (mSE) and Spearman correlation coefficient ρ.

3 Results

3.1 Segmentation Performance

The segmentation performance on BraTS 2019 validation dataset is reported in Table 3, in terms of Dice score and Hausdorff distance. Among each individual models, model 1 (UNet with cross entropy loss and CRF) achieves the highest Dice scores of whole tumour and enhanced tumour. For Dice score of tumour core, model 2 (UNet with soft Dice loss and CRF) gives the best result. The ensemble of all six models gives the best result across all metrics. The ensemble of models significantly improves the overall performance of the models. Although the performance of UNet-AT (models 5, 6) is not as good as UNet (models 1–4), but adding UNet-AT improves the overall performance of the ensemble.

An example showing improvement of using ensemble of models is given in Fig. 5.

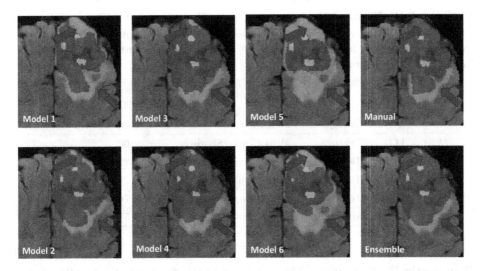

Fig. 5. Segmentation result of the brain tumour from a training image in BraTS 2019 segmented by 6 models and the ensemble model, where green depicts oedema, red depicts tumour core, and yellow depicts enhancing tumour. (Color figure online)

The performance of the ensemble of models on test dataset is given in Table 2.

Table 2. Segmentation result on validation set.

Model	Dice_ET	Dice_WT	Dice_TC	Hausdorff_ET	Hausdorff_WT	Hausdorff_TC
1	0.74	0.90	0.78	6.56	5.84	8.66
2	0.71	0.89	0.79	6.08	5.37	8.19
3	0.73	0.89	0.80	6.68	6.93	7.50
4	0.73	0.89	0.80	7.06	7.05	7.89
5	0.70	0.89	0.73	6.14	8.61	13.13
6	0.72	0.89	0.77	6.00	4.81	7.99
Ensemble	0.75	0.90	0.81	4.99	**4.70**	7.11
Post-processing	**0.79**	**0.90**	**0.83**	**3.37**	5.04	**5.56**

Table 3. Segmentation result on test set.

Label	Dice_ET	Dice_WT	Dice_TC	Hausdorff_ET	Hausdorff_WT	Hausdorff_TC
Mean	0.82	0.88	0.82	2.55	5.49	4.80
StdDev	0.18	0.12	0.26	4.53	7.19	8.38
Median	0.85	0.92	0.92	1.73	3.16	2.24

3.2   Survival Prediction

Subset of Radiomics Features. One morphological feature and four texture features were selected through RFE. The selected radiomics features were ranked according to feature importance (Table 4), which were included into the radiomics prognostic model.

Table 4. Selected radiomics features.

Rank	Region	Categories	Feature name	Importance (%)
1	TC	Texture	glcm_ClusterShade	100
2	WT	Texture	glcm_MaximumProbability	87
3	TC	Texture	glcm_SumSquares	79
4	TC	Texture	glszm_MaximumProbability	70
5	TC	Morphology	shape_center_Z	55

Prognostic Performance. The training set includes 101 patients with Gross Tumour Resection (GTR) and the validation set includes 29 patients. We first trained the models on the full training set and evaluated on the training set. This usually overestimates the real performance so we also repeated ten-fold cross-validation to assess the generalisation performance on unseen data. The training set was split into ten folds where models were trained on nine folds and evaluated on the hold-out fold. The training performance and cross-validation performance are reported in Table 5. Finally, we evaluated the performance of the trained models on the independent validation set (Table 6).

Table 5. Prognostic model performance on the training set.

Model	Training				Cross-validation			
	Accuracy	MSE	mSE	ρ	Accuracy	MSE	mSE	ρ
Baseline	0.48	88822	21135	0.48	0.48	**93203**	21923	0.47
Radiomics	**0.73**	**22249**	**6174**	**0.93**	0.47	103896	31265	0.37
Invasiveness	0.51	95728	17165	0.49	**0.50**	99707	**18218**	**0.47**

The radiomics model performed the best when tested on the same training data, reaching an accuracy of 0.73. However, the performance dropped significantly on the cross-validation results and independent validation set, highlighting the over-fitting problem. In contrast, the baseline model and invasiveness model show good capabilities of generalisation on unseen datasets. The linear fitting result of the baseline model and prediction error are shown in Fig. 6. Large errors are found for patients survived more than 1,000 days. The invasiveness model outperformed the baseline model with an accuracy of 0.50 on the cross-validation and 0.59 on the independent validation set. We chose the invasiveness model to submit for test set evaluation. The model won the second place of the MICCAI 2019 BraTS Challenge for the overall survival prediction task with an accuracy of 0.56. The leader board is available at https://www.med.upenn.edu/cbica/brats2019/rankings.html.

Table 6. Prognostic model performance on the independent validation set.

Model	Validation set			
	Accuracy	MSE	mSE	ρ
Baseline	0.45	90109	36453	0.27
Radiomics	0.48	97883	**29535**	0.28
Invasiness*	**0.59**	**89724**	36121	**0.36**

*This model is submitted for evaluation on test set.

4   Discussion

For tumour segmentation task, [6] demonstrated that the generic UNet with a set of carefully selected hyperparameters and a pre-processing method achieved the state-of-the-art performance on brain tumour segmentation. Our experiments results further confirm this. However, [6] adopted co-training scheme to further improve the UNet performance on segmenting enhancing tumour and our models also encountered some difficulties when segmenting enhancing tumour for some subjects. Including networks with different or more complex structure in our ensembles has been demonstrated to be effective to alleviate this issue to some extent.

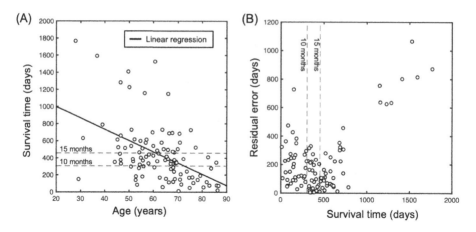

Fig. 6. Age-only linear regression results (A) and the distribution of residual error on the training set (B). The circle dots and straight line are the training data and prediction results respectively. Two dash lines represents the two thresholds for short-, mid- and long-survival categories.

The performance of our method on test dataset did not get into top three in the end. Judging from the Dice score and Hausdorff distance we achieved, we suspect including 6 models in the ensemble may not necessary and may introduce a higher false negative rate in the prediction of whole tumour.

For survival prediction, our invasiveness model achieved the best prognostic performance using age and RIC as predictors. Although more complex features were integrated in the radiomics model, it did not outperform the invasiveness model or even the baseline model on validation sets. This highlights the over-fitting risk of data-driven model and demonstrates the advantage of features from biophysics modelling. The RIC feature is designed from the diffusion equation of tumour growth, reflecting the physiological information of tumour infiltration. We note that RIC can be calculated from the 161 radiomics features but relevant features were not selected in the data-driven approach. This suggested that a better feature selection algorithm may be needed. However, it is a challenging task to select radiomics features due to several difficulties. First, the 'large n small p' problem is underdetermined and poses a risk of taking none-sense features by chance. On the other hand, high-order radiomics features are usually sensitive to the intensity distribution and image noise, limiting the performance on unseen data set. The reproducibility and robustness of radiomics feature should be examined in future studies. Moreover, the radiomics features are difficult to explain, which prevents the application in clinical practice. The biophysics modelling and other prior knowledge could be integrated in the feature design and selection schemes.

In this study, we assume that the boundary of WT represents the tumour infiltration front while the boundary of TC represents the active proliferation. Prognostic value of this image-derived feature was verified in the regression of survival time. However, the four structural modalities of MRI (e.g. T1-Gd, T1, T2 and T2-FLAIR) are not capable to reflect the physiological process. Advanced MRI modalities such as perfusion weighted imaging (PWI) and diffusion tensor imaging (DTI) could be used for better assessment of tumour heterogeneity and invasiveness [11].

It is also noted that all the three prognostic models achieved lower mSE than MSE, which indicates the skewed distribution of prediction error. The univariate linear regression fits well for the mid-survival patients, while large errors were observed for short- and long-survival patients (Fig. 6). Weighted loss or appropriate data transformation could be used to reduce the influence of long-tail survival distribution in future studies.

5 Conclusion

We have developed a deep learning framework for automatic brain tumour segmentation and a biophysics-guided prognostic model that performs well for overall survival prediction of patients.

Acknowledgement. This work was supported by the SmartHeart EPSRC Programme Grant (EP/P001009/1) and the NIHR Imperial Biomedical Research Centre (BRC).

References

1. Bakas, S., et al.: Segmentation labels and radiomic features for the pre-operative scans of the TCGA-GBM collection. The cancer imaging archive (2017)
2. Bakas, S., et al.: Segmentation labels and radiomic features for the pre-operative scans of the TCGA-LGG collection. The Cancer Imaging Archive 286/2017 (2017)
3. Bakas, S.: Advancing the cancer genome atlas Glioma MRI collections with expert segmentation labels and radiomic features. Sci. Data **4**, 170117 (2017)
4. Bakas, S., Reyes, M., Jakab, A., Bauer, S., Rempfler, M., et al.: Identifying the Best Machine Learning Algorithms for Brain Tumor Segmentation, Progression Assessment, and Overall Survival Prediction in the BraTS Challenge. arXiv:1811.02629 (2018)
5. Baldock, A.L., Ahn, S., Rockne, R., Johnston, S., Neal, M., et al.: Patient-specific metrics of invasiveness reveal significant prognostic benefit of resection in a predictable subset of gliomas. PLoS One **9**(10), e99057 (2014)
6. Isensee, F., Kickingereder, P., Wick, W., Bendszus, M., Maier-Hein, K.H.: No new-net. In: Crimi, A., Bakas, S., Kuijf, H., Keyvan, F., Reyes, M., van Walsum, T. (eds.) BrainLes 2018. LNCS, vol. 11384, pp. 234–244. Springer, Cham (2019). https://doi.org/10.1007/978-3-030-11726-9_21
7. Kamnitsas, K., et al.: Ensembles of multiple models and architectures for robust brain tumour segmentation. In: Crimi, A., Bakas, S., Kuijf, H., Menze, B., Reyes, M. (eds.) BrainLes 2017. LNCS, vol. 10670, pp. 450–462. Springer, Cham (2018). https://doi.org/10.1007/978-3-319-75238-9_38
8. Kuhn, M.: Building predictive models in R using the caret package. J. Stat. Softw. **28**(5), 1–26 (2008)

9. Li, C., Wang, S., Liu, P., Torheim, T., Boonzaier, N.R., et al.: Decoding the inter-dependence of multiparametric magnetic resonance imaging to reveal patient subgroups correlated with survivals. Neoplasia **21**(5), 442–449 (2019)

10. Li, C., Wang, S., Serra, A., Torheim, T., Yan, J.L., et al.: Multi-parametric and multi-regional histogram analysis of MRI: modality integration reveals imaging phenotypes of glioblastoma. Eur. Radiol. **29**, 1–12 (2019)

11. Li, C., et al.: Intratumoral heterogeneity of glioblastoma infiltration revealed by joint histogram analysis of diffusion tensor imaging. Neurosurgery **85**, 524–534 (2018)

12. Li, C., Wang, S., Yan, J.L., Torheim, T., Boonzaier, N.R., et al.: Characterizing tumor invasiveness of glioblastoma using multiparametric magnetic resonance imaging. J. Neurosurg. **1**, 1–8 (2019)

13. Menze, B.H., et al.: The multimodal brain tumor image segmentation benchmark (brats). IEEE Trans. Med. Imaging **34**(10), 1993–2024 (2014)

14. Myronenko, A.: 3D MRI brain tumor segmentation using autoencoder regularization. In: Crimi, A., Bakas, S., Kuijf, H., Keyvan, F., Reyes, M., van Walsum, T. (eds.) BrainLes 2018. LNCS, vol. 11384, pp. 311–320. Springer, Cham (2019). https://doi.org/10.1007/978-3-030-11726-9_28

15. Ricard, D., Idbaih, A., Ducray, F., Lahutte, M., Hoang-Xuan, K., Delattre, J.Y.: Primary brain tumours in adults. Lancet **379**(9830), 1984–1996 (2012)

16. Scialpi, M., Bianconi, F., Cantisani, V., Palumbo, B.: Radiomic machine learning: is it really a useful method for the characterization of prostate cancer? Radiology **291**(1), 269 (2019)

17. van Griethuysen, J.J.M., Fedorov, A., Parmar, C., Hosny, A., Aucoin, N., et al.: Computational radiomics system to decode the radiographic phenotype. Cancer Res. **77**(21), e104–e107 (2017)

18. Wang, F., et al.: Residual attention network for image classification. In: Proceedings of the IEEE Conference on Computer Vision and Pattern Recognition, pp. 3156–3164 (2017)

19. Wang, G., Li, W., Ourselin, S., Vercauteren, T.: Automatic brain tumor segmentation using cascaded anisotropic convolutional neural networks. In: Crimi, A., Bakas, S., Kuijf, H., Menze, B., Reyes, M. (eds.) BrainLes 2017. LNCS, vol. 10670, pp. 178–190. Springer, Cham (2018). https://doi.org/10.1007/978-3-319-75238-9_16

20. Wu, C., Zou, Y., Zhan, J.: DA-U-Net: densely connected convolutional networks and decoder with attention gate for retinal vessel segmentation. In: IOP Conference Series: Materials Science and Engineering, vol. 533, p. 012053. IOP Publishing (2019)

Multimodal Brain Tumor Segmentation and Survival Prediction Using Hybrid Machine Learning

Linmin Pei, Lasitha Vidyaratne, M. Monibor Rahman[(✉)],
Zeina A. Shboul, and Khan M. Iftekharuddin

Vision Lab, Electrical and Computer Engineering, Old Dominion University,
Norfolk, VA 23529, USA
{lxpei001,lvidy001,mrahm006,zshbo001,
kiftekha}@odu.edu

Abstract. In this paper, we propose a UNet-VAE deep neural network architecture for brain tumor segmentation and survival prediction. UNet-VAE architecture has shown great success in brain tumor segmentation in the multimodal brain tumor segmentation (BraTS) 2018 challenge. In this work, we utilize the UNet-VAE to extract high dimension features, then fuse with handcrafted texture features to perform survival prediction. We apply the proposed method to the BraTS 2019 validation dataset for both tumor segmentation and survival prediction. The tumor segmentation result shows dice score coefficient (DSC) of 0.759, 0.90, and 0.806 for enhancing tumor (ET), whole tumor (WT), and tumor core (TC), respectively. For the feature fusion-based survival prediction method, we achieve 56.4% classification accuracy with mean square error (MSE) 101577, and 51.7% accuracy with MSE 70590 for training and validation, respectively. In testing phase, the proposed method for tumor segmentation achieves average DSC of 0.81328, 0.88616, and 0.84084 for ET, WT, and TC, respectively. Moreover, the model offers accuracy of 0.439 with MSE of 449009.135 for overall survival prediction in testing phase.

Keywords: Deep neural network · Tumor segmentation · Survival prediction · Feature fusion

1 Introduction

Gliomas are the most primary brain tumors in the central nervous system (CNS) in adults. The overall average annual age-adjusted incidence rate for all primary brain and other CNS tumors was 23.03 per 100,000 population [1]. The estimated five- and ten-year relative survival rates are 35.0% and 29.3% for patients with a malignant brain tumor, respectively [1]. According to degrees of aggressiveness, variable prognosis and various heterogeneous histological sub-regions, gliomas are categorized into four grades [2–4]. Survival in days of glioma patients are highly related to the tumor grade. In general, high-grade patients have less survival period. The median survival period of patients with glioblastoma remains two years or less [4]. Brain tumor segmentation is very critical for brain tumor prognosis, treatment planning, and follow-up evaluation.

© Springer Nature Switzerland AG 2020
A. Crimi and S. Bakas (Eds.): BrainLes 2019, LNCS 11993, pp. 73–81, 2020.
https://doi.org/10.1007/978-3-030-46643-5_7

Manual brain tumor segmentation is low efficiency and error-prone to an observer. Therefore, computer-aided semi-/full- automatic brain tumor segmentation is needed. Structural Magnetic resonance imaging (MRI) is widely used for brain tumor study because of non-invasive and good resolution for soft tissues. Multimodal MRI (mMRI) offers more information for brain tumor segmentation. The mMRI sequences include T1-weighted MRI (T1), T1-weighted MRI with contrast enhancement (T1ce), T2-weighted MRI (T2), and T2-weighted MRI with fluid-attenuated inversion recovery (T2-FLAIR). With the advanced mMRI, intensity-based brain tumor segmentation is still a challenging task because of the impact of signal noise, multi-image co-registration, and intensity inhomogeneity, etc.

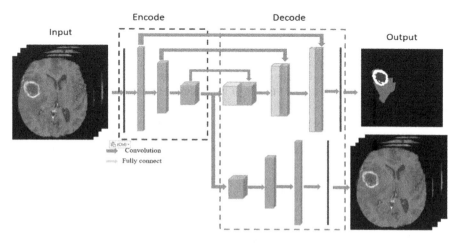

Fig. 1. The framework of the proposed method for brain tumor segmentation. The decoder has two parts. On the top part, it is a regular 3D-UNet for tumor segmentation. The bottom part shows variational auto-encoder (VAE) for image reconstruction. The VAE is used for network regularization.

Brain tumor segmentation has been studied using different techniques such as threshold-based, region-based, to conventional machine learning-based method [5–10]. However, those methods are limited because of the complex mathematic model or difficult hand-crafted feature extraction. Recently, deep learning has great success in computer vision, medical imaging analysis, etc. It overcomes the drawbacks of conventional machine learning, and it achieves great success in many fields [11–15].

In this work, we use a 3D UNet with variational auto-encoder (VAE) architecture for brain tumor segmentation known as UNet-VAE. VAE is used to regularize the training network. We believe that the features are most dense and valuable at end of the encoder of UNet. These features are then fused with hand-crafted features obtained by using a machine learning method, followed by linear regression for overall survival prediction.

2   Method

2.1   Brain Tumor Segmentation

Accurate brain tumor segmentation is very important for brain tumor prognosis and
following-up evaluation for patients with gliomas. We use a 3D-UNet with variational
auto-encoder (UNet-VAE) based deep neural network for tumor segmentation. The
UNet-VAE has two parts: UNet and VAE. UNet is used to for tumor segmentation, and
the VAE reconstructs images, and works to regularize the network. The framework of
the proposed method is shown in Fig. 1.

2.2   Survival Prediction

First, we extract texture features from the four mMRI modalities. These features
include texture, volumetric and area-related features. There are total 1107 texture
features, eight volumetric textures, and fractal texture features extracted from raw
mMRI sequences. In addition, there are six histogram based statistics features are
extracted (more details in [16]). Along with the patients' age, we fuse these texture
features with the high dimensional UNet-VAE extracted features. From our texture
feature extraction, we get 1702 features and UNet-VAE extraction techniques gives us
7,37,280 features. We believe the two types of extracted features are important in
predicting the overall survival. After fusing all the features, we select significant features using least absolute shrinkage and selection operator (LASSO) on these hybrid
features. Finally, linear regression is applied on the selected features for survival
prediction. Our proposed fusion-based survival prediction model is illustrated in Fig. 2.

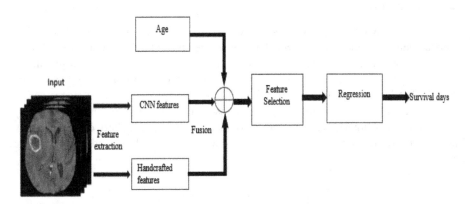

Fig. 2. The framework of our proposed fusion-based survival prediction method.

3   Materials and Pre-processing

3.1   Data

The training data is obtained from the Multimodal Brain Tumor Segmentation Challenge 2019 (BraTS 2019) [2–4, 17, 18], with a total of 335 cases which have 259 high-grade gliomas (HGG) and 76 low-grade gliomas (LGG). Each patient case has T1-weighted MRI (T1), T1-weighted MRI with contrast enhancement (T1ce), T2-weighted MRI (T2), and T2-weighted MRI with fluid-attenuated inversion recovery (T2-FLAIR). All modality sequences are co-registered, skull-stripped, denoised, and bias-corrected [3]. Image size is 240 × 240 × 155 for each image modality. Tumors have different sub-tissues: necrotic (NC), peritumoral edema (ED), and enhancing tumor (ET). The rank of the BraTS 2019 is based on the dice similarity score (DSC) of tumor sub-region evaluation: enhancing tumor (ET), tumor core (TC), and whole tumor (WT). TC consists of ET and TC. WT is made up of all sub-tumors (ED, NC, and EN). Ground truth is created and/or corrected by radiologists.

For survival prediction task, 210 cases are available with overall survival (in days), age, and resection status. Cases with resection status Gross Total Resection (GTR) is evaluated in BraTS 2019 Challenge. One-hundred-one cases are with GTR status. The overall survival is categorized into three classes/groups based on survival days; Long-term, Short-term, and Mid-term survival groups. Long-term group contains cases with overall survival greater than 15 months. Short-term group contain cases with overall survival less than 10 months. Mid-term group contain cases with overall survival between 15 and 10 months. The distribution of 210 cases with overall survival are 82, 56, 72 cases for short-term, mid-term, and long-term, respectively.

3.2   Pre-processing

Even though all MRIs are pre-processed by the organizer with the co-registered, skull-stripped, denoised and bias-corrected process, the intensities of MRI are still very different across cases. The intensity inhomogeneity can result in tumor misclassification. It hampers accurate brain tumor segmentation, especially for intensity-based deep neural network methods. In order to minimize the impact of intensity variance across cases and modalities, we perform z-score intensity normalization for brain region only in MRIs. The normalized image has zero mean and unit standard deviation (std). Figure 3 illustrates an example of image comparison before and after z-score normalization.

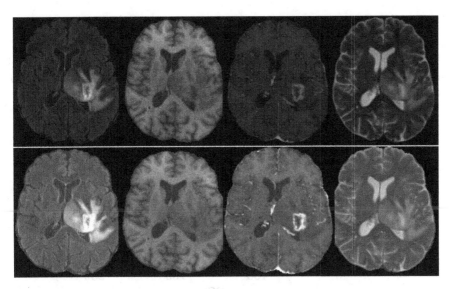

Fig. 3. An example of using z-score intensity normalization. Top row: original images. Bottom row: normalized images. From left to right: T2-FLAIR, T1, T1ce, and T2 image.

4 Experiments and Results

4.1 Hyper-parameter Setting

Due to the limited capacity of the graphics processing unit (GPU), we empirically crop all MRIs with size as $160 \times 192 \times 128$. Within the cropped size image, it has all brain information. The batch size is set as 1. The loss function consists of 2 terms:

$$L = L_{dice} + 0.1 \times L_{vae}, \tag{1}$$

where L_{dice} is a L1 loss applied to the decoder prediction output to compare with the ground truth and L_{vae} represents the loss function that computes the difference between input image and reconstructed image. Dice loss is computed as:

$$L_{dice} = 1 - DSC, \tag{2}$$

where $DSC = \frac{2TP}{FP + 2TP + FN}$ is dice similarity coefficient [19]. TP, FP, and FN are the numbers of true positive, false positive and false negative, respectively.

We use Adam [20] optimizer with an initial learning rate of $lr_0 = 0.001$ in training phase, and the learning rate (lr_i) is gradually reduced by the following:

$$lr_i = lr_0 * \left(1 - \frac{i}{N}\right)^{0.9}, \tag{3}$$

where i is epoch counter, and N is a total number of epochs in training.

4.2   Training Stage

For brain tumor segmentation task, we randomly split 80% data for training, and 20% for validation based on HGG and LGG. The training performance is shown in Fig. 4. Figure 5 shows a comparison between segmentation and ground truth in multiple views.

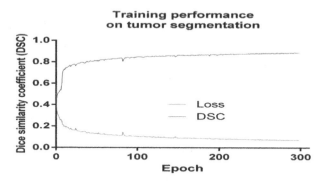

Fig. 4. The DSC and loss changes in the training stage.

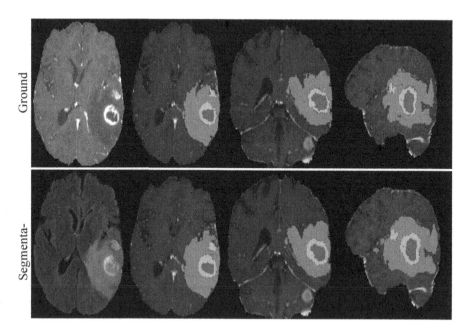

Fig. 5. Comparison between tumor segmentation using the proposed method and ground truth. Top row: T1ce, T1ce image overlaid with ground truth, and bottom row: T2-FLAIR, T1ce image overlaid with our segmentation. From left to right: axial view, axial view, sagittal view, and coronal view, respectively.

Our fused-based survival prediction model achieves an average of 10-fold cross-validation accuracy of 56.0% ± 9.1% and a mean square error of 101577, using the 101 GTR cases with overall survival.

4.3 Online Evaluation

We apply the proposed method to BraTS 2019 validation dataset and evaluate the performance online on the Challenge web. With the 125 unknown tumor grade cases, our result achieves average DSC as 0.726, 0.90, and 0.806 for ET, WT, and TC, respectively. After post-processing steps, the performance increases up to DSC as 0.759, 0.90, and 0.806 for ET, WT, and TC, respectively. The tumor segmentation performance is shown in Table 1. In addition, we also applied the proposed method to BraTS 2019 testing dataset consisting of 166 cases with unknown tumor grade. The post-process steps include small object removal, and replacing ET with NC if the number of ET is less than a threshold. The online evaluation offers average DSC of 0.8128, 0.88616, and 0.84084 for ET, WT, and TC, respectively. According to the performance comparison in Table 1, DSC of WT is 1% lower than that of in validation, however, DSC of ET and TC shows 6% and 4% improvement in testing phase. Note that the Hausdorff distance in testing phase is constantly lower than that of validation. These results suggest that our model offers stable and reliable tumor segmentation results.

Table 1. Brain tumor segmentation performance using the online evaluation of BraTS 2019 validation dataset.

Phase	Dice_ET	Dice_WT	Dice_TC	Hausdorff95_ET	Hausdorff95_WT	Hausdorff95_TC
Validation	0.75946	0.90036	0.8066	4.21057	5.46384	8.02428
Testing	0.81328	0.88616	0.84084	2.24108	4.82841	4.14012

In BraTS 2019 survival prediction, there are 101 GTR cases in training dataset, and 29 GTR cases in validation dataset, respectively. Our proposed fused-based survival prediction model achieves an accuracy of 56.4%, and an accuracy of 51.5%, as illustrated in Table 2, using the training and validation datasets respectively. In testing phase (107 cases), the proposed method offers accuracy of 0.439 with MSE of 449009.785.

Table 2. Performance of our fused-based survival prediction model using the online evaluation.

Phase	Accuracy	MSE	medianSE	stdSE	SpearmanR
Training	0.564	101577	23914	225717	0.54
Validation	0.515	70590	31510	98549	0.436
Testing	0.439	449009	44604	1234471	0.279

5   Conclusion

In the paper, we propose a deep learning-based method, namely UNet-VAE for brain tumor segmentation, and also a hybrid feature-based method for survival prediction. For segmentation, we use variational auto-encoder for regularization as this offers better performance comparing to pure UNet method. In addition, we use hybrid hand-crafted features and high dimensional features from UNet-VAE for survival prediction. The overall hybrid feature-based ML method in this work offers a better survival prediction result.

Acknowledgements. This work was partially funded through NIH/NIBIB grant under award number R01EB020683. This work is also partially supported in part by NSF under grant CNS-1828593.

References

1. Ostrom, Q.T., Gittleman, H., Truitt, G., Boscia, A., Kruchko, C., Barnholtz-Sloan, J.S.: CBTRUS statistical report primary brain and other central nervous system tumors diagnosed in the United States in 2011–2015. Neuro-oncology. **20 suppl. 4**, iv1–iv86 (2018)
2. Bakas, S. et al.: Identifying the best machine learning algorithms for brain tumor segmentation, progression assessment, and overall survival prediction in the BRATS challenge. arXiv preprint arXiv:1811.02629 (2018)
3. Bakas, S., et al.: Advancing the cancer genome atlas glioma MRI collections with expert segmentation labels and radiomic features. Sci. data **4**, 170117 (2017)
4. Menze, B.H., et al.: The multimodal brain tumor image segmentation benchmark (BRATS). IEEE Trans. Med. Imaging **34**(10), 1993–2024 (2014)
5. Mustaqeem, A., Javed, A., Fatima, T.: An efficient brain tumor detection algorithm using watershed & thresholding based segmentation. Int. J. Image Graph. Signal Process. 4(10), 34 (2012)
6. Pei, L., Bakas, S., Vossough, A., Reza, S.M., Davatzikos, C., Iftekharuddin, K.M.: Longitudinal brain tumor segmentation prediction in MRI using feature and label fusion. Biomed. Signal Process. Control **55**, 101648 (2020)
7. Pei, L., Reza, S.M., Li, W., Davatzikos, C., Iftekharuddin, K.M.: Improved brain tumor segmentation by utilizing tumor growth model in longitudinal brain MRI. In: Medical Imaging 2017: Computer-Aided Diagnosis, vol. 10134, p. 101342L. International Society for Optics and Photonics (2017)
8. Pei, L., Reza, S.M., Iftekharuddin, K.M.: Improved brain tumor growth prediction and segmentation in longitudinal brain MRI. In: 2015 IEEE International Conference on Bioinformatics and Biomedicine (BIBM), pp. 421–424. IEEE (2015)
9. Prastawa, M., Bullitt, E., Ho, S., Gerig, G.: A brain tumor segmentation framework based on outlier detection. Med. Image Anal. **8**(3), 275–283 (2004)
10. Ho, S., Bullitt, E., Gerig, G.: Level-set evolution with region competition: automatic 3-D segmentation of brain tumors. In: null, p. 10532. Citeseer (2002)
11. LeCun, Y., Bengio, Y., Hinton, G.: Deep learning. Nature **521**, 436 (2015)
12. Goodfellow, I., Bengio, Y., Courville, A.: Deep Learning. MIT press, Cambridge (2016)

13. Mohsen, H., El-Dahshan, E.-S.A., El-Horbaty, E.-S.M., Salem, A.-B.M.: Classification using deep learning neural networks for brain tumors. Future Comput. Inform. J. **3**(1), 68–71 (2018)
14. He, K., Zhang, X., Ren, S., Sun, J.: Deep residual learning for image recognition. In: Proceedings of the IEEE Conference on Computer Vision and Pattern Recognition, pp. 770–778 (2016)
15. Ronneberger, O., Fischer, P., Brox, T.: U-Net: convolutional networks for biomedical image segmentation. In: Navab, N., Hornegger, J., Wells, W.M., Frangi, A.F. (eds.) MICCAI 2015. LNCS, vol. 9351, pp. 234–241. Springer, Cham (2015). https://doi.org/10.1007/978-3-319-24574-4_28
16. Shboul, Z.A., Alam, M., Vidyaratne, L., Pei, L., Elbakary, M.I., Iftekharuddin, K.M.: Feature-guided deep radiomics for glioblastoma patient survival prediction (in English). Front. Neurosci. **13**(966) (2019). Original Research
17. Bakas, S. et al.: Segmentation labels and radiomic features for the pre-operative scans of the TCGA-GBM collection. The Cancer Imaging Archive (2017)
18. Bakas, S. et al.: Segmentation labels and radiomic features for the pre-operative scans of the TCGA-LGG collection. The Cancer Imaging Archive, vol. 286 (2017)
19. Dice, L.R.: Measures of the amount of ecologic association between species. Ecology **26**(3), 297–302 (1945)
20. Kingma, D.P., Ba, J.: Adam: a method for stochastic optimization. arXiv preprint arXiv: 1412.6980 (2014)

Robust Semantic Segmentation of Brain Tumor Regions from 3D MRIs

Andriy Myronenko[✉] and Ali Hatamizadeh

NVIDIA, Santa Clara, CA, USA
{amyronenko,ahatamizadeh}@nvidia.com

Abstract. Multimodal brain tumor segmentation challenge (BraTS) brings together researchers to improve automated methods for 3D MRI brain tumor segmentation. Tumor segmentation is one of the fundamental vision tasks necessary for diagnosis and treatment planning of the disease. Previous years winning methods were all deep-learning based, thanks to the advent of modern GPUs, which allow fast optimization of deep convolutional neural network architectures. In this work, we explore best practices of 3D semantic segmentation, including conventional encoder-decoder architecture, as well combined loss functions, in attempt to further improve the segmentation accuracy. We evaluate the method on BraTS 2019 challenge.

1 Introduction

Brain tumors are categorized into primary and secondary tumor types. Primary brain tumors originate from brain cells, whereas secondary tumors metastasize into the brain from other organs. The most common type of primary brain tumors are gliomas, which arise from brain glial cells. Gliomas can be of low-grade (LGG) and high-grade (HGG) subtypes. High grade gliomas are an aggressive type of malignant brain tumor that grow rapidly, usually require surgery and radiotherapy and have poor survival prognosis. Magnetic Resonance Imaging (MRI) is a key diagnostic tool for brain tumor analysis, monitoring and surgery planning. Usually, several complimentary 3D MRI modalities are acquired - such as T1, T1 with contrast agent (T1c), T2 and Fluid Attenuation Inversion Recover (FLAIR) - to emphasize different tissue properties and areas of tumor spread. For example the contrast agent, usually gadolinium, emphasizes hyperactive tumor subregions in T1c MRI modality.

Automated segmentation of 3D brain tumors can save physicians time and provide an accurate reproducible solution for further tumor analysis and monitoring. Recently, deep learning based segmentation techniques surpassed traditional computer vision methods for dense semantic segmentation. Convolutional neural networks (CNN) are able to learn from examples and demonstrate state-of-the-art segmentation accuracy both in 2D natural images [5,7] and in 3D medical image modalities [15].

Multimodal Brain Tumor Segmentation Challenge (BraTS) aims to evaluate state-of-the-art methods for the segmentation of brain tumors by providing a 3D

© Springer Nature Switzerland AG 2020
A. Crimi and S. Bakas (Eds.): BrainLes 2019, LNCS 11993, pp. 82–89, 2020.
https://doi.org/10.1007/978-3-030-46643-5_8

MRI dataset with ground truth tumor segmentation labels annotated by physicians [1–4,14]. This year, BraTS 2019 training dataset included 335 cases, each with four 3D MRI modalities (T1, T1c, T2 and FLAIR) rigidly aligned, resampled to $1 \times 1 \times 11$ mm isotropic resolution and skull-stripped. The input image size is $240 \times 240 \times 155$. The data were collected from multiple institutions, using various MRI scanners. Annotations include 3 tumor subregions: the enhancing tumor, the peritumoral edema, and the necrotic and non-enhancing tumor core. The annotations were combined into 3 nested subregions: whole tumor (WT), tumor core (TC) and enhancing tumor (ET), as shown in Fig. 1. Two additional datasets without the ground truth labels were provided for validation and testing. These datasets required participants to upload the segmentation masks to the organizers' server for evaluations. The validation dataset (125 cases) allowed multiple submissions and was designed for intermediate evaluations. The testing dataset allowed only a single submission, and is used to calculate the final challenge ranking.

In this work, we describe our semantic segmentation approach for volumetric 3D brain tumor segmentation from multimodal 3D MRIs and participate in BraTS 2019 challenge.

2   Related Work

Previous year, BraTS 2018 top submissions included Myronenko [16], Isensee et al. [11], McKinly et al. [13] and Zhou et al. [19]. In our previous work [16], we explored how an additional decoder for a secondary task get impose additional structure on the network. Isensee et al. [11] demonstrated that a generic U-net architecture with a few minor modifications is enough to achieve competitive performance. McKinly et al. [13] proposed a segmentation CNN in which a DenseNet [9] structure with dilated convolutions was embedded in U-net-like network. Finally, Zhou et al. [19] proposed to use an ensemble of different networks: taking into account multi-scale context information, segmenting 3 tumor subregions in cascade with a shared backbone weights and adding an attention block.

Here, we generally follow the previous year submission [16], but instead of secondary task decoder we explore various architecture design choices and complimentary loss functions. We also utilize multi-gpu systems for data parallelism to be able to use larger batch sizes.

3   Methods

Our segmentation approach generally follows [16] with encoder-decoder based CNN architecture.

3.1 Encoder Part

The encoder part uses ResNet [8] blocks, where each block consists of two convolutions with normalization and ReLU, followed by additive identity skip connection. For normalization, we experimented with Group Normalization (GN) [18], Instance Normalization [17] and Batch Normalization [10]. We follow a common CNN approach to progressively downsize image dimensions by 2 and simultaneously increase feature size by 2. For downsizing we use strided convolutions. All convolutions are $3 \times 13 \times 3$ with initial number of filters equal to 32. The encoder part structure is shown in Table 1. The encoder endpoint has size $256 \times 120 \times 24 \times 116$, and is 8 times spatially smaller than the input image. We decided against further downsizing to preserve more spatial content.

Table 1. Encoder structure, where GN stands for group normalization (with group size of 8), Conv - $3 \times 13 \times 3$ convolution, AddId - addition of identity/skip connection. Repeat column shows the number of repetitions of the block. We refer to the final output of the encoder, as the encoder endpoint

Name	Ops	Repeat	Output size
Input			$4 \times 160 \times 192 \times 128$
InitConv	Conv	1	$32 \times 160 \times 192 \times 128$
EncoderBlock0	GN,ReLU,Conv,GN,ReLU,Conv, AddId	1	$32 \times 160 \times 192 \times 128$
EncoderDown1	Conv stride 2	1	$64 \times 80 \times 96 \times 64$
EncoderBlock1	GN,ReLU,Conv,GN,ReLU,Conv, AddId	2	$64 \times 80 \times 96 \times 64$
EncoderDown2	Conv stride 2	1	$128 \times 40 \times 48 \times 32$
EncoderBlock2	GN,ReLU,Conv,GN,ReLU,Conv, AddId	2	$128 \times 40 \times 48 \times 32$
EncoderDown3	Conv stride 2	1	$256 \times 20 \times 24 \times 16$
EncoderBlock3	GN,ReLU,Conv,GN,ReLU,Conv, AddId	4	$256 \times 20 \times 24 \times 16$

3.2 Decoder Part

The decoder structure is similar to the encoder one, but with a single block per each spatial level. Each decoder level begins with upsizing: reducing the number of features by a factor of 2 (using $1 \times 1 \times 1$ convolutions) and doubling the spatial dimension (using 3D bilinear upsampling), followed by an addition of encoder output of the equivalent spatial level. The end of the decoder has the same spatial size as the original image, and the number of features equal to the initial input feature size, followed by $1 \times 1 \times 1$ convolution into 3 channels and a sigmoid function. The decoder structure is shown in Table 2.

3.3 Loss

We use a hybrid loss function that consists of the following terms:

$$\mathbf{L} = \mathbf{L}_{dice} + \mathbf{L}_{focal} + \mathbf{L}_{acl} \tag{1}$$

Table 2. Decoder structure, where GN stands for group normalization (with group size of 8), Conv - $3 \times 3 \times 3$ convolution, Conv1 - $1 \times 1 \times 1$ convolution, AddId - addition of identity/skip connection, UpLinear - 3D linear spatial upsampling

Name	Ops	Repeat	Output size
DecoderUp2	Conv1, UpLinear, +EncoderBlock2	1	$128 \times 40 \times 48 \times 32$
DecoderBlock2	GN,ReLU,Conv,GN,ReLU,Conv, AddId	1	$128 \times 40 \times 48 \times 32$
DecoderUp1	Conv1, UpLinear, +EncoderBlock1	1	$64 \times 80 \times 96 \times 64$
DecoderBlock1	GN,ReLU,Conv,GN,ReLU,Conv, AddId	1	$64 \times 80 \times 96 \times 64$
DecoderUp0	Conv1, UpLinear, +EncoderBlock0	1	$32 \times 160 \times 192 \times 128$
DecoderBlock0	GN,ReLU,Conv,GN,ReLU,Conv, AddId	1	$32 \times \times 160 \times 192 \times 128$
DecoderEnd	Conv1, Sigmoid	1	$1 \times 160 \times 192 \times 144$

$\mathbf{L}_{dice}$ is a soft dice loss [15] applied to the decoder output p_{pred} to match the segmentation mask p_{true}:

$$\mathbf{L}_{dice} = 1 - \frac{2 * \sum p_{true} * p_{pred}}{\sum p_{true}^2 + \sum p_{pred}^2 + \epsilon} \tag{2}$$

where summation is voxel-wise, and ϵ is a small constant to avoid zero division. Since the output of the segmentation decoder has 3 channels (predictions for each tumor subregion), we simply add the three dice loss functions together. $\mathbf{L}_{acl}$ is the 3D extension of supervised active contour loss [6] that consists of volumetric and length terms:

$$\mathbf{L}_{acl} = \mathbf{L}_{vol} + \mathbf{L}_{length} \tag{3}$$

in which:

$$\mathbf{L}_{vol} = \mid \sum p_{pred}(c_1 - p_{true})^2 \mid + \mid \sum (1 - p_{pred})(c_2 - p_{true})^2 \mid \tag{4}$$

$$\mathbf{L}_{length} = \sum \sqrt{\mid (\nabla p_{pred,x})^2 + (\nabla p_{pred,y})^2 + (\nabla p_{pred,z})^2 \mid + \epsilon} \tag{5}$$

Where c_1 and c_2 represent the energy of the foreground and background. $\mathbf{L}_{focal}$ is a focal loss function [12] defined as:

$$\mathbf{L}_{focal} = -\frac{1}{N} \sum (1 - p_{pred})^\gamma p_{true} \log (p_{pred} + \epsilon) \tag{6}$$

Where N is the total number of voxels, and γ is set to 2.

3.4 Optimization

We use Adam optimizer with initial learning rate of $\alpha_0 = 1e-4$ and progressively decrease it according to:

$$\alpha = \alpha_0 * \left(1 - \frac{e}{N_e}\right)^{0.9} \tag{7}$$

where e is an epoch counter, and N_e is a total number of epochs (300 in our case). We draw input images in random order (ensuring that each training image is drawn once per epoch).

3.5 Regularization

We use L2 norm regularization on the convolutional kernel parameters with a weight of $1e - 5$. We also use the spatial dropout with a rate of 0.2 after the initial encoder convolution.

3.6 Data Preprocessing and Augmentation

We normalize all input images to have zero mean and unit std (based on non-zero voxels only). We apply a random (per channel) intensity shift ($-0.1..0.1$ of image std) and scale ($0.9..1.1$) on input image channels. We also apply a random axis mirror flip (for all 3 axes) with a probability 0.5.

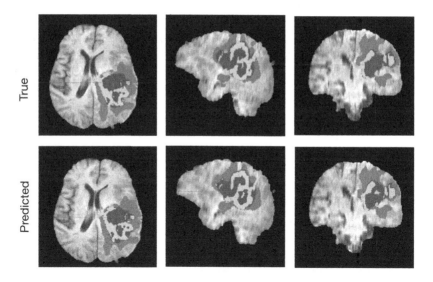

Fig. 1. A typical segmentation example with true and predicted labels overlaid over T1c MRI axial, sagittal and coronal slices. The whole tumor (WT) class includes all visible labels (a union of green, yellow and red labels), the tumor core (TC) class is a union of red and yellow, and the enhancing tumor core (ET) class is shown in yellow (a hyperactive tumor part). The predicted segmentation results match the ground truth well. (Color figure online)

4   Results

We implemented our network in PyTorch[1] and trained it on NVIDIA Tesla V100 32 GB GPUs using BraTS 2019 training dataset (335 cases) without any additional in-house data. During training we used a random crop of size $160 \times 192 \times 128$, which ensures that most image content remains within the crop area. We concatenated 4 available 3D MRI modalities into the 4 channel image as an input. The output of the network is 3 nested tumor subregions (after the sigmoid).

We report the results of our approach on BraTS 2019 validation (125 cases). We uploaded our segmentation results to the BraTS 2019 server for evaluation of per class dice, sensitivity, specificity and Hausdorff distances.

The results of our model on the BratTS 2019 data are shown in Table 3 for the validation dataset and in Table 4 for the testing dataset.

Table 3. BraTS 2019 validation dataset results. Mean Dice and Hausdorff measurements of the proposed segmentation method. EN - enhancing tumor core, WT - whole tumor, TC - tumor core.

	Dice			Hausdorff (mm)		
Validation dataset	ET	WT	TC	ET	WT	TC
Single model (batch 8)	0.800	0.894	0.834	3.921	5.89	6.562

Table 4. BraTS 2019 testing dataset results. Mean Dice and Hausdorff measurements of the proposed segmentation method. EN - enhancing tumor core, WT - whole tumor, TC - tumor core.

	Dice			Hausdorff (mm)		
Testing dataset	ET	WT	TC	ET	WT	TC
Ensemble	0.826	0.882	0.837	2.203	4.713	3.968

Time-wise, each training epoch (335 cases) on a single GPU (NVIDIA Tesla V100 32 GB) takes 10 min. Training the model for 300 epochs takes 2 days. We trained the model on NVIDIA DGX-1 server (that includes 8 V100 GPUs interconnected with NVLink); this allowed to train the model in 8 hours. The inference time is 0.4 sec for a single model on a single V100 GPU.

5   Discussion and Conclusion

In this work, we described a semantic segmentation network for brain tumor segmentation from multimodal 3D MRIs for BraTS 2019 challenge. We have

---

[1] https://pytorch.org/.

experimented with various normalization functions, and found groupnorm and instancenorm to perform equivalent, whereas batchnorm was always inferior, which could be due the fact of the largest batch size attempted being only 16. Since instancenorm is simpler to understand and implement, we used it for normalization by default. Multi-gpu systems, such as DGX-1 server, contains 8 GPU, which allows data-parallel implementation of batch size of 1 (where each each GPU get a batch of 1). We found the performance of multi-gpu system to be equivalent to a single gpu (batch 1) case, thus we used a batch of 8 by default, since it is almost 8 times faster to train. We have also experimented with more sophisticated data augmentation techniques, including random histogram matching, affine image transforms, rotations, random image filtering, which did not demonstrate any additional improvements. Increasing the network depth further did not improve the performance, but increasing the network width (the number of features/filters) consistently improved the results. Our BraTS 2019 final testing dataset results were 0.826, 0.882 and 0.837 average dice for enhanced tumor core, whole tumor and tumor core, respectively.

References

1. Bakas, S., et al.: Segmentation labels and radiomic features for the pre-operative scans of the TCGA-GBM collection. The Cancer Imaging Archive (2017). https://doi.org/10.7937/K9/TCIA.2017.KLXWJJ1Q
2. Bakas, S., et al.: Segmentation labels and radiomic features for the pre-operative scans of the TCGA-LGG collection. The Cancer Imaging Archive (2017). https://doi.org/10.7937/K9/TCIA.2017.GJQ7R0EF
3. Bakas, S., Akbari, H., et al.: Advancing the cancer genome atlas glioma MRI collections with expert segmentation labels and radiomic features. Sci. Data **4**, 170117 (2017)
4. Bakas, S., Reyes, M., et Int, Menze, B.: Identifying the best machine learning algorithms for brain tumor segmentation, progression assessment, and overall survival prediction in the BRATS challenge. arXiv:1811.02629 (2018)
5. Chen, L.C., Zhu, Y., Papandreou, G., Schroff, F., Adam, H.: Encoder-decoder with atrous separable convolution for semantic image segmentation. arXiv:1802.02611 (2018)
6. Chen, X., Williams, B.M., Vallabhaneni, S.R., Czanner, G., Williams, R., Zheng, Y.: Learning active contour models for medical image segmentation. In: Proceedings of the IEEE Conference on Computer Vision and Pattern Recognition, pp. 11632–11640 (2019)
7. Hatamizadeh, A., Sengupta, D., Terzopoulos, D.: End-to-end deep convolutional active contours for image segmentation. arXiv preprint arXiv:1909.13359 (2019)
8. He, K., Zhang, X., Ren, S., Sun, J.: Identity mappings in deep residual networks. In: Leibe, B., Matas, J., Sebe, N., Welling, M. (eds.) ECCV 2016. LNCS, vol. 9908, pp. 630–645. Springer, Cham (2016). https://doi.org/10.1007/978-3-319-46493-0_38
9. Huang, G., Liu, Z., van der Maaten, L., Weinberger, K.Q.: Densely connected convolutional networks. In: Proceedings of the IEEE Conference on Computer Vision and Pattern Recognition, pp. 2261–2269 (2017)
10. Ioffe, S., Szegedy, C.: Batch normalization: Accelerating deep network training by reducing internal covariate shift. In: International Conference on Machine Learning (ICML), pp. 448–456 (2015)

11. Isensee, F., Kickingereder, P., Wick, W., Bendszus, M., Maier-Hein, K.H.: No new-net. In: Crimi, A., Bakas, S., Kuijf, H., Keyvan, F., Reyes, M., van Walsum, T. (eds.) BrainLes 2018. LNCS, vol. 11384, pp. 234–244. Springer, Cham (2019). https://doi.org/10.1007/978-3-030-11726-9_21

12. Lin, T.Y., Goyal, P., Girshick, R., He, K., Dollár, P.: Focal loss for dense object detection. In: Proceedings of the IEEE international conference on computer vision, pp. 2980–2988 (2017)

13. McKinley, R., Meier, R., Wiest, R.: Ensembles of densely-connected CNNs with label-uncertainty for brain tumor segmentation. In: Crimi, A., Bakas, S., Kuijf, H., Keyvan, F., Reyes, M., van Walsum, T. (eds.) BrainLes 2018. LNCS, vol. 11384, pp. 456–465. Springer, Cham (2019). https://doi.org/10.1007/978-3-030-11726-9_40

14. Menze, B.H., et al.: The multimodal brain tumor image segmentation benchmark (BRATS). IEEE Trans. Med. Imaging **34**(10), 1993–2024 (2015)

15. Milletari, F., Navab, N., Ahmadi, S.A.: V-net: fully convolutional neural networks for volumetric medical image segmentation. In: Fourth International Conference on 3D Vision (3DV) (2016)

16. Myronenko, A.: 3D MRI brain tumor segmentation using autoencoder regularization. In: Crimi, A., Bakas, S., Kuijf, H., Keyvan, F., Reyes, M., van Walsum, T. (eds.) BrainLes 2018. LNCS, vol. 11384, pp. 311–320. Springer, Cham (2019). https://doi.org/10.1007/978-3-030-11726-9_28

17. Ulyanov, D., Vedaldi, A., Lempitsky, V.S.: Instance normalization: the missing ingredient for fast stylization. In: CVPR (2016)

18. Wu, Y., He, K.: Group normalization. In: European Conference on Computer Vision (ECCV) (2018)

19. Zhou, C., Chen, S., Ding, C., Tao, D.: Learning contextual and attentive information for brain tumor segmentation. In: Crimi, A., Bakas, S., Kuijf, H., Keyvan, F., Reyes, M., van Walsum, T. (eds.) BrainLes 2018. LNCS, vol. 11384, pp. 497–507. Springer, Cham (2019). https://doi.org/10.1007/978-3-030-11726-9_44

Brain Tumor Segmentation with Cascaded Deep Convolutional Neural Network

Ujjwal Baid[1]([✉]), Nisarg A. Shah[2], and Sanjay Talbar[1]

[1] SGGS Institute of Engineering and Technology, Nanded, India
{baidujjwal,sntalbar}@sggs.ac.in
[2] Department of Electrical Engineering, Indian Institute of Technology Jodhpur,
Jodhpur, Rajasthan, India
shah.2@iitj.ac.in

Abstract. Cancer is the second leading cause of death globally and is responsible for an estimated 9.6 million deaths in 2018. Approximately 70% of deaths from cancer occur in low and middle-income countries. One defining feature of cancer is the rapid creation of abnormal cells that grow uncontrollably causing tumor. Gliomas are brain tumors that arises from the glial cells in brain and comprise of 80% of all malignant brain tumors. Accurate delineation of tumor cells from healthy tissues is important for precise treatment planning. Because of different forms, shapes, sizes and similarity of the tumor tissues with rest of the brain segmentation of the Glial tumors is challenging. In this study we have proposed fully automatic two step approach for Glioblastoma (GBM) brain tumor segmentation with Cascaded U-Net. Training patches are extracted from 335 cases from Brain Tumor Segmentation (BraTS) Challenge for training and results are validated on 125 patients. The proposed approach is evaluated quantitatively in terms of Dice Similarity Coefficient (DSC) and Hausdorff95 distance.

Keywords: Brain tumor segmentation · Convolutional Neural Networks · Deep learning · U-Net · GPU

1 Introduction

Cancer is a term used for diseases in which abnormal cells divide without control and can invade other tissues. It spreads beyond their usual boundaries, and which can then invade adjoining parts of the body. It can affect almost any part of the body. Lack of awareness, suboptimal medical infrastructure, fewer chances of screening, and low doctor-patient ratio are the prime reasons for the rise in various cancers [3].

According to Central Brain Tumor Registry of the United States (CBTRUS), 86,970 new cases of primary malignant and non-malignant brain tumors are expected to be diagnosed in the United States in 2019 [1]. An estimated 16,830

A. Crimi and S. Bakas (Eds.): BrainLes 2019, LNCS 11993, pp. 90–98, 2020.
https://doi.org/10.1007/978-3-030-46643-5_9

deaths will be attributed to primary malignant brain tumors in the US in 2018. In developing country India alone, total deaths due to cancer in 2018 are 7,84,821 [2]. More than 50% of the cases in India are diagnosed in stage 3 or 4, which decreases the patient's chances of survival. Reports say India has highest mortality-to-incidence ratio in the whole world.

Late-stage presentation and inaccessible diagnosis and treatment are common in various types of cancer. In 2017, only 26% of low-income countries reported having pathology services generally available in the public sector. More than 90% of high-income countries reported treatment services are available compared to less than 30% of low-income countries. Only 1 in 5 low- and middle-income countries have the necessary data to drive cancer policy. Annotation of brain tumors in MRI images is time consuming task for the radiologists. It has been observed that there is high inter-rater variation when the same tumor is marked by different radiologists [16]. Thus, we aimed at developing a fully automatic algorithm which accurately segment the Glial brain tumor without any manual intervention. Also, on average, segmentation of the intra-tumor parts for a single patient is completed in 60 seconds. An automatic segmentation algorithm will be useful as a reference and will save the radiologist time to attend more patients. This is of most importance in developing countries with large population.

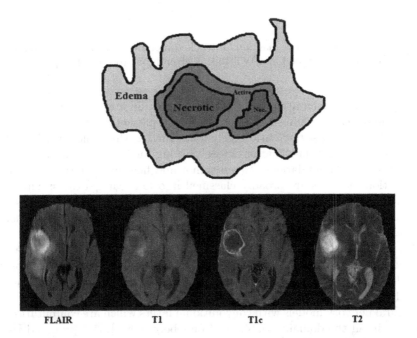

Fig. 1. Top row: sub-tumor parts like Edema, Necrotic and Enhancing tumor. Bottom row: appearance of Sub tumor parts in different MR modalities: FLAIR, T1, T2, T1ce [8]

Gliomas are the most frequent primary brain tumors in adults and account for 70% of adult malignant primary brain tumors. Glioma arises from glial cells and infiltrate the surrounding tissues such as white matter fibre tracts with very rapid growth [16]. Patients with High Grade Gliomas, also known as Glioblastoma tumors have average survival time of one year. Patient undergoes MRI scan for imaging of the brain tumor and treatment planning. Various intra-tumor parts like Enhancing Tumor (ET), Tumor Core (TC)/Necrosis and Edema appears differently in different MR modalities as shown in Fig. 1.

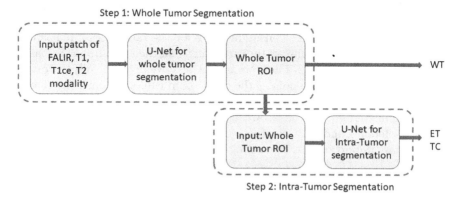

Fig. 2. Proposed two step flow diagram for whole tumor and Intra-tumor segmentation. In first step WT is segmented and WT ROI is given input to second step.

Brain tumor segmentation is a challenging task because of its non-rigid and complex shape, variation in size and position from patient to patient. These challenges make classical segmentation techniques such as thresholding, edge detection, region growing, classification and clustering ineffective at accurate delineation of complex boundaries between tumor and healthy tissues. Brain tumor segmentation methods are broadly classified into four categories as: Threshold based, Region based, Pixel classification based and Model based techniques with pros and cons over each other [4,6]. Many approaches to brain tumor segmentation have been implemented over decades but there is no winning theory.

Recently, methods based on Deep Convolutional Neural Networks are outperforming over all traditional machine learning methods in various domains like medical image segmentation, image classification, object detection and tracking etc. [7,18]. The computational power of GPUs has enabled researchers to design deep neural network models with convolutional layers which are computationally expensive in all the domains [5,14,15]. Ronneberger et al. [17] proposed U-Net architecture for biomedical image segmentation with limited images. This paper is major breakthrough in the field of medical image segmentation like liver, brain, lung etc. Inspired from the above literature we developed two step approach based on Deep Convolutional Neural Network 2D U-Net model. Various heterogeneous histologic sub-regions like peritumoral edema, enhancing tumor and necrosis were accurately segmented in spite of thin boundaries between intra-tumor parts.

2 Patients and Method

2.1 Database

In this study, we trained model on popular Brain Tumor Segmentation Challenge (BraTS) 2019 dataset [9–13]. BraTS is the popular challenge organised at Medical Image Computing and Computer Assisted Intervention (MICCAI) conference since 2012. Organisers collected dataset from various hospitals all over the Globe with various scanners and acquisition protocols which made the task challenging[1]. The BraTS dataset comprised of 335 training patients out of which 259 are with High Grade Glioma (HGG) and 76 with Low Grade Glioma (LGG). MRI data of each patient was provided with four channels as FLAIR, T1, T2, T1ce with volume size $240 \times 240 \times 155$. Also, for each case annotations marked by expert radiologists were provided for Whole Tumor (WT), Enhancing Tumor (ET), Tumor Core (TC) and Edema. Validation dataset of 125 patients was provided without ground truths. In addition to this 166 patients data was provided for testing purpose. Evaluation of the proposed method is done by submitting the segmentation results on online evaluation portal[2].

2.2 Preprocessing

BraTS dataset is already skull-stripped and registered to $1\,\mathrm{mm} \times 1\,\mathrm{mm} \times 1\,\mathrm{mm}$ isotropic resolution. Bias fields were corrected with N4ITK tool [19]. All four MR channel is normalised to zero mean and unit variance.

2.3 Patch Extraction

Researchers have proposed several ways of training U-Net for biomedical image segmentation like training with complete multi-modal images and with images patches. In our database there was high class imbalance among the different intra-tumor labels to be segmented along with the non-tumor region. In order to address this, we extracted patches from all the four channels of MRI data. We explicitly augmented patches of underrepresented class by applying various affine transformation like scaling, rotation, translation. Training patches of size 64×64 were extracted from the FLAIR, T1, T2, T1ce for training.

---

[1] https://www.med.upenn.edu/cbica/brats2019.html.

[2] https://ipp.cbica.upenn.edu.

2.4   Cascaded U-Net

The proposed two step approach is shown in Fig. 2. We have cascaded two U-Net architecture for WT segmentation and intra-tumor segmentation respectively. In step one first U-Net is trained to segment the Whole tumor. Four 64×64 patches from T1, T2, T1ce, FLAIR are concatenated together and given input along with the corresponding WT mask to the first layer of U-Net as shown in Fig. 3. Second U-Net is trained with WT patch along with corresponding intra-tumor patch as input to give segmentation output for enhancing tumor and tumor core. It should be noted that the area without ET and TC in whole tumor is Edema.

We modified 2D U-Net with large number of feature channels in down-sampling and up-sampling layers as shown in Fig. 3. The architecture consists of a contracting path to capture context and a symmetric expanding path that enables precise localization. At the first layer four 64×64 multichannel MR volume data was given input for training along with the corresponding ground truth. The number of features maps increases in the subsequent layers to learn the deep tumor features. These are followed by Leaky ReLU activation function and the features were down-sampled in encoding layer. Similarly, in decoding layer after convolution layers and Leaky ReLU activation function, features maps were up-sampled by factor of 2. Features maps from encoding layers were concatenated to corresponding decoding layer in the architecture. At the output layer, the segmentation map predicted by the model was compared with the corresponding ground truth and the error was back propagated in the intermediate U-Net layers. The output layer is a 1×1 convolution with one filter for the first stage i.e. WT segmentation, and three filters for the second stage i.e. ET, TC and Edema segmentation. The learning rate (α) was initialised to 0.001. After every epoch learning rate was decreased linearly by the factor of 10^{-1} which avoid convergence of the model to local minima. The model was trained for 100 epochs since validation loss did not improved beyond that. Further, for better optimization a momentum strategy was included in the implementation. This used a temporally averaged gradient to damp the optimization velocity.

3   Result and Discussion

The quantitative evaluation of the proposed model was done on BraTS 2019 challenge dataset. No ground truths are provided for validation dataset and predicted results are to be uploaded on online evaluation portal for fair evaluation. The sample result on BraTS challenge dataset is shown in Fig. 4 with Glial tumors. Edema, Enhancing Tumor and Tumor Core segmented by our approach are shown with Green, Blue and Red colours respectively. The performance in terms of dice similarity index is shown in Table 1.

The proposed architecture was implemented using Keras and Tensorflow library which supported the use of GPUs. GPU implementation greatly accelerated the implementation of deep learning algorithms. The approximate time to train the model was 48 hours on 16 GB NVIDIA P100 GPU using cuDNN v5.0 and CUDA 8.0 with 128 GB RAM. The prediction on validation data took

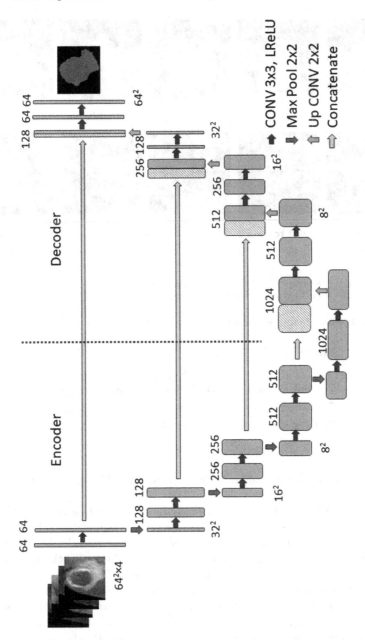

Fig. 3. U-Net architecture for WT segmentation

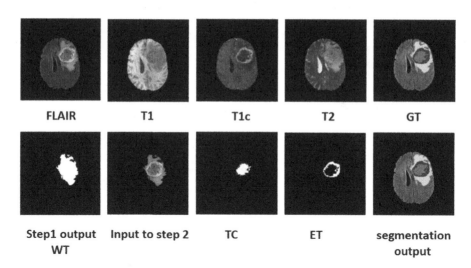

| FLAIR | T1 | T1c | T2 | GT |

| Step1 output WT | Input to step 2 | TC | ET | segmentation output |

Fig. 4. Sample segmentation results with proposed approach. Green: Edema, Blue: Enhancing tumor (ET), Red: Tumor Core (TC) (Color figure online)

Table 1. Performance evaluation of proposed algorithm on BraTS challenge

Brats Dataset	Dice			Hausdorff95		
	WT	ET	TC	WT	ET	TC
Training Dataset	0.94	0.85	0.94	6.62	2.06	5.13
Validation Dataset	0.87	0.70	0.77	13.36	6.45	12.71
Test Dataset	0.89	0.78	0.82	4.92	4.45	4.66

less than 60 seconds for a single patient with four MR channels data, each of dimension $240 \times 240 \times 155$.

There was high class imbalance in the dataset. Around more than 98% pixels belonged to background/healthy class, and rest were labelled as Edema, enhancing tumor and necrotic tumor. Training of the model was challenging because of this class imbalance as there were chances of overfitting to the healthy class. This overfitting of the model to healthy tissue would lead to misclassification of necrotic pixels to healthy pixels. This problem was overcome by augmenting the data with under represented regions. We increased the training data with augmentation techniques like rotation, scaling, shifting etc. This had improved the performance of the model with better class balance. Patches from the boundary region of the tumor were added explicitly for better training of the model. Thus, the segmentation accuracy at the tumor boundaries improved because of additional patches.

4   Conclusion

In conclusion, we presented two step cascaded brain tumor segmentation approach with 2D U-Net architecture based on Deep Convolutional Neural Networks. The encoder-decoder type ConvNet model for pixel-wise segmentation is proposed. We incorporated information from all four MR channels for segmentation which could delineate the tumor boundaries more accurately. We considered different training schems with variable patch sizes, data augmentation methods, activation functions, loss functions and optmizers. This automated approach will definitely provide second opinion to the radiologists for automatic segmentation of brain tumor within minutes. This will enable radiologists for quick reporting and deal with more patients where there is poor patient to doctor ratio.

Acknowledgment. This publication is an outcome of the R & D work undertaken project under the Visvesvaraya PhD Scheme funded by Ministry of Electronics & Information Technology, Government of India, being implemented by Digital India Corporation with reference number: PhD-MLA/4(67/2015-16).

References

1. Central brain tumor registry of the united states (2018). http://www.cbtrus.org/factsheet/factsheet.html
2. Central brain tumor registry of the united states (2018). http://cancerindia.org.in/cancer-statistics/
3. World health organization fact-sheets (2018). https://www.who.int/news-room/fact-sheets/detail/cancer
4. Angulakshmi, M., Lakshmi Priya, G.: Automated brain tumour segmentation techniques a review. Int. J. Imaging Syst. Technol. **27**(1), 66–77. https://doi.org/10.1002/ima.22211
5. Baheti, B., Gajre, S., Talbar, S.: Detection of distracted driver using convolutional neural network. In: 2018 IEEE/CVF Conference on Computer Vision and Pattern Recognition Workshops (CVPRW), pp. 1145–11456, June 2018. https://doi.org/10.1109/CVPRW.2018.00150
6. Baid, U., Talbar, S.: Comparative study of k-means, gaussian mixture model, fuzzy c-means algorithms for brain tumor segmentation. In: International Conference on Communication and Signal Processing 2016 (ICCASP 2016). Atlantis Press (2016)
7. Baid, U., et al.: Deep learning radiomics algorithm for gliomas (drag) model: a novel approach using 3D UNET based deep convolutional neural network for predicting survival in gliomas. In: Crimi, A., Bakas, S., Kuijf, H., Keyvan, F., Reyes, M., van Walsum, T. (eds.) Brainlesion: Glioma, Multiple Sclerosis, Stroke and Traumatic Brain Injuries, pp. 369–379. Springer, Cham (2019). https://doi.org/10.1007/978-3-030-11726-9_33
8. Baid, U., Talbar, S., Talbar, S.: Brain tumor segmentation based on non negative matrix factorization and fuzzy clustering. In: Proceedings of the 10th International Joint Conference on Biomedical Engineering Systems and Technologies, BIOIMAGING, BIOSTEC 2017, Porto, Portugal, 21–23 February 2017, vol. 2, pp. 134–139 (2017). https://doi.org/10.5220/0006250701340139

9. Bakas, S., et al.: Advancing the cancer genome atlas glioma MRI collections with expert segmentation labels and radiomic features. Nat. Sci. Data **4**, 170117 (2017). https://doi.org/10.1038/sdata.2017.117

10. Bakas, S., et al.: Segmentation labels and radiomic features for pre operative scans of the TCGA-GBM collection. Cancer Imaging Arch. **170117** (2017). https://doi.org/10.1038/sdata.2017.117

11. Bakas, S., et al.: Segmentation labels and radiomic features for pre operative scans of the TCGA-LGG collection. Cancer Imaging Arch. **170117**, (2017). https://doi.org/10.1038/sdata.2017.117

12. Bakas, S., et al.: Identifying the best machine learning algorithms for brain tumor segmentation, progression assessment, and overall survival prediction in the BRATS challenge. CoRR abs/1811.02629 (2018). http://arxiv.org/abs/1811.02629

13. Bakas, S., et al.: Identifying the best machine learning algorithms for brain tumor segmentation, progression assessment, and overall survival prediction in the brats challenge. ArXiv abs/1811.02629 (2018)

14. Hariharan, B., Arbelaez, P., Girshick, R., Malik, J.: Object instance segmentation and fine-grained localization using hypercolumns. IEEE Trans. Pattern Anal. Mach. Intell. **39**(4), 627–639 (2017)

15. Kamnitsas, K., et al.: Efficient multi-scale 3D CNN with fully connected crf for accurate brain lesion segmentation. Med. Image Anal. **36**, 61–78 (2017)

16. Menze, B.H., et al.: The multimodal brain tumor image segmentation benchmark (brats). IEEE Trans. Med. Imaging **34**(10), 1993–2024 (2015). https://doi.org/10.1109/TMI.2014.2377694

17. Ronneberger, O., Fischer, P., Brox, T.: U-Net: convolutional networks for biomedical image segmentation. In: Navab, N., Hornegger, J., Wells, W.M., Frangi, A.F. (eds.) MICCAI 2015. LNCS, vol. 9351, pp. 234–241. Springer, Cham (2015). https://doi.org/10.1007/978-3-319-24574-4_28

18. Smistad, E., Falch, T.L., Bozorgi, M., Elster, A.C., Lindseth, F.: Medical image segmentation on GPUs - a comprehensive review. Med. Image Anal. **20**(1), 1–18 (2015). https://doi.org/10.1016/j.media.2014.10.012

19. Tustison, N.J., et al.: N4ITK: improved N3 bias correction. IEEE Trans. Med. Imaging **29**(6), 1310–1320 (2010). https://doi.org/10.1109/TMI.2010.2046908

Fully Automated Brain Tumor Segmentation and Survival Prediction of Gliomas Using Deep Learning and MRI

Chandan Ganesh Bangalore Yogananda[1(✉)], Ben Wagner[1],
Sahil S. Nalawade[1], Gowtham K. Murugesan[1], Marco C. Pinho[1],
Baowei Fei[2], Ananth J. Madhuranthakam[1], and Joseph A. Maldjian[1]

[1] Advanced Neuroscience Imaging Research Lab, Department of Radiology,
University of Texas Southwestern Medical Center, Dallas, TX, USA
ChandanGanesh.BangaloreYogananda@utsouthwestern.edu
[2] Department of Bioengineering, University of Texas at Dallas,
Richardson, TX, USA

Abstract. Tumor segmentation of magnetic resonance images is a critical step in providing objective measures of predicting aggressiveness and response to therapy in gliomas. It has valuable applications in diagnosis, monitoring, and treatment planning of brain tumors. The purpose of this work was to develop a fully-automated deep learning method for tumor segmentation and survival prediction. Well curated brain tumor cases with multi-parametric MR Images from the BraTS2019 dataset were used. A three-group framework was implemented, with each group consisting of three 3D-Dense-UNets to segment whole-tumor (WT), tumor-core (TC) and enhancing-tumor (ET). Each group was trained using different approaches and loss-functions. The output segmentations of a particular label from their respective networks from the three groups were ensembled and post-processed. For survival analysis, a linear regression model based on imaging texture features and wavelet texture features extracted from each of the segmented components was implemented. The networks were tested on both the BraTS2019 validation and testing datasets. The segmentation networks achieved average dice-scores of 0.901, 0.844 and 0.801 for WT, TC and ET respectively on the validation dataset and achieved dice-scores of 0.877, 0.835 and 0.803 for WT, TC and ET respectively on the testing dataset. The survival prediction network achieved an accuracy score of 0.55 and mean squared error (MSE) of 119244 on the validation dataset and achieved an accuracy score of 0.51 and MSE of 455500 on the testing dataset. This method could be implemented as a robust tool to assist clinicians in primary brain tumor management and follow-up.

Keywords: Brain tumor segmentation · Deep learning · BraTS · Dense-UNet · MRI · Survival prediction · Imaging features · Radiomics features · Pyradiomics

© Springer Nature Switzerland AG 2020
A. Crimi and S. Bakas (Eds.): BrainLes 2019, LNCS 11993, pp. 99–112, 2020.
https://doi.org/10.1007/978-3-030-46643-5_10

1   Introduction

Gliomas account for the most common malignant primary brain tumors in both pediatric and adult populations [1]. They arise from glial cells and are divided into low grade and high grade gliomas with significant differences in patient survival. Patients with aggressive high grade gliomas have life expectancies of less than 2 years [2]. Glioblastoma (GBM) are aggressive brain tumors classified by the world health organization (WHO) as stage IV brain cancer [3, 4]. The overall survival for GBM patient is poor and is in the range of 12 to 15 months [5–7]. These tumors are typically treated by surgery, followed by radiotherapy and chemotherapy. Gliomas often consist of active tumor tissue, necrotic tissue and surrounding edema. Magnetic Resonance Imaging (MRI) is the most commonly used modality to assess brain tumors because of its superior soft tissue contrast. It is routinely used in the clinical work-up of patients for brain tumor diagnosis, monitoring progression and treatment planning [8, 9]. Each MR sequence provides specific information about different tissue subcomponents of gliomas. For instance, T1-weighted images with intravenous contrast highlight the most vascular regions of the tumor, called 'enhancing tumor' (ET), along with the 'tumor core' (TC) that does not involve peri-tumoral edema. Conventional T2-weighted (T2W) and T2W-Fluid Attenuation Inversion Recovery (FLAIR) images are used to evaluate the tumor and peri-tumoral edema together defined as the 'whole tumor' (WT) [10].

MRI tumor segmentation is used to identify the subcomponents as enhancing, necrotic or edematous tissue. Due to heterogeneity and tissue relaxation differences in these subcomponents, multi-parametric (or multi-contrast) MRI are often used simultaneously for accurate segmentation [11]. Manual brain tumor segmentation is a challenging and tedious task for human experts due to the variability of tumor appearance, unclear borders of the tumor and the need to evaluate multiple MR images with different contrasts simultaneously [12]. In addition, manual segmentation is often prone to significant intra- and inter-rater variability [12, 13]. Hence machine learning algorithms have been developed for tumor segmentation with high reproducibility and efficiency [12–14]. Following the early success of CNNs [14, 15], they are used as one of the major machine learning methods to achieve great success in clinical applications [16, 17]. Furthermore, underlying molecular heterogeneity in gliomas makes it difficult to predict the overall survival (OS) of GBM patients based on MR imaging alone [16, 17]. Clinical features [5], along with MR imaging based texture features [18–20] have been used to predict the OS in GBM patients. In this work, we utilized designed a 3D Dense-Unet for segmenting brain tumors into subcomponents and used MRI based texture features from each of these subcomponents for survival prediction in GBM patients. The purpose of this work was to develop a deep learning method with high prediction accuracy for brain tumor segmentation and survival prediction that can be easily incorporated into the clinical workflow.

2   Material and Methods

2.1   Data and Pre-processing

2.1.1   Brain Tumor Segmentation

335 well curated multi-parametric brain MR images including T2w, T2w-FLAIR, T1w and T1C (post contrast) from the BraTS2019 dataset were used [10, 21–24]. The dataset consisted of 259 high grade glioma (HGG) cases and 76 low grade glioma (LGG) cases. The dataset also included three ground truth labels for (a) enhancing tumor, (b) non-enhancing tumor including necrosis and (c) edema. 125 cases from the BraTS2019 validation dataset was used for evaluating the network's performance. Pre-processing steps included N4BiasCorrection to remove the RF inhomogeneity [25] and intensity normalization to zero-mean and unit variance.

2.1.2   Survival Analysis

259 HGG subjects were provided for the BraTS2019 Survival challenge. Information regarding age, survival days and resection status (Gross Total Resection (GTR), Subtotal Resection (STR) or not available (NA)) were also provided. 210 subjects out of 259 subjects were selected, as the other 49 subjects were either alive or survival days were not available (NA). The results were evaluated on 29 GTR subjects from the validation dataset. Pre-processing steps were similar to the segmentation task including N4BiasCorrection and intensity normalization.

2.2   Network Architecture

Brain tumors contain a complex structure of sub-components and are challenging for automated tumor segmentation. Specifically, the appearance of enhancing tumor and non-enhancing tumor are often different between HGG and LGG. In order to simplify the complex segmentation problem, first, a simple convolutional neural network (CNN) was developed to classify HGG and LGG cases. Next, to maintain consistency with the output results provided by the BraTS challenge, separate networks were trained to recognize and predict whole tumor (consisting of enhancing + non-enhancing + necrosis + edema; Whole-net), tumor core (consisting of enhancing + non-enhancing + necrosis; Core-net), and enhancing tumor (enhance-net) as binary features using 3D Dense UNets. Each of these three networks was designed separately for HGG and LGG cases.

The networks were designed to predict the local structures for tumors and sub-components in multi-parametric brain MR images. The architecture of each network is shown in Fig. 1. On the encoder side of the network, multi-parametric images passed through initial convolution to generate 64 feature maps to be used in subsequent dense blocks. Each dense block consisted of five layers as shown in Fig. 1. Each layer included four sublayers, BatchNormalization, Rectified Linear Unit (ReLu), 3D

Convolution and 3D Spatial dropout that were connected sequentially. At each layer, the input was used to generate k feature maps (referred to as growth rate and set to 16) that were subsequently concatenated to the next input layer. The next layer was then applied to create another k feature maps. To generate the final dense block output, inputs from each layer were concatenated with the output of the last layer. At the end of each dense block, the input to the dense block was also concatenated to the output of that dense block. The output of each dense block followed a skip connection to the adjacent decoder. In addition, each dense block output went through a transition down block until the bottleneck block. With this connecting pattern, all feature maps were reused such that every layer in the architecture received a direct supervision signal [26]. On the decoder side, a transition up block preceded each dense block until the final convolution layer followed by a sigmoid activation layer. Two techniques were used to circumvent the problem of maintaining a high number of convolution layers. A) The bottleneck block (Dense block 4 in Fig. 1 was used such that if the total number of feature maps from a layer exceeded the initial number of convolution maps (i.e. 64) then it was reduced to $1/4^{th}$ of the total generated feature maps in that layer. B) A compression factor of 0.75 was used to reduce the total number of feature maps after every block in the architecture. In addition, due to a large number of high resolution feature maps, a patch based 3D Dense-Unet approach was implemented where higher resolution information was passed through the standard skip connections.

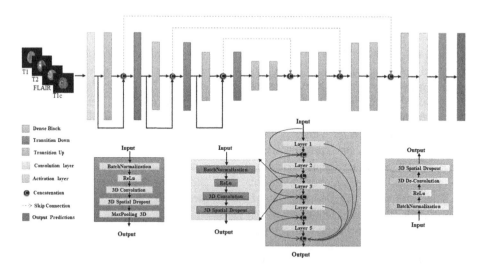

Fig. 1. Network Architecture. Three such networks were designed separately for whole tumor (WT), tumor core (TC) and Enhancing tumor (EN).

2.3   Training

2.3.1   Brain Tumor Segmentation

Three groups of three Dense-UNets were designed. Each group consisted of networks designed to segment whole tumor, tumor core and enhancing tumor. Group 1 was trained using the dice-coefficient [27] as the loss function while group 2 was trained with binary focal loss [28] as the loss function. Group 3 consisted of 2 sub-groups namely, HGG group and LGG group. As the first step of group 3, a simple convolutional neural network was developed to separate HGG from LGG. The networks in the HGG group were trained using HGG cases only and the networks in the LGG group used the LGG cases only.

All 335 cases from the BraTS2019 dataset were used for training the networks from group 1 and group 2. 259 HGG cases were used to train the networks from the HGG group and 76 LGG cases were used to train the networks from the LGG group. 75% overlapping patches were extracted from the multi-parametric brain MR images that had at least one non-zero pixel on the corresponding ground truth patch. 20% of the extracted patches were used for in-training validation. Data augmentation steps included horizontal flipping, vertical flipping, random rotation, and translational rotation. Down sampled data ($128 \times 128 \times 128$) was also provided in the training as an additional data augmentation step. To circumvent the problem of data leakage, no patch from the same subject was mixed between training and in-training validation [29, 30].

Dice loss: The Dice co-efficient determines the amount of spatial overlap between the ground truth segmentation (X) and the network segmentation (Y)

$$dice\ loss = \frac{2|X_1 \bigcap Y_1|}{|X_1| + |Y_1|}$$

Focal loss: The focal loss was designed to address the problem of class imbalance between the foreground and background classes during training.

$$p_t = \begin{cases} p & if\ y = 1 \\ 1 - p & otherwise \end{cases} \quad 'y'\ specifies\ the\ ground\ truth\ and\ 'p'$$
$$\in [0, 1]\ is\ the\ model's\ predictio$$

$$Focal\ Loss = FL(p_t) = -\alpha_t(1 - p_t)^\gamma \log(p_t)$$

HGG and LGG Classifier: 335 cases from the BraTS2019 dataset including 259 HGG cases and 76 LGG cases were used for training the classifier network. Data augmentation steps included horizontal flipping, vertical flipping, random and translational rotation. The dataset was randomly shuffled and split into 60% (155 HGG and 46 LGG) for training, 20% (52 HGG and 15 LGG) for in-training validation and 20% (52 HGG and 15 LGG) for testing (Fig. 2).

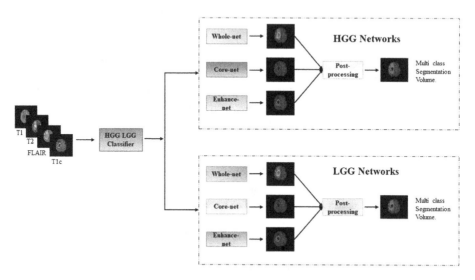

Fig. 2. Overview of the 3rd approach using the HGG/LGG classifier.

2.3.2 Survival Analysis

Pyradiomics (a python package) was used to extract imaging (or radiomics) features [31]. Multi contrast MR images along with ground truth labels for each of the subcomponents were used to extract features from the BraTS2019 training dataset. For the BraTS2019 validation and testing datasets, the segmented labels from the three 3D Dense UNets were used to extract features from the multi contrast MR images. 318 features were extracted using T1-pre contrast with enhancing, edema and non-enhancing tumor with necrosis. Similar procedure was followed using the 3 other MR image sequence. Thus a total of 1272 features were extracted from 4 MR imaging sequences. Pywavelet, a python toolbox, was also used to extract wavelet based features [32]. 8 wavelet components for each MR image were extracted from level 1 of wavelet transform using coiflets (order = 1) [33]. These 8 components (extracted from pywavelet toolbox) for the four MR imaging sequences in combination with 3 tumor subcomponents were used to extract 10,176 features using the pyradiomics package. These imaging based features were combined with additional features including surface area, volume of tumor core, volume of whole tumor, ratio of tumor subcomponents to tumor core, ratio of tumor subcomponents to tumor volume, ratio of tumor core to tumor volume, and variance of enhancing tumor with T1C. The degree of angiogenesis was calculated by subtracting T1w and T1C in the tumor ROI, followed by a threshold of 50%. Finally, age and resection status were added to the feature set [34]. A total of 11,468 features were extracted combining the above features including imaging, texture and wavelet based features. Feature reduction was used to reduce the large number of features based on the feature importance determined by the gradient boost model. This reduced the feature space from 11,468 to 17 features. Table 1 shows a list of the selected 17 features for survival analysis. The 17 features were a combination of 10 Imaging features, 6 wavelet-imaging features and one non-imaging feature. All networks were trained using Tensorflow [35] backend engines, the Keras [36] python package and Pycharm IDEs on Tesla V100s and/or P40 NVIDIA-GPUs.

Table 1. Selected 17 features for survival analysis

Feature No.	MR Imaging sequence	Tumor Mask	Pyradiomics Feature Name	Wavelet Component	Feature Category
1	T1C	Necrosis	original_glszm_SizeZoneNonUniformity	NA	Imaging
2	T2w-FLAIR	Enhancing	original_firstorder_Skewness	NA	Imaging
3	T2w-FLAIR	Necrosis	original_glszm_LargeAreaLowGrayLevelEmphasis	NA	Imaging
4	T2w-FLAIR	Edema	original_glszm_SmallAreaLowGrayLevelEmphasis	NA	Imaging
5	T2w	Enhancing	original_firstorder_Kurtosis	NA	Imaging
6	T2w	Enhancing	original_firstorder_Maximum	NA	Imaging
7	T2w	Necrosis	original_shape_Maximum2DDiameterRow	NA	Imaging
8	T1w	Enhancing	original_firstorder_Range	NA	Imaging
9	T1w	Enhancing	original_firstorder_Skewness	NA	Imaging
10	T1w	Edema	original_firstorder_Minimum	NA	Imaging
11	T2w-FLAIR	Enhancing	original_firstorder_Skewness	Component 1	Wavelet- Imaging
12	T1C	Necrosis	original_glszm_HighGrayLevelZoneEmphasis	Component 2	Wavelet- Imaging
13	T1C	Edema	original_firstorder_Minimum	Component 2	Wavelet- Imaging
14	T1C	Edema	original_glszm_GrayLevelNonUniformity	Component 4	Wavelet- Imaging
15	T1C	Edema	original_firstorder_Minimum	Component 4	Wavelet- Imaging
16	T1w	Edema	original_glszm_SizeZoneNonUniformity	Component 5	Wavelet- Imaging
17	Age	NA	NA	NA	Non Imaging Feature

2.4   Testing

2.4.1   HGG and LGG Classifier

The classifier was evaluated on 67 cases including 52 HGGs and 15 LGGs. To generalize the network's performance, a 3 fold cross-validation was also performed. While we used the HGG and LGG classifier as the first step for brain tumor segmentation, the BraTS2019 validation dataset did not include labels for HGG and LGG precluding evaluation of classification accuracy for this initial step on the validation dataset.

2.4.2   Brain Tumor Segmentation

All the networks were tested on 125 cases from the BraTS2019 validation dataset and 166 cases from BraTS2019 testing dataset. Patches of size $32 \times 32 \times 32$ were provided to the networks for testing. The prediction patches were then used to reconstruct a full segmentation volume. Each group was tested in 3 different ways including (a) non-overlapping patches, (b) 25% overlapping patches and (c) 50% overlapping patches. At the end of testing, each group produced 3 segmentation volumes for a particular label resulting in 9 segmentation volumes across the 3 groups. The 9 segmentation volumes were assigned with equal weights, averaged and thresholded at 0.5 for enhancing tumor, 0.8 for whole tumor and 0.7 for tumor core for every voxel. The same procedure was performed on whole tumor, tumor core and enhancing tumor labels. The ensembled output of whole tumor, tumor core and enhancing tumor were fused in a post-processing step that included the 3D connected components algorithm to improve prediction accuracy by removing false positives.

2.4.3   Survival Analysis

The network was evaluated on 29 cases from the BraTS2019 validation dataset and 107 cases from BraTS2019 testing dataset. The Survival analysis was evaluated only for the GTR cases. The predicted overall survival (OS) task classified the subjects as long-survivors (greater than 15 months), mid- survivors (between 10 to 15 months) and short-survivors (less than 10 months).

2.4.4   Uncertainty Task

Uncertainty masks were also created for the 125 cases from the BraTS2019 validation dataset. The predicted probability maps for a particular label from each group were assigned with equal weights, averaged and scaled from 0 to 100 such that 0 represents the most certain prediction and 100 represents the most uncertain. In this task, uncertain voxels were removed at multiple predetermined threshold points. The performance of the networks was assessed based on the dice score of the remaining voxels.

3   Results

3.1   HGG and LGG Classifier

The network achieved a testing accuracy of 89% and an average cross-validation accuracy of 90%.

3.2   Brain Tumor Segmentation

This method achieved average dice scores of 0.95, 0.93 and 0.90 on WT, TC and ET on the training dataset (Table 1). It achieved average dice scores of 0.901, 0.844, 0.801 and sensitivity of 0.924, 0.846, 0.796 specificity of 0.991, 0.999, 0.998 for WT, TC and ET respectively on the BraTS2019 validation cases along with Hausdorff distances of 7.60, 8.31 and 5.49 mm respectively. The network also achieved average dice scores of 0.877, 0.835, and 0.803 for WT, TC and ET respectively on the BraTS2019 testing cases (Table 2).

Table 2. Mean dice scores on BraTS2019 datasets

	Whole tumor	Tumor core	Enhancing tumor
BraTS2019 training dataset	0.951	0.930	0.900
BraTS2019 validation dataset	0.901	0.844	0.801
BraTS2019 testing dataset	0.877	0.835	0.803

3.3   Survival Task

The linear regression model achieved an accuracy of 55.2% and mean squared error (MSE) of 119244.6 () on the BraTS2019 validation dataset. It also achieved an accuracy of 51% and MSE of 455500 on the BraTS2019 testing dataset.

3.4   Uncertainty Task

The networks achieved average dice_AUC of 0.894, 0.887, 0.876 and FTP_AUC_ratio of 0.012, 0.061 and 0.046 for WT, TC and ET respectively on the BraTS2019 validation dataset (Figs. 3, 4 and Table 3).

Table 3. Uncertainty task results

	Dice AUC whole tumor	Dice AUC tumor core	Dice AUC enhancing tumor	FTP ratio AUC whole tumor	FTP ratio AUC tumor core	FTP ratio AUC enhancing
BraTS2019 Validation dataset	0.894	0.887	0.876	0. 012	0. 061	0. 046

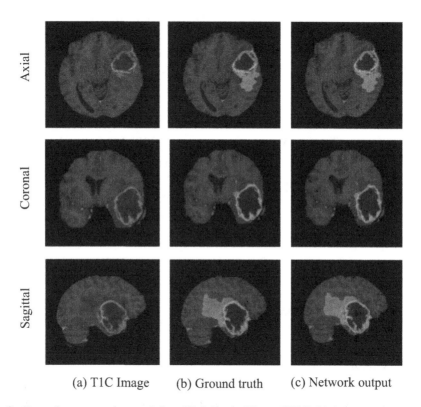

(a) T1C Image   (b) Ground truth   (c) Network output

Fig. 3. Example segmentation result for a High Grade Glioma (HGG) (a) A post-contrast image. (b) Ground truth (c) Network output. Color Code: Red = Enhancing tumor, Blue = tumor core (Enhancing tumor + non-enhancing tumor + necrosis), Green = Edema, Whole tumor = Green + Blue + Red. (Color figure online)

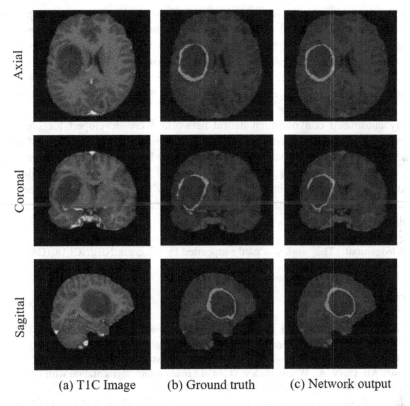

(a) T1C Image   (b) Ground truth   (c) Network output

Fig. 4. Example segmentation result for a Low Grade Glioma (HGG) (a) A post-contrast image. (b) Ground truth (c) Network output. Color Code: Red = Enhancing tumor, Blue = tumor core (Enhancing tumor + non-enhancing tumor + necrosis), Green = Edema, Whole tumor = Green + Blue + Red. (Color figure online)

4 Discussion and Conclusion

Accurate, efficient, and reliable tumor segmentation algorithms have the potential to improve management of GBM patients. Currently the vast majority of clinical and research efforts to evaluate response to therapy utilize gross geometric measurements. MRI-based glioma segmentation algorithms represent a method to reduce subjectivity and provide accurate quantitative analysis to assist in clinical decision making and improve patient outcomes. Although the underlying task can be simply stated as a voxel-level classification, a wide variety of automated and semi-automated tumor segmentation algorithms have been proposed. In this work, we developed a fully automated deep learning method to classify gliomas as high grade and low-grade, segment brain tumors into subcomponents and predict overall survival. This method was tested on 125 cases from the BraTS2019 validation dataset and 166 cases from the BraTS2019 testing dataset for the tumor segmentation task. It was also tested 29 cases from the BraTS2019 validation dataset and 107 cases from the BraTS2019 testing dataset for the survival analysis task. A three group framework was used, providing several advantages compared to the currently existing methods. Using three binary segmentation networks for segmenting the tumor into its sub-components allowed us to use a simpler network for each task [14]. The networks were easier to train with reduced over-fitting [14]. Furthermore, since all three networks were trained separately as binary segmentation problems, misclassification was greatly reduced, as demonstrated by the results from the uncertainty task.

Fully automated convolutional neural networks were developed for segmenting brain tumors into their subcomponents. This algorithm reached high performance accuracy on the BraTS2019 validation dataset. High Dice scores, accuracy and speed of this network allows for large scale application in brain tumor segmentation. This method can be implemented in the clinical workflow for reliable tumor segmentation, survival prediction and for providing clinical guidance in diagnosis, surgical planning and follow up assessments.

Acknowledgement. This work was partly supported by the grant, NIH/NCI U01CA207091.

References

1. Ostrom, Q.T., Gittleman, H., Truitt, G., Boscia, A., Kruchko, C., Barnholtz-Sloan, J.S.: CBTRUS statistical report: primary brain and other central nervous system tumors diagnosed in the United States in 2011–2015. Neuro Oncol. **20**(suppl_4), iv1–iv86 (2018)
2. Havaei, M., Davy, A., Warde-Farley, D., et al.: Brain tumor segmentation with Deep Neural Networks. Med. Image Anal. **35**, 18–31 (2017)
3. Louis, D.N., Ohgaki, H., Wiestler, O.D., et al.: The 2007 WHO classification of tumours of the central nervous system. Acta Neuropathol. **114**(2), 97–109 (2007). https://doi.org/10.1007/s00401-007-0243-4
4. Kleihues, P., Cavenee, W.K.: Pathology and genetics of tumours of the nervous system, vol 2. International Agency for Research on Cancer (2000)
5. Lacroix, M., Abi-Said, D., Fourney, D.R., et al.: A multivariate analysis of 416 patients with glioblastoma multiforme: prognosis, extent of resection, and survival. J. Neurosurg. **95**(2), 190–198 (2001)
6. Hakin-Smith, V., Jellinek, D., Levy, D., et al.: Alternative lengthening of telomeres and survival in patients with glioblastoma multiforme. Lancet **361**(9360), 836–838 (2003)
7. Johnson, D.R., O'Neill, B.P.: Glioblastoma survival in the United States before and during the temozolomide era. J. Neurooncol. **107**(2), 359–364 (2012)
8. Myronenko, A.: 3D MRI brain tumor segmentation using autoencoder regularization. In: MICCAI_BraTS_2018_Proceedings_shortpapers (2018)
9. Holland, E.C.: Progenitor cells and glioma formation. Curr. Opin. Neurol. **14**(6), 683–688 (2001)
10. Menze, B.H., Jakab, A., Bauer, S., et al.: The multimodal brain tumor image segmentation benchmark (BRATS). IEEE Trans. Med. Imaging **34**(10), 1993–2024 (2015)
11. Shreyas, V., Pankajakshan, V.: A deep learning architecture for brain tumor segmentation in MRI images. In: 2017 IEEE 19th International Workshop on Multimedia Signal Processing (MMSP), Luton, pp. 1–6 (2017)

12. Pei, L., Reza, S.M.S., Li, W., Davatzikos, C., Iftekharuddin, K.M.: Improved brain tumor segmentation by utilizing tumor growth model in longitudinal brain MRI. Proc. SPIE Int. Soc. Opt. Eng. **10134** (2017)

13. Kamnitsas, K., et al.: Ensembles of multiple models and architectures for robust brain tumour segmentation. In: Crimi, A., Bakas, S., Kuijf, H., Menze, B., Reyes, M. (eds.) BrainLes 2017. LNCS, vol. 10670, pp. 450–462. Springer, Cham (2018). https://doi.org/10. 1007/978-3-319-75238-9_38

14. Wang, G., Li, W., Ourselin, S., Vercauteren, T.: Automatic brain tumor segmentation using cascaded anisotropic convolutional neural networks. In: Crimi, A., Bakas, S., Kuijf, H., Menze, B., Reyes, M. (eds.) BrainLes 2017. LNCS, vol. 10670, pp. 178–190. Springer, Cham (2018). https://doi.org/10.1007/978-3-319-75238-9_16

15. Funke, J., Martel, J.N., Gerhard, S., et al.: Candidate sampling for neuron reconstruction from anisotropic electron microscopy volumes. Med. Image Comput. Comput. Assist. Interv. **17**(Pt 1), 17–24 (2014)

16. Soeda, A., Hara, A., Kunisada, T., Yoshimura, S.-I., Iwama, T., Park, D.M.: The evidence of glioblastoma heterogeneity. JSR **5**, 7979 (2015)

17. Shboul, Z.A., Vidyaratne, L., Alam, M., Iftekharuddin, K.M.: Glioblastoma and survival prediction. Paper presented at International MICCAI Brainlesion Workshop (2017)

18. Yang, D., Rao, G., Martinez, J., Veeraraghavan, A., Rao, A.: Evaluation of tumor-derived MRI-texture features for discrimination of molecular subtypes and prediction of 12-month survival status in glioblastoma. JMP **42**(11), 6725–6735 (2015)

19. Lee, J., Jain, R., Khalil, K., et al.: Texture feature ratios from relative CBV maps of perfusion MRI are associated with patient survival in glioblastoma. Am. J. Neuroradiol. **37**(1), 37–43 (2016)

20. Sanghani, P., Ang, B.T., King, N.K.K., Ren, H.: Overall survival prediction in glioblastoma multiforme patients from volumetric, shape and texture features using machine learning. JSO **27**(4), 709–714 (2018)

21. Bakas, S., Akbari, H., Sotiras, A., et al.: Advancing the cancer genome atlas glioma MRI collections with expert segmentation labels and radiomic features. Sci. Data. **4**, 170117 (2017)

22. Bakas, S., et al.: Identifying the best machine learning algorithms for brain tumor segmentation, progression assessment, and overall survival prediction in the BRATS challenge. https://doi.org/10.17863/CAM.38755

23. Bakas, S., Akbari, H., Sotiras, A., Bilello, M., Rozycki, M., Kirby, J., et al.: Segmentation labels and radiomic features for the pre-operative scans of the TCGA-GBM collection. Cancer Imaging Arch. (2017). https://doi.org/10.7937/k9/tcia.2017.klxwjj1q

24. Bakas, S., Akbari, H., Sotiras, A., Bilello, M., Rozycki, M., Kirby, J., et al.: Segmentation labels and radiomic features for the pre-operative scans of the TCGA-LGG collection. Cancer Imaging Arch. (2017). https://doi.org/10.7937/k9/tcia.2017.gjq7r0ef

25. Tustison, N.J., Cook, P.A., Klein, A., et al.: Large-scale evaluation of ANTs and FreeSurfer cortical thickness measurements. Neuroimage **99**, 166–179 (2014)

26. Jegou, S., Drozdzal, M., Vazquez, D., Romero, A., Bengio, Y.: The One Hundred Layers Tiramisu: Fully Convolutional DenseNets for Semantic Segmentation (2017)

27. Taha, A.A., Hanbury, A.: Metrics for evaluating 3D medical image segmentation: analysis, selection, and tool. BMC Med. Imaging **15**, 29 (2015)

28. Lin, T.Y., Goyal, P., Girshick, R., He, K., Dollar, P.: Focal loss for dense object detection. IEEE Trans. Pattern Anal. Mach. Intell. **42**(2), 318–327 (2020)

29. Wegmayr, V., AS, B.J., Petrick, N., Mori, K. (eds.): Classification of brain MRI with big data and deep 3D convolutional neural networks. In: Published in SPIE Proceedings, Medical Imaging 2018: Computer-Aided Diagnosis, p. 1057501 (2018)

30. Feng, X., Yang, J., Lipton, Z.C., Small, S.A., Provenzano, F.A.: Deep learning on MRI affirms the prominence of the hippocampal formation in Alzheimer's disease classification. bioRxiv. 2018; 2018:456277

31. Van Griethuysen, J.J., Fedorov, A., Parmar, C., et al.: Computational radiomics system to decode the radiographic phenotype. Cancer Res. **77**(21), e104–e107 (2017)

32. Lee, G.R., Gommers, R., Waselewski, F., Wohlfahrt, K., O'Leary, A.: PyWavelets: a Python package for wavelet analysis. J. Open Source Softw. **4**(36), 1237 (2019)

33. Winger, L.L., Venetsanopoulos, A.N.: Biorthogonal nearly coiflet wavelets for image compression. JSPIC **16**(9), 859–869 (2001)

34. Feng, X., Tustison, N., Meyer, C.: Brain tumor segmentation using an ensemble of 3d U-Nets and overall survival prediction using radiomic features. Paper presented at International MICCAI Brainlesion Workshop (2018)

35. Abadi, M., et al.: Tensorflow: a system for large-scale machine learning. In: OSDI, pp. 265–284 (2016)

36. Charles, P.W.D.: Keras. GitHub repository (2013)

3D Automatic Brain Tumor Segmentation Using a Multiscale Input U-Net Network

S. Rosas González[1], T. Birgui Sekou[2], M. Hidane[2],
and C. Tauber[1(✉)]

[1] UMR U1253 iBrain, Université de Tours, Inserm, Tours, France
sarahi.rosasgonzalez@etu.univ-tours.fr,
clovis.tauber@univ-tours.fr
[2] INSA CVL, Université de Tours, 6300 LIFAT EA, Tours, France

Abstract. Quantitative analysis of brain tumors is crucial for surgery planning, follow-up and subsequent radiation treatment of glioma. Finding an automatic and reproducible solution may save time to physicians and contribute to improve overall poor prognosis of glioma patients. In this paper, we present our current BraTS contribution on developing an accurate and robust tumor segmentation algorithm. Our network architecture implements a multiscale input module which has been thought to maximize the extraction of features associated to the multiple image modalities before they are merged in a modified U-Net network avoiding the loss of specific information provided by each modality and improving brain tumor segmentation performance. Our method's current performance on the BraTS 2019 test set is dice scores of 0.775 ± 0.212, 0.865 ± 0.133 and 0.789 ± 0.266 for enhancing tumor, whole tumor and tumor core, respectively with and overall dice of 0.81.

Keywords: Tumor segmentation · Deep-learning · BraTS

1 Introduction

Glioma is the most frequent primary brain tumor that originates from glial cells [1]. The main treatment is surgical resection followed by radiation therapy and/or chemotherapy. Gliomas can be classified into grade I to IV based on the phenotypic cell characteristics. In this grading system, Low Grade Gliomas (LGG) correspond to grades I and II whereas High Grade Gliomas (HGG) are grades III and IV. Compared with HGG, LGG are less aggressive and have a better prognosis with a survival range from 1 to 15 years in contrast with 1 to 2 years of HGG [2].

Magnetic Resonance Imaging (MRI) is a non-invasive imaging technique commonly used for diagnosis, surgery planning and follow-up of brain tumors due to its high-resolution on brain structures.

Commonly, multiple 3D MRI modalities are acquired to obtain complementary information from tumor regions based on the difference of magnetic tissue properties.

S. Rosas González and T. Birgui Sekou—Authors stand for equal contribution.

A. Crimi and S. Bakas (Eds.): BrainLes 2019, LNCS 11993, pp. 113–123, 2020.
https://doi.org/10.1007/978-3-030-46643-5_11

Conventional protocols for glioma imaging assessment generally include T2, Fluid Attenuation Inversion Recovery (FLAIR), T1 and post-contrast T1 (T1c).

Currently, tumor regions are segmented manually from MRI images by radiologists, but due to variability in image appearance, the process is time consuming and inter-observer reproducibility is considerably low [3].

Since accurate tumor segmentation is determinant in surgery planning, follow-up and subsequent radiation treatment of glioma, finding an automatic and reproducible solution may save time to physicians and contribute to improve overall poor prognosis of glioma patients.

The aim of Multimodal Brain Tumor Segmentation Challenge (BraTS) is to evaluate state of the art segmentation algorithms by making available an extensive pre-operative multimodal MRI dataset with ground truth tumor segmentation labels annotated by multiple experts and approved by experienced neuro-radiologist [4–8].

In this work, we implemented a multiscale input module which has been thought to maximize the extraction of features associated to the multiple image modalities before they are merged in a modified 3D U-Net network [9] avoiding the loss of specific information provided by each modality to better segment brain tumors.

1.1 Related Work

One popular neural network approach for image segmentation models is to proceed from an encoder/decoder structure where the encoder module consists of multiple connected convolution layers that aim to gradually reduce the spatial dimension of feature maps and capture more high-level semantic features learned to be highly efficient at discriminating between classes. The decoder module uses up sampling layers to recover the spatial dimension and object representation.

An intrinsic dilemma between extracting semantic and spatial information was described by J. Long et al. [10] due to the difficulty of obtaining a model able to make local predictions that respect global structure. The authors address this problem by adding "skip connections" between intermediate up sampled layers from the encoded representation and lower layers, and summing both feature maps. The main idea is to combine local and spatial information. The benefits of skip connections have been investigated by Drozdzal et al. [11] in medical image context.

Ronneberger et al. [12] improved previous architectures through expanding the capacity of the decoder module of the network. They proposed a model called U-Net which is also a Fully Convolutional Network (FCN) encoder/decoder architecture. The main contribution of U-Net is that while upsampling and going deeper in the network the model concatenates the higher resolution features from encoder module with the upsampled features in a symmetric decoder path to better localize and learn representations in following convolutions. U-Net has been widely adapted for a variety of segmentation problems and has gained much attention [13–16].

Many modifications have been proposed based on U-Net architecture. The standard model consists of a series of blocks containing convolution operations. Some variants include replacing these by residual [11] and dense blocks [17]. Residual block introduces short skip connections over a single block alongside the existing long skip connections between the corresponding feature maps of encoder and decoder modules.

The authors report that the short skip connections allow for faster convergence when training and are beneficial in creating deep architectures.

Many contributions from previous editions focused on training their networks to segment the brain regions evaluated (whole tumor, core tumor and enhancing tumor) instead of training on segmenting the regions associated to the labels 1 to 4 which suggests helping to get better scores. Moreover, many competitors used ensemble of models to improve their results. Top performing submissions of BraTS 2018 challenge included the first-place of the competition A. Myronenko [18] who proposed the implementation of a variational autoencoder (VAE) branch to an encoder/decoder model to reconstruct the input image simultaneously with segmentation, it seems to have a regularization effect on the encoder part of the network, forcing earlier layers in the network to extract more relevant features, the second-place contribution Isensee et al. [19] proposed the use of a 3D U-Net architecture with minor but important modifications, this include the implementation of instance normalization [20] and leaky ReLU, demonstrating that with a well-trained U-Net network it is possible to obtain competitive results. McKinly et al. [21] proposed an ensemble of DenseNet structures with dilated convolutions [22] embedded in a U-Net like network with encoder/decoder structure and skip connections. Finally, Zhou et al. [23] proposed and ensemble of different models to segment the three different regions in cascade considering multi-scale context information.

2   Method

2.1   Network Architecture

Our proposed network consists of two components: a modified 3D U-Net responsible for segmentation, and a multiscale-input module which has been thought to maximize learning from independent features associated to each image modality before they are merged in the encoding/decoding architecture, avoiding loss of specific information provided by each modality. Despite the images provided were already preprocessed to homogenize data and they were resampled to an isotropic resolution, the images can still contain inhomogeneities in spatial information distribution either in the same plane or between axial, sagittal and coronal planes. We start by dividing the input in 4 pathways one for each image modality, then we applied a 3D convolution followed by a multiscale module. This module has two main characteristics, the first one is that it contains three branches, each with two convolution blocks with asymmetric filter size to emphasize a different orthogonal plane, i.e. axial, sagittal and coronal planes, as shown in Fig. 2a. The second characteristic is that each branch has kernel filters of different size of three, five and seven.

The choice of multiscale blocks is motivated by the fact that different kernel sizes allow to extract different spatial information, which can be globally or locally located in the image. After each convolution, an instance normalization [20] and ReLU activation are applied. We call this structure Multiscale input module.

The network architecture is illustrated in Fig. 1. In the second component of our network (a modified U-Net), each level of the encoding pathway consists of two

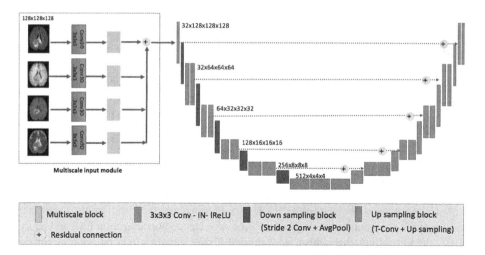

Fig. 1. Proposed Multiscale Input U-Net. A modified 3D U-Net architecture using instance normalization [20], transpose convolutions for up sampling and residual connections between encoding and decoding corresponding feature maps instead of concatenation operation used in the conventional U-Net. (Color figure online)

residual blocks (green blocks) followed by a down sampling block (blue blocks), which includes a 3D convolution with a kernel size of $3 \times 3 \times 3$ and a stride of 2 (see Fig. 2b). As a result, the spatial dimension of the input tensor is reduced by a factor of 8 in the encoding path. In the decoding pathway, we proposed a symmetrical operation to that of the encoding module, using a 3D transpose convolution with a kernel size of $3 \times 3 \times 3$ and a stride of 2 join with upsample operation (Fig. 2b) which progressively doubles the spatial dimension on each level to eventually recover the original spatial dimension. We added residual connections inside each block since these have been shown to be beneficial for faster convergence.

The final activation layer uses a sigmoid as activation function to output three prediction maps corresponding to the regions evaluated (enhancing tumor, whole tumor and tumor core).

2.2   Dataset

The preoperative MRI images used in this work were provided by BraTS 2019 Challenge [4–8]. The training dataset contains four modalities T2, FLAIR, T1 and T1c for each of the 335 glioma patients, of which 259 are HGG and 76 are LGG.

All the subjects in the training dataset are provided with ground truth labels for 3 regions: enhancing tumor (ET - label 4), the peritumoral edema (ED - label 2), and the necrotic and non-enhancing tumor core (NCR/NET - label 1). The validation dataset includes the same MRI modalities for 125 subjects which have no expert segmentation annotations and the grading information.

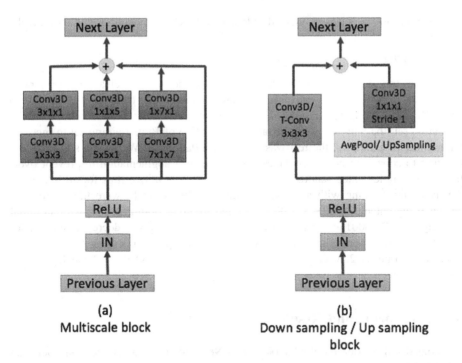

Fig. 2. Three different blocks used inside the proposed model. The block (a) is used inside the Multiscale input module. (b) Summarize two symmetric blocks, down sampling and up sampling blocks (blue and orange blocks in Fig. 1). IN represents Instance Normalization. (Color figure online)

2.3 Preprocessing

The multimodal scans in BraTS challenge were acquired from multiple institutions, employing different clinical protocols, resulting in a non-standardized distribution.

Even though the images provided were already preprocessed to homogenize data as described in [5], image intensity variability is still a problem, contrarily to other imaging techniques such as computed tomography, magnetic resonance imaging does not have a standard intensity scale. Therefore, image intensity normalization is a necessary stage for model convergence.

We chose to normalize the MRI images first dividing each modality by the maximum value and then using the z-scores due to its simplicity and qualitative good performance. It uses the brain mask of each image to determine the mean of image intensities inside the brain mask and subtracting it from individual intensities and then dividing the difference by the standard deviation of the image intensities (Eq. 1).

$$z = \frac{X - \mu}{\sigma} \tag{1}$$

where X is the image intensities, μ is mean of the image intensities, and σ is the standard deviation of the image intensities.

The preprocessing was conducted independently across modalities and individuals.

2.4   Training

During the training phase of the competition and before the release of the BraTS validation set, the data was split into training, validation and test subsets with 80, 10 and 10% of the original training set respectively. After the official release of the validation set, the networks were retrained using 85% of Brats training set to increase the number of examples, the rest was used as validation set. We used a soft dice loss function to cope with class imbalances and some data augmentation to prevent over-fitting. We implemented random crops and random axis mirror flips for all 3 axes on the fly during training with a probability of 0.5. The networks were trained for 100 epochs using patches of $128 \times 128 \times 128$ and Adam as optimizer [24] with an initial learning rate of 0.0001 and batch size of 1. The learning rate was decreased by a factor of 5 if no improvement is seen within 10 epochs. All experiments were conducted on a workstation Intel-i7 2.20 GHz CPU, 48 GB RAM and an NVIDIA Titan Xp 12 GB GPU.

3   Experiments and Results

We considered as our baseline model a U-Net with residual connections between encoding and decoding corresponding feature maps instead of concatenation operation usually used in conventional U-Net. We experimented with different variants of U-Net, using residual blocks, multiscale blocks and dense blocks instead of using simple convolutional blocks included in conventional U-Net. We compared the performance using multiscale blocks inside U-Net (U-Net + MSb) which improves overall dice with respect to Conventional U-Net (Table 1). We then experimented replacing multiscale blocks with residual blocks i.e. convolutional blocks with short residual connections within each block (U-Net + ResCon) that helped reducing memory demanding by the model and allowing to increment the number of filters on each level, which also improved the overall performance compared with U-Net + MSb.

Afterwards, we experimented by incorporating an input module to the previously adapted U-Net (U-Net + ResCon) that allowed to enter the four different modalities of magnetic resonance imaging independently and extract some features before concate-nating the images to be entered the adapted U-Net. We have continued to explore different input module configurations until we found a configuration that extracts as much information as possible from each image modality, we called this network: Multiscale Input U-Net (MIU-Net).

We also experimented with using different loss functions, soft dice, weighted soft dice, binary cross entropy (bce) and adding soft dice with bce. We found similar behavior using the addition of soft dice and bce and soft dice alone, both better than training with bce, then we choose to train with soft dice.

Finally, we implemented a post processing of the final predictions maps to reduce the number of false positives on enhancing tumor detection by replacing all the voxels values of the enhancing tumor region with the values corresponding to necrosis region

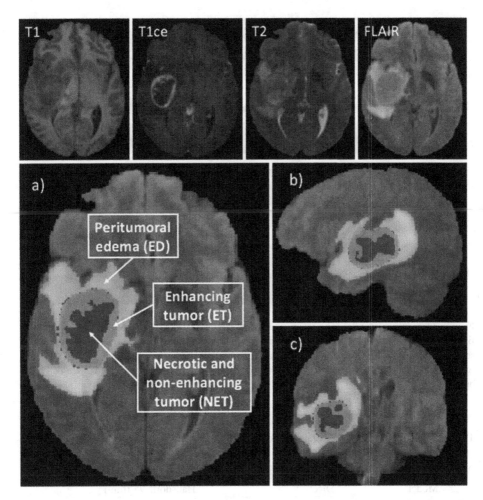

Fig. 3. Qualitative results. The case shown is patient BraTS19_TCIA01_147_1 from the BRATS 2019 validation dataset. Up the four MRI modalities: T1, T1ce, T2, Flair. Down the predicted segmentation over the flair modality in a) axial, b) sagittal and c) coronal view. The whole tumor (WT) class includes the union of the three regions (ED, ET, and NET), the tumor core (TC) class is the union of red and yellow (ET and NET). (Color figure online)

when the total number of predicted voxel is less than a threshold value found as the minimum number of voxels on annotated enhancing tumor region in the training data.

We report the results of our proposed benchmark on BraTS 2019 validation dataset (125 cases) (Table 1) and in the test dataset (166 cases) (Table 2). The best performance was obtained when using the Multiscale Input U-Net combined with postprocessing (MIU-Net + PostP).

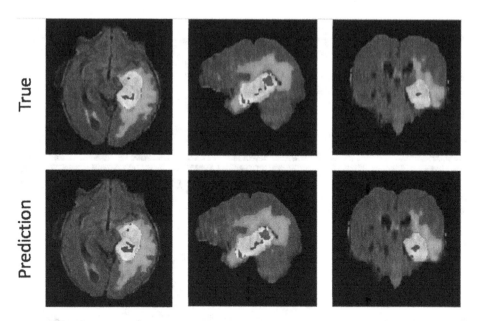

Fig. 4. Qualitative results. The case shown is patient BraTS19_TCIA01_147_1 from the BRATS 2019 training dataset. Up the Ground Truth segmentation (True) and down the predicted segmentation over flair MRI axial, sagittal and coronal view depicting the regions: Edema (green), enhancing tumor (yellow), and necrotic and non-enhancing tumor (red). (Color figure online)

Table 1. Results on BraTS 2019 validation data (125 cases). Metrics were computed by the online evaluation platform. The mean Dice coefficient and mean Hausdorff distance are reported. ET – enhancing tumor, WT – whole tumor, TC – tumor core.

Model	Dice			Hausdorff 95 (mm)		
	ET	WT	TC	ET	WT	TC
Conventional U-Net	0.63064	0.88401	0.77782	7.39285	8.91082	8.86673
U-Net + MSb	0.66249	0.88288	0.74763	6.77522	6.9740	8.89163
U-Net + ResCon	0.69308	0.88803	0.78079	6.34973	7.41774	8.77463
MIU-Net	0.71193	**0.89096**	**0.79217**	5.73875	**7.40938**	7.76064
MIU-Net + PostP	**0.72312**	0.88816	0.78334	**4.91319**	8.12305	**7.56179**

An example of image segmentation predicted by our model is shown in Fig. 3. In this image it can be verified that the information obtained from the different MRI modalities is complementary, since each image provides different contrast from the different tumor regions. In Fig. 4 the predicted segmentation and the ground truth are compared, qualitatively they seem to match well.

Table 2. Results on BraTS 2019 test data (166 cases) using our final model. Metrics were computed by the online evaluation platform. ET – enhancing tumor, WT – whole tumor, TC – tumor core.

	Dice			Hausdorff 95 (mm)		
	ET	WT	TC	ET	WT	TC
Mean	**0.77479**	**0.86505**	**0.78931**	**3.08938**	**7.42054**	**6.23278**
StdDev	0.21235	0.13252	0.26564	3.53525	10.89553	8.50941
Median	0.83309	0.90487	0.89361	2	4.12311	3.503
25 quantile	0.74319	0.84705	0.79877	1.41421	2.23607	2
75 quantile	0.89199	0.93734	0.94046	3.16228	7.07107	7.0533

4 Discussion

We have experimented using different models, in all of them the most challenging region to segment was enhancing tumor. The proposed method improved our overall performance by improving segmenting this region. We consider that some preprocessing and post processing techniques would help to improve predictions in this region. We improved slightly the dice on the enhancing tumor region by applying a post-processing to modify our final predictions maps and reduce the number of false positives in this region, nevertheless the performance in the other two regions was affected in the same proportion, maintaining the same overall dice value of 0.80 which is the average of the three regions on the validation dataset.

Comparing the values obtained in the validation and test datasets we can conclude that our model generalized well since very close values were obtained in both sets even when we did not perform any assembly of models and we did not use more complicated data augmentation methods. The post-processing worked better on the test dataset since the performance on the enhancing and tumor core regions was improved with respect to the values on the validation set, and the overall dice increased to 0.81. The improvement in Hausdorff distance was also significant.

We attribute the good generalization of our model to the fact that we have performed a validation of our model using the training data set and we used the official validation data set as a first test set. Therefore, if the distributions of validation and testing datasets are similar, the results had to be similar in both.

We have obtained higher median values than average, which means that our model has a good performance in most of the images but that there are few in which a bad segmentation is obtained, causing the average to remain below the median.

Even when we did not observe overfitting behavior we could improve the performance of our model by adding more complex data augmentation methods, such as affine and image intensity transformations. We could also improve the results obtained by implementing an ensemble of models either an ensemble that averages different models or that uses the voting system method to select the most repeated predictions by the models. In the future work we intend to investigate further the effect of our multiscale module in the improvement of segmentation and implement different methods of data augmentation and ensemble of models methods.

5 Conclusions

We have proposed an end-to-end FCN architecture for preoperative MRI tumor segmentation. To demonstrate the effect of the proposed multiscale input module we have compared the results of trained models without this module. We can conclude that our model generalized well since values obtained on validation and test datasets are very close. We obtained dice scores of 0.775 ± 0.212, 0.865 ± 0.133 and 0.789 ± 0.266 for enhancing tumor, whole tumor and tumor core, respectively with and overall dice of 0.81. We have experimented incorporating different modifications into U-Net architecture, results are promising but is still possible to improve them. Further study is needed to demonstrate the usefulness of this module over other improvements.

Acknowledgements. This work was supported by the Mexican Council for Science and Technology CONACYT (Grant 494208), INSA CVL and the Inserm unit 1253.

References

1. Kleihues, P., Burger, P.C., Scheithauer, B.W.: The new WHO classification of brain tumours. Brain Pathol. **3**(3), 255–268 (1993)
2. Hoshide, R., Jandial, R.: 2016 world health organization classification of central nervous system tumors: an era of molecular biology. World Neurosurg. **94**, 561–562 (2016)
3. Kubben, P.L., Postma, A.A., Kessels, A.G.H., van Overbeeke, J.J., van Santbrink, H.: Intraobserver and interobserver agreement in volumetric assessment of glioblastoma multiforme resection. Neurosurgery **67**(5), 1329–1334 (2010)
4. Menze, B.H., et al.: The multimodal brain tumor image segmentation benchmark (BRATS). IEEE TMI **34**(10), 1993–2024 (2015)
5. Bakas, S., et al.: Advancing The Cancer Genome Atlas glioma MRI collections with expert segmentation labels and radiomic features. Nat. Sci. Data (2017, in press)
6. Bakas, S., et al.: Segmentation labels and radiomic features for the pre-operative scans of the TCGA-GBM collection. TCIA (2017)
7. Bakas, S., et al.: Segmentation labels and radiomic features for the pre-operative scans of the TCGA-LGG collection. TCIA (2017)
8. Bakas, S., et al.: Identifying the best machine learning algorithms for brain tumor segmentation, progression assessment, and overall survival prediction in the BRATS challenge. arXiv preprint arXiv:1811.02629 (2018)
9. Çiçek, Ö., Abdulkadir, A., Lienkamp, S.S., Brox, T., Ronneberger, O.: 3D U-Net: learning dense volumetric segmentation from sparse annotation. In: Ourselin, S., Joskowicz, L., Sabuncu, M., Unal, G., Wells, W. (eds.) MICCAI 2016. LNCS, vol. 9901, pp. 424–432. Springer, Cham (2016). https://doi.org/10.1007/978-3-319-46723-8_49
10. Long, J., Shelhamer, E., Darrell, T.: Fully convolutional networks for semantic segmentation. In: Proceedings of the IEEE Conference on Computer Vision and Pattern Recognition, pp. 3431–3440 (2015)
11. Drozdzal, M., Vorontsov, E., Chartrand, G., Kadoury, S., Pal, C.: The importance of skip connections in biomedical image segmentation. In: Carneiro, G., et al. (eds.) LABELS/DLMIA -2016. LNCS, vol. 10008, pp. 179–187. Springer, Cham (2016). https://doi.org/10.1007/978-3-319-46976-8_19

12. Ronneberger, O., Fischer, P., Brox, T.: U-Net: convolutional networks for biomedical image segmentation. In: Navab, N., Hornegger, J., Wells, W., Frangi, A.F. (eds.) MICCAI 2015. LNCS, vol. 9351, pp. 234–241. Springer, Cham (2015). https://doi.org/10.1007/978-3-319-24574-4_28

13. Norman, B., Pedoia, V., Majumdar, S.: Use of 2D U-Net convolutional neural networks for automated cartilage and meniscus segmentation of knee MR imaging data to determine relaxometry and morphometry. Radiology **288**, 177–185 (2018). 172322

14. Sevastopolsky, A.: Optic disc and cup segmentation methods for glaucoma detection with modification of U-Net convolutional neural network. Pattern Recogn. Image Anal. **27**(3), 618–624 (2017). https://doi.org/10.1134/S1054661817030269

15. Roy, A.G., et al.: ReLayNet: retinal layer and fluid segmentation of macular optical coherence tomography using fully convolutional networks. Biomed. Opt. Express **8**(8), 3627–3642 (2017)

16. Skourt, B.A., El Hassani, A., Majda, A.: Lung CT image segmentation using deep neural networks. Proc. Comput. Sci. **127**, 109–113 (2018)

17. Jégou, S., Drozdzal, M., Vazquez, D., Romero, A., Bengio, Y.: The one hundred layers Tiramisu: fully convolutional DenseNets for semantic segmentation. In: 2017 IEEE Conference on Computer Vision and Pattern Recognition Workshops (CVPRW), pp. 1175–1183. IEEE (2017)

18. Myronenko, A.: 3D MRI brain tumor segmentation using autoencoder regularization. In: Crimi, A., Bakas, S., Kuijf, H., Keyvan, F., Reyes, M., van Walsum, T. (eds.) BrainLes 2018. LNCS, vol. 11384, pp. 311–320. Springer, Cham (2019). https://doi.org/10.1007/978-3-030-11726-9_28

19. Isensee, F., Kickingereder, P., Wick, W., Bendszus, M., Maier-Hein, Klaus H.: No new-net. In: Crimi, A., Bakas, S., Kuijf, H., Keyvan, F., Reyes, M., van Walsum, T. (eds.) BrainLes 2018. LNCS, vol. 11384, pp. 234–244. Springer, Cham (2019). https://doi.org/10.1007/978-3-030-11726-9_21

20. Ulyanov, D., Vedaldi, A., Lempitsky, V.: Instance normalization: the missing ingredient for fast stylization. arXiv preprint arXiv:1607.08022 (2016)

21. McKinley, R., Meier, R., Wiest, R.: Ensembles of CNNs with label-uncertainty for brain tumor segmentation. In: Crimi, A., Bakas, S., Kuijf, H., Keyvan, F., Reyes, M., van Walsum, T. (eds.) BrainLes 2018. LNCS, vol. 11384, pp. 456–465. Springer, Cham (2019). https://doi.org/10.1007/978-3-030-11726-9_40

22. Yu, F., Koltun, V.: Multi-scale context aggregation by dilated convolutions. In: ICLR (2016)

23. Zhou, C., Chen, S., Ding, C., Tao, D.: Learning contextual and attentive information for brain tumor segmentation. In: Crimi, A., Bakas, S., Kuijf, H., Keyvan, F., Reyes, M., van Walsum, T. (eds.) BrainLes 2018. LNCS, vol. 11384, pp. 497–507. Springer, Cham (2019). https://doi.org/10.1007/978-3-030-11726-9_44

24. Kingma, D.P., Ba, L.J.: Adam: a method for stochastic optimization. In: International Conference on Learning Representations ICLR (2015)

Semi-supervised Variational Autoencoder for Survival Prediction

Sveinn Pálsson[1(✉)], Stefano Cerri[1], Andrea Dittadi[2],
and Koen Van Leemput[1,3]

[1] Department of Health Technology, Technical University of Denmark,
Lyngby, Denmark
svpa@dtu.dk

[2] Department of Applied Mathematics and Computer Science,
Technical University of Denmark, Lyngby, Denmark

[3] Athinoula A. Martinos Center for Biomedical Imaging,
Massachusetts General Hospital, Harvard Medical School, Boston, USA

Abstract. In this paper we propose a semi-supervised variational autoencoder for classification of overall survival groups from tumor segmentation masks. The model can use the output of any tumor segmentation algorithm, removing all assumptions on the scanning platform and the specific type of pulse sequences used, thereby increasing its generalization properties. Due to its semi-supervised nature, the method can learn to classify survival time by using a relatively small number of labeled subjects. We validate our model on the publicly available dataset from the Multimodal Brain Tumor Segmentation Challenge (BraTS) 2019.

Keywords: Survival time · Deep generative models · Semi-supervised VAE

1 Introduction

Brain tumor prognosis involves forecasting the future disease progression in a patient, which is of high potential value for planning the most appropriate treatment. Glioma is the most common primary brain tumor and patients suffering from its most aggressive form, glioblastoma, have generally very poor prognosis. Glioblastoma patients have a median overall survival (OS) of less than 15 months, and a 5-year OS rate of only 10% even when they receive treatment [1]. Automatic prediction of overall survival of glioblastoma patients is an important but unsolved problem, with no established method available in clinical practice.

The last few years have seen an increased interest in brain tumor survival time prediction from magnetic resonance (MR) images, often using discriminative

---

S. Pálsson, S. Cerri and A. Dittadi—Contributed equally.

© Springer Nature Switzerland AG 2020
A. Crimi and S. Bakas (Eds.): BrainLes 2019, LNCS 11993, pp. 124–134, 2020.
https://doi.org/10.1007/978-3-030-46643-5_12

methods that directly encode the relationship between image intensities and prediction labels [2]. However, due to the flexibility of MR imaging, such methods do not generalize well to images acquired at different centers and with different scanners, limiting their potential applicability in clinical settings. Furthermore, being supervised methods, they require "labeled" training data where for each training subject both imaging data and ultimate survival time are available. Although public imaging databases with survival information have started to be collected [3–6], the requirement of such labeled data fundamentally limits the number of subjects available for training, severely restricting the prediction performance attainable with current methods.

In this paper, we explore whether the aforementioned issues with supervised intensity-based methods can be ameliorated by using a semi-supervised approach instead, using only segmentation masks as input. In particular, we adapt a semi-supervised variational autoencoder model [7] to predict overall survival from a small amount of labeled training subjects, augmented with *unlabeled* subjects in which only imaging data is available. The method only takes segmentation masks as input, thereby removing all assumptions on the image modalities and scanners used.

The Multimodal Brain Tumor Segmentation Challenge (BraTS) [3] has been held every year since 2012, and focuses on the task of segmenting three different brain tumors structures ("enhancing tumor", "tumor core" and "whole tumor") and "background" from multimodal MR images. Since 2017, BraTS has also included the task of OS prediction. In this paper we focus on the latter, classifying the scans into three prognosis groups: **long-survivors** (>15 months), **short-survivors** (<10 months), and **mid-survivors** (between 10 and 15 months), all relative to the time of diagnosis.

2   Model

We begin by formally describing the problem we aim to solve. The available training data consists of a set of N_l labeled pairs $\{(\boldsymbol{x}_1, y_1), ..., (\boldsymbol{x}_{N_l}, y_{N_l})\}$, possibly augmented with a set of N_u *unlabeled* data points $\{\boldsymbol{x}_{N_l+1}, ..., \boldsymbol{x}_{N_l+N_u}\}$, where $\boldsymbol{x}_i \in \{1, ..., M_x\}^D$ is the i-th subject's image data in the form of a segmentation map with D voxels, and the target variable $y_i \in \{1, ..., M_y\}$ denotes the survival group the subject belongs to. In our case we have the segmentation of $M_x = 4$ different tumor structures as input to the model, and $M_y = 3$ different survival groups. For convenience, we will omit the index i when possible in the remainder.

We assume that the data is generated by a random process, illustrated in Fig. 1, that involves some latent variables $\boldsymbol{z} \in \mathcal{R}^L$, assumed to be independent of y, where $L \ll D$. These latent variables encode high-level tumor shape and location features shared across survival groups. Specifically, we assume a generative model of the form

$$p_\theta(\boldsymbol{x}, y, \boldsymbol{z}) = p_\theta(\boldsymbol{x}|y, \boldsymbol{z})p(\boldsymbol{z})p(y), \tag{1}$$

where $p(z) = \mathcal{N}(z|\mathbf{0}, \mathbf{I})$ is a zero-mean isotropic multivariate Gaussian, $p(y) \propto 1$ is a flat categorical prior distribution over y, and $p_\theta(x|y, z)$ is a conditional distribution parameterized by θ.

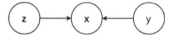

Fig. 1. Probabilistic graphical model of the generative process.

Our task is to find the maximum likelihood parameters, i.e., the parameter values θ that maximize the probability of the training data under the model. This is equivalent to maximizing

$$\sum_{i=1}^{N_l} \log p_\theta(x_i, y_i) + \sum_{i=N_l+1}^{N_l+N_u} \log p_\theta(x_i) \tag{2}$$

with respect to θ, where

$$p_\theta(x, y) = \int_z p_\theta(x, y, z) dz \tag{3}$$

and

$$p_\theta(x) = \sum_y p_\theta(x, y). \tag{4}$$

Once suitable parameter values are found, the survival group of a new subject with image data x can be predicted by assessing $p_\theta(y|x) = p_\theta(x, y)/p_\theta(x)$.

2.1   Semi-supervised Variational Autoencoder

Maximizing Eq. (2) for θ directly is not feasible due to intractability of the integral over the latent variables in Eq. (3). We therefore use an Expectation-Maximization (EM) [8] algorithm to exploit the fact that the optimization would be easier if the latent variables were known. The algorithm iteratively constructs and maximizes a lower bound to Eq. (2) in a process that involves "filling in" the missing latent variables using their posterior distribution. Since this posterior distribution is intractable, we follow [7] and approximate $p_\theta(z, y|x)$ using a specific functional form $q_\phi(z|x, y)$ with parameters ϕ:

$$q_\phi(z, y|x) = q_\phi(z|x, y)q_\phi(y|x),$$

where $q_\phi(z|x, y)$ is a multivariate Gaussian distribution with diagonal covariance matrix, and $q_\phi(y|x)$ is a categorical distribution. This approximation can be used to obtain a lower bound to Eq. (2) as follows. The probability of each *labeled* data point (first term in Eq. (2)) can be rewritten as:

$$\log p_\theta(\boldsymbol{x}, y) = \mathbb{E}_{q_\phi(\boldsymbol{z}|\boldsymbol{x},y)}[\log p_\theta(\boldsymbol{x}, y)]$$

$$= \mathbb{E}_{q_\phi(\boldsymbol{z}|\boldsymbol{x},y)}\left[\log\left[\frac{p_\theta(\boldsymbol{x}, y, \boldsymbol{z})}{p_\theta(\boldsymbol{z}|\boldsymbol{x}, y)}\right]\right]$$

$$= \mathbb{E}_{q_\phi(\boldsymbol{z}|\boldsymbol{x},y)}\left[\log\left[\frac{p_\theta(\boldsymbol{x}, y, \boldsymbol{z})}{q_\phi(\boldsymbol{z}|\boldsymbol{x}, y)}\frac{q_\phi(\boldsymbol{z}|\boldsymbol{x}, y)}{p_\theta(\boldsymbol{z}|\boldsymbol{x}, y)}\right]\right]$$

$$= \underbrace{\mathbb{E}_{q_\phi(\boldsymbol{z}|\boldsymbol{x},y)}\left[\log\left[\frac{p_\theta(\boldsymbol{x}, y, \boldsymbol{z})}{q_\phi(\boldsymbol{z}|\boldsymbol{x}, y)}\right]\right]}_{=\mathcal{L}_{\theta,\phi}(\boldsymbol{x},y)} + \underbrace{\mathbb{E}_{q_\phi(\boldsymbol{z}|\boldsymbol{x},y)}\left[\log\left[\frac{q_\phi(\boldsymbol{z}|\boldsymbol{x}, y)}{p_\theta(\boldsymbol{z}|\boldsymbol{x}, y)}\right]\right]}_{=D_{KL}(q_\phi(\boldsymbol{z}|\boldsymbol{x},y)\|p_\theta(\boldsymbol{z}|\boldsymbol{x},y))}$$

where D_{KL} denotes the Kullback-Leibler (KL) divergence. Since the KL divergence is always non-negative, we have that

$$\log p_\theta(\boldsymbol{x}, y) \geq \mathcal{L}_{\theta,\phi}(\boldsymbol{x}, y). \tag{5}$$

Using a similar derivation, the probability of each *unlabeled* data point can be bounded as follows:

$$\log p_\theta(\boldsymbol{x}) \geq \mathbb{E}_{q_\phi(y,\boldsymbol{z}|\boldsymbol{x})}\left[\log\frac{p_\theta(\boldsymbol{x}, y, \boldsymbol{z})}{q_\phi(\boldsymbol{z}|y, \boldsymbol{x})} - \log q_\phi(y|\boldsymbol{x})\right]$$

$$= \sum_y q_\phi(y|\boldsymbol{x})(\mathcal{L}_{\theta,\phi}(\boldsymbol{x}, y)) + \mathcal{H}(q_\phi(y|\boldsymbol{x})) = \mathcal{U}_{\theta,\phi}(\boldsymbol{x}), \tag{6}$$

where $\mathcal{H}(\cdot)$ denotes the entropy of a probability distribution.

By combining (5) and (6), a lower bound to Eq. (2) is finally obtained as:

$$\mathcal{J}_{\theta,\phi} = \sum_{i=1}^{N_l}\mathcal{L}_{\theta,\phi}(\boldsymbol{x}_i, y_i) + \sum_{i=N_l+1}^{N_l+N_u}\mathcal{U}_{\theta,\phi}(\boldsymbol{x}_i), \tag{7}$$

which we optimize with respect to both the variational parameters ϕ and the generative parameters θ. We use stochastic gradient ascent for the optimization, approximating gradients of the expectations in (7) as described in [9]. Implementation details are discussed in Sect. 4.

From a information theory point of view, the latent unobserved variables $\boldsymbol{z}$ can be interpreted as a code. Therefore, we can refer to the distributions $q_\phi(\boldsymbol{z}|\boldsymbol{x}, y)$ and $p_\theta(\boldsymbol{x}|y, \boldsymbol{z})$ as a probabilistic *encoder* and *decoder*, respectively [9]. The label predictive distribution $q_\phi(y|\boldsymbol{x})$ has the form of a discriminative *classifier*, and can be used as an approximation to $p_\theta(y|\boldsymbol{x})$ for classifying new cases after training.

2.2   Model Modifications

Here we describe a few model modifications for making the parameter learning process faster and less prone to overfitting.

Classification Objective. Note that in the objective function (7), the label predictive approximation $q_\phi(y|x)$ only appears in the bound for unlabeled data. To let $q_\phi(y|x)$ also learn from labeled data, we follow [7] and add a weak classification loss, resulting in the modified objective

$$\mathcal{J}_{\theta,\phi}^\alpha = \mathcal{J}_{\theta,\phi} + \alpha \sum_{i=1}^{N_l} \log q_\phi(y_i|x_i) \qquad (8)$$

where α controls the relative weight between generative and purely discriminative learning.

Gumbel-Softmax. One of the issues of training a semi-supervised VAE is that the marginalization over $q_\phi(y|x)$ in Eq. (6) can be computationally expensive. This marginalization can be avoided by using Gumbel-Softmax [10,11], a continuous distribution on the probability simplex that approximates a categorical sample and can be smoothly annealed (through a temperature parameter) to the categorical distribution. Gumbel-Softmax is reparameterizable so that the gradient of the loss function can be propagated back through the sampling step $y \sim q_\phi(y|x)$ for single-sample gradient estimation.

Regularization. The lower bound for labeled data can be rewritten as

$$\mathcal{L}_{\theta,\phi}(x, y) = \mathbb{E}_{q_\phi(z|x,y)} \left[\log \frac{p_\theta(x, y, z)}{q_\phi(z|x, y)} \right]$$

$$= \mathbb{E}_{q_\phi(z|x,y)} \left[\log p_\theta(x|z, y) \right] + \log p(y) - D_{KL}(q_\phi(z|x, y)||p(z))$$

where $\log p(y)$ is a constant, the first term can be interpreted as expected negative reconstruction error, and the last term is the negative KL divergence from the prior to the approximate posterior. Similarly, we can express the bound for unlabeled data as follows:

$$\mathcal{U}_{\theta,\phi}(x) = \mathbb{E}_{q_\phi(z,y|x)} \left[\log p_\theta(x|z, y) \right] - D_{KL}(q_\phi(z, y|x)||p(z, y))$$

In both cases, the KL divergence acts as a regularization term that encourages the approximate posterior to be close to the prior, thereby constraining the amount of information encoded in the latent variables. The overall lower bound (7) thus trades off reconstruction error with this regularization term. When training a VAE, we can control such trade-off in order to favor more accurate reconstructions or more constrained latent space, by simply multiplying the KL term by a factor $\beta > 0$ as proposed in [12]. Similarly, we found it beneficial in practice to scale the entropy of $q_\phi(y|x)$ in Eq. (6) by a factor $\gamma > 1$. Intuitively, the entropy term acts as a regularizer in the classifier by encouraging $q_\phi(y|x)$ to have high entropy: the amplification of this term helps to further reduce overfitting in the classifier.

3   Data

The BraTS 2019 challenge is composed of a training, a validation and a test set. The training set is composed of 335 delineated tumor images, in which 210 images have survival labels. The validation set is composed of 125 non-delineated images without survival labels, in which only 29 images with resection status of GTR (i.e., Gross Total Resection) are part of the online evaluation platform (CBICA's Image Processing Portal). Finally, the test set will be made available to the challenge participants during a limited time window, and the results will be part of the BraTS 2019 workshop.

In all our experiments we performed 3-fold cross-validation by randomly splitting the BraTS 2019 training set with survival labels into a "private" training (75%) and validation set (25%) in each fold, in order to have an alternative to the online evaluation platform. This help us having a more informative indication of the model performance, since the online evaluation platform includes just 29 cases (vs. 53 cases in our private validation sets). With this set-up, which we call **S0** in the remainder, we effectively trained the model on a training set of $N_l = 157$ and $N_u = 125$ for each of the three cross-validation folds. These models were subsequently tested on their corresponding private validation sets of 53 subjects, as well as on the standard BraTS 2019 validation set of 29 subjects.

In order to evaluate just how much the proposed method is able to learn from *unlabeled* data (i.e., subjects with tumor delineations but no survival time information), we used three open-source methods [13–15] to automatically segment both the entire BraTS 2019 training and validation sets in order to have many more unlabeled training subjects available. We further augmented these unlabeled data sets by flipping the images in the coronal plane. With this new set-up, which we call **S1**, we then trained the model on an "augmented" private training set of $N_l = 157$ and $N_u = 2268$ for each of the three cross-validation folds. Ideally, dramatically increasing the set of unlabeled data points this way should help the model learn to better encode tumor representations, thereby increasing classification accuracy.

4   Implementation

We implemented the encoder $q_\phi(z|x, y)$, the decoder $p_\theta(x|z, y)$ and the classifier $q_\phi(y|x)$ all as deep convolutional networks using PyTorch [16]. The segmentation volumes provided in the BraTS challenge have size $240 \times 240 \times 155$, but since large parts of the volume are always zero, we cropped the volumes to $146 \times 188 \times 128$ without losing any tumor voxels. We further reduced the volume by a factor of 2 in all dimensions, resulting in a shape of $73 \times 94 \times 64$, roughly a 95% overall reduction in input image size. This leads to faster training and larger batches fitting in memory, while losing minimal information.

We optimized the model end-to-end with Adam optimizer [17], using a batch size of 32, learning rate $2 \cdot 10^{-5}$, latent space size 32, $\alpha = 10^{-5} \cdot D \approx 4.4$ with D the data dimensionality (number of voxels), β from 0 to $6 \cdot 10^3$ in $3 \cdot 10^4$ steps,

$\gamma = 50$, and exponentially annealing the Gumbel-Softmax sampling temperature from 1.0 to 0.2 in $5 \cdot 10^4$ steps. Hyperparameters were found by grid search, although not fine-tuned because of the computational cost. The total number of parameters in the model is around 2.7×10^6.

4.1 Network Architecture

The three networks consist of 3D convolutional layers, with the exception of a few fully connected layers in the classifier. There are nonlinearities (Scaled Exponential Linear Units, [18]) and dropout [19] after each layer, except when noted. What follows is a high-level description of the network architecture, represented in diagrams in Fig. 2. For more details, the code is available at https://github. com/sveinnpalsson/semivaebrats.

The inference network consists of a convolutional layer (B1_e) with large kernel size and stride (7 and 4, respectively), followed by two residual blocks [20] (B2_e and B3_e). The input to each block is processed in parallel in two branches, one consisting of two convolutional layers, the other of average pooling followed by a linear transformation (without nonlinearities). The results of the two branches are added together. The output of the first layer is also fed into the classifier network, which outputs the class scores (these will be used to compute the classification loss for labeled data). A categorical sample from $q_\phi(y|\mathbf{x})$ is drawn using the Gumbel-Softmax reparameterization given the class scores, and is embedded by a fully connected layer into a real vector space. Such embedding is then concatenated to the output of the two encoder blocks, so that the means and variances of the approximate posterior $q_\phi(\mathbf{z}|\mathbf{x}, y)$, that are computed

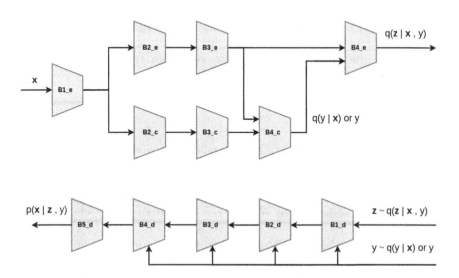

Fig. 2. Networks architectures: encoder, decoder and classifier architectures.

by a final convolutional layer, are conditioned on the sampled label. The classifier consists of two residual blocks similar to the ones in the encoder (B2_c and B3_c), followed by two fully connected layers (B4_c).

The decoder network consists of two convolutional layers (B1_d and B2_d), two residual blocks similar to those in the encoder (B3_d and B4_d), and a final convolution followed by a sigmoid nonlinearity (B5_d). In the decoder, most convolutions are replaced by transposed convolutions (for upsampling), and pooling in the residual connections is replaced by nearest neighbour interpolation. The input to the decoder network is a latent vector z sampled from the approximate posterior. The embedding of y, computed as in the final stage of the inference network, is also concatenated to the input of each layer (except the ones in the middle of a block) to express the conditioning of the likelihood function on the label. Here, the label is either the ground truth (for labeled examples) or a sample from the inferred posterior (for unlabeled examples).

5   Results

5.1   Conditional Generation

We visually tested whether the decoder $p_\theta(x|y, z)$ is able to generate tumor-like images after training, and whether it can disentangle the classes. For this purpose we sampled z from $\mathcal{N}(z|0, I)$ and varied y between the three classes, namely, short survivor, mid survivor and long survivor. Figure 3 shows the three shapes generated accordingly by one of the models trained in set-up **S0**. From the images we can see that the generated tumor for the short survivor class has an irregular shape with jagged edges while the long survivor generated tumor has a more compact shape with rounded edges.

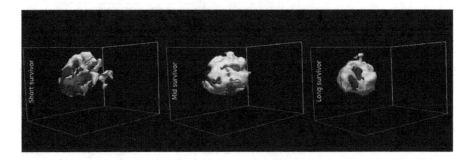

Fig. 3. Generated tumor from $p_\theta(x|y, z)$ where we sampled z from $\mathcal{N}(z|0, I)$ and we varied y between short survivor, mid survivor and long survivor.

5.2   Quantitative Evaluation

All the classification accuracies are reported with binomial confidence interval with normal approximation [21], defined as

$$a \pm z^* \sqrt{\frac{a(1-a)}{n}}$$

where a is the classification accuracy, $z^* = 1.96$ is the critical value with confidence level at 95% and n is the number of subjects. In Table 1 we show the classification accuracy of the proposed method on the "private" validation set of 53 subjects for each of the three cross-validation folds, both for the set-up with fewer (**S0**) and more (**S1**) unlabeled training subjects. The corresponding results based on the online evaluation platform (29 validation subjects) are summarized in Table 2, where we submitted the majority vote for survival group prediction across the three models trained in the cross-validation folds. The online evaluation platform takes the estimated number of days as input and returns the accuracy along with mean- and median squared error and Spearman's rank correlation coefficient. To make these predictions we input the average survival from each class. Our scores on the challenge leaderboard for set-up **S0** are as follows: 37.9% accuracy, 111214.828 mean squared error, 51076.0 median squared error and a correlation of 0.36. When testing the models we found that they are insensitive to the segmentation method used to produce the input.

Table 1. Classification accuracies [%] for both set-ups on the "private" validation set for each of the three cross-validation folds.

Set-up	Fold 1	Fold 2	Fold 3	Avg
S0	42.18 ± 13.30	35.90 ± 12.91	39.53 ± 13.16	39.20 ± 7.59
S1	47.55 ± 13.45	41.13 ± 13.40	42.91 ± 13.32	43.86 ± 7.71

Table 2. Classification accuracies [%] for both set-ups on the BraTS 2019 online evaluation platform.

Set-up	Majority voting
S0	37.90 ± 17.57
S1	31.00 ± 16.83

The results show that in none of the experiments our model achieved a significant improvement over always predicting the largest class, which constitutes around 40% of the labeled cases.

6   Discussion and Conclusions

In this paper we evaluated the potential of a semi-supervised deep generative model for classifying brain tumor patients into three overall survival groups, based only on tumor segmentation masks. The main potential advantages of this approach are (1) its in-built invariance to MR intensity variations when different scanners and protocols are used, enabling wide applicability across clinics; and (2) its ability to learn from unlabeled data, which is much more widely available than fully-labeled data.

We compared two different set-ups: one where fewer unlabeled subjects were available for training, and one where their number was (largely artificially) increased using automatic segmentation and data augmentation. Although the latter set-up increased classification performance in our "private" experiments, this increase did not reach statistically significant levels and was not replicated on the small BraTS 2019 validation set. We demonstrated visually that the proposed model effectively learned class-specific information, but overall failed to achieve classification accuracies significantly higher than predicting always the largest class.

The results described here are only part of a preliminary analysis. More real unlabeled data, obtained from truly different subjects pooled across treatment centers, and more clinical covariates of the patients, such as age and resection status, may be necessary to reach better classification accuracies. Future work may also involve stacking hierarchical generative models to further increase the classification performance of the model [7].

Acknowledgements. This project was funded by the European Union's Horizon 2020 research and innovation program under the Marie Sklodowska-Curie project TRABIT (agreement No 765148).

References

1. Poulsen, S.H., et al.: The prognostic value of fet pet at radiotherapy planning in newly diagnosed glioblastoma. Eur. J. Nucl. Med. Mol. Imaging **44**(3), 373–381 (2017)
2. Bakas, S., et al.: Identifying the best machine learning algorithms for brain tumor segmentation, progression assessment, and overall survival prediction in the BRATS challenge. CoRR, abs/1811.02629 (2018)
3. Menze, B.H., et al.: The multimodal brain tumor image segmentation benchmark (BRATs). IEEE Trans. Med. Imaging **34**(10), 1993–2024 (2015)
4. Sotiras, A., et al.: Advancing the cancer genome atlas glioma MRI collections with expert segmentation labels and radiomic features (2017)
5. Bakas, S., et al.: Segmentation labels and radiomic features for the pre-operative scans of the TCGA-GBM collection (2017)
6. Bakas, S., et al.: Segmentation labels and radiomic features for the pre-operative scans of the TCGA-LGG collection (2017)
7. Kingma, D.P., Rezende, D.J., Mohamed, S., Welling, M.: Semi-supervised learning with deep generative models. arXiv e-prints arXiv:1406.5298, June 2014

8. Dempster, A.P., Laird, N.M., Rubin, D.B.: Maximum likelihood from incomplete data via the EM algorithm. J. R. Stat. Soc.: Ser. B (Methodol.) **39**(1), 1–22 (1977)
9. Kingma, D.P., Welling, M.: Auto-encoding variational Bayes. arXiv e-prints arXiv:1312.6114, December 2013
10. Jang, E., Gu, S., Poole, B.: Categorical reparameterization with Gumbel-softmax. arXiv e-prints arXiv:1611.01144, November 2016
11. Maddison, C.J., Mnih, A., Teh, Y.W.: The concrete distribution: a continuous relaxation of discrete random variables. arXiv preprint arXiv:1611.00712 (2016)
12. Higgins, I., et al.: beta-VAE: Learning basic visual concepts with a constrained variational framework
13. Wang, G., Li, W., Ourselin, S., Vercauteren, T.: Automatic brain tumor segmentation using cascaded anisotropic convolutional neural networks. In: Crimi, A., Bakas, S., Kuijf, H., Menze, B., Reyes, M. (eds.) BrainLes 2017. LNCS, vol. 10670, pp. 178–190. Springer, Cham (2018). https://doi.org/10.1007/978-3-319-75238-9_16
14. Isensee, F., Kickingereder, P., Wick, W., Bendszus, M., Maier-Hein, K.H.: No new-net. In: Crimi, A., Bakas, S., Kuijf, H., Keyvan, F., Reyes, M., van Walsum, T. (eds.) BrainLes 2018. LNCS, vol. 11384, pp. 234–244. Springer, Cham (2019). https://doi.org/10.1007/978-3-030-11726-9_21
15. Nuechterlein, N., Mehta, S.: 3D-ESPNet with pyramidal refinement for volumetric brain tumor image segmentation. In: Crimi, A., Bakas, S., Kuijf, H., Keyvan, F., Reyes, M., van Walsum, T. (eds.) BrainLes 2018. LNCS, vol. 11384, pp. 245–253. Springer, Cham (2019). https://doi.org/10.1007/978-3-030-11726-9_22
16. Paszke, A., et al.: Automatic differentiation in PyTorch. In: NIPS-W (2017)
17. Kingma, D.P., Ba, J.: Adam: a method for stochastic optimization. arXiv preprint arXiv:1412.6980 (2014)
18. Klambauer, G., Unterthiner, T., Mayr, A., Hochreiter, S.: Self-normalizing neural networks. In: Advances in Neural Information Processing Systems, pp. 971–980 (2017)
19. Srivastava, N., Hinton, G., Krizhevsky, A., Sutskever, I., Salakhutdinov, R.: Dropout: a simple way to prevent neural networks from overfitting. J. Mach. Learn. Res. **15**(1), 1929–1958 (2014)
20. He, K., Zhang, X., Ren, S., Sun, J.: Deep residual learning for image recognition. In: Proceedings of the IEEE Conference on Computer Vision and Pattern Recognition, pp. 770–778 (2016)
21. Brown, L.D., Tony Cai, T., DasGupta, A.: Interval estimation for a binomial proportion. Stat. Sci. **16**(2), 101–133 (2001)

Multi-modal U-Nets with Boundary Loss and Pre-training for Brain Tumor Segmentation

Pablo Ribalta Lorenzo[1,2], Michal Marcinkiewicz[3], and Jakub Nalepa[2(✉)]

[1] NVIDIA, Warsaw, Poland
pribalta@nvidia.com
[2] Silesian University of Technology, Gliwice, Poland
jnalepa@ieee.org
[3] Netguru, Poznan, Poland
michal.marcinkiewicz@netguru.com

Abstract. Gliomas are the most common primary brain tumors, and their manual segmentation is a time-consuming and user-dependent process. We present a two-step multi-modal U-Net-based architecture with unsupervised pre-training and surface loss component for brain tumor segmentation which allows us to seamlessly benefit from all magnetic resonance modalities during the delineation. The results of the experimental study, performed over the newest release of the BraTS test set, revealed that our method delivers accurate brain tumor segmentation, with the average DICE score of 0.72, 0.86, and 0.77 for the enhancing tumor, whole tumor, and tumor core, respectively. The total time required to process one study using our approach amounts to around 20 s.

Keywords: Brain tumor · Segmentation · Boundary loss · U-Net

1 Introduction

Gliomas are the most common primary brain tumors in humans. They are characterized by different levels of aggressiveness which directly influences prognosis. Due to the gliomas' heterogeneity (in terms of shape and appearance) manifested in multi-modal magnetic resonance imaging (MRI), their accurate delineation is an important yet challenging medical image analysis task. However, manual segmentation of such brain tumors is very time-consuming and prone to human errors and bias. It also lacks reproducibility which adversely affects the effectiveness of patient's monitoring, and can ultimately lead to inefficient treatment.

Automatic brain tumor *detection* (i.e., which pixels in an input image are tumorous) and *classification* (what is a type of a tumor and/or which part of a tumor, e.g., edema, non-enhancing solid core, or enhancing structures a given

© Springer Nature Switzerland AG 2020
A. Crimi and S. Bakas (Eds.): BrainLes 2019, LNCS 11993, pp. 135–147, 2020.
https://doi.org/10.1007/978-3-030-46643-5_13

pixel belongs to) from MRI are vital research topics. A very wide practical applicability of such techniques encompasses computer-aided diagnosis, prognosis, staging, and monitoring of a patient. In this paper, we propose a deep learning-powered pipeline to segment gliomas from MRI. First, we detect the whole tumor (WT) area, while the second part of the pipeline segments it into the enhancing tumor (ET), tumor core (TC), and the peritumoral edema (ED).

There is an apparent downside of using deep neural networks for image segmentation—they require significant quantities of data to obtain high generalization capabilities. The training of (especially large) models on such great amount of data is computationally expensive and can take a very long time. A remedy to this problem was established by Erhan et al. [10]—an unsupervised pre-training phase was used, which accelerates the convergence and results in better generalization of deep models. Here, we train the model to compress the input data in an encoder-decoder manner. Reducing the reconstruction loss allows us to iteratively update the weights of the model so that they better match the input data. It, in turn, makes the final training faster.

As a base for our model, we selected a variant of a U-Net [38] adapted for multi-modal data and multi-class brain tumor segmentation, inspired by our recent work [26]. Our experiments, performed over the training and validation BraTS 2019 sets, revealed that exploiting multiple modalities allows us to obtain accurate delineation, and that our models can be seamlessly trained in an end-to-end fashion without any additional overhead of dealing multiple modalities.

In Sect. 2, we discuss the current state of the art in brain tumor delineation. The proposed multi-modal U-Nets are presented in Sect. 3. The results of our experiments are analyzed in Sect. 4. Section 5 concludes the paper.

2   Related Literature

Approaches for automated brain tumor delineation can be divided into *atlas-based*, *unsupervised*, *supervised*, and *hybrid* techniques (Fig. 1). In the *atlas-based* algorithms, manually segmented images (*atlases*) are used to segment incoming scans [36]. These atlases model the anatomical variability of the brain tissue [34]. Atlas images are extrapolated to new frames by warping and applying non-rigid registration techniques. Within the atlas-based algorithms, we distinguish different segmentation strategies, with the single- and multi-atlas label propagation and the probabilistic atlas-based segmentation being the most popular [7]. An important drawback of the atlas-based techniques is the necessity of creating large reference sets. It is time-consuming and error-prone in practice, and may lead to atlases which cannot be applied to other tumors [1,6].

Unsupervised algorithms search for hidden structures within the input (unlabeled) data [11,28]. In various meta-heuristic approaches, e.g., in evolutionary algorithms [43], brain segmentation is understood as an optimization problem, in which pixels (or voxels) of similar characteristics are searched. It is tackled in a biologically-inspired manner, in which a population of candidate solutions

Fig. 1. Automated delineation of brain tumors from MRI—a taxonomy [26].

(being the pixel or voxel labels) evolves in time [8]. Other unsupervised algorithms encompass clustering-based techniques [20,39,44], and Gaussian modeling [41]. Clustering-based approaches partition an input image (or volume) into consistent groups of pixels manifesting similar characteristics [20,39,41,44]. In hard clustering, each pixel belongs to a single cluster, and the clusters do not overlap. On the other hand, soft clustering methods assign a probability of a pixel being in a given cluster, and such clusters *can* overlap [45]. The clustering methods, albeit being pixel-wise, can also benefit from additional spatial information.

In *supervised* techniques, manually segmented image sets are utilized to train a model. Such algorithms include, among others, decision forests [13,49], conditional random fields [46], support vector machines [24], and extremely randomized trees [35]. Appearance and context-based features have also been also utilized in [35], in their extremely randomized forests. The major drawback of both unsupervised and supervised conventional machine-learning techniques is the necessity of extracting hand-crafted features. In the pixel-wise classification, such features are extracted for each pixel (or voxel) separately [42]. Unfortunately, this process can lead to extracting features which do not benefit from the full information contained in an input image, as some features may be easily omitted in the extraction process. Additionally, if a large number of features are determined, we may easily face the curse of dimensionality problem for smaller training sets [40], hence it is often coupled with feature selection.

Deep neural networks, which established the state of the art in a plethora of image-processing and image-recognition tasks, have been successful in segmentation of different kinds of brain tissue as well [15,23,31]. Holistically nested neural nets for MRI were introduced in [48]. White matter was segmented in [14], and convolutional neural networks were applied to segment tumors in [17]. Interestingly, the winning BraTS'17 algorithm used deep neural nets ensembles [21]. Although the last year's edition (2018) of the BraTS competition was won by a deep network which benefits from the encoder-decoder architecture [5,32], the other top-performing methods included U-Net-based networks with minor modifications [19], and the U-Net-inspired ensembles [27]. Specifically, Isensee et al. [19] showed that U-Nets are extremely powerful deep architectures which can easily outperform more sophisticated and deeper networks. It is also worth mentioning that the U-Net-based architectures have been widely researched in this context, and constituted the mainstream in the BraTS challenge for multi-class tumor segmentation from multiple MRI modalities [9]. In [26], we exploited a cascaded U-Net for tumor detection and further segmentation. A three pathways U-Net structure has been utilized by Fang and He [12], where each modality is analyzed in a separate pathway, and they are finally fused. Such multi-modal approaches can fully benefit from all available modalities, but they require having them co-registered which is a challenging image-analysis task [16]. *Hybrid* algorithms couple together methods from other categories [37,42,47].

3   Methods

We propose a multi-modal U-Net architecture and apply it to brain tumor (BT) segmentation. In the following sections, we discuss the pre-processing (Sect. 3.1), pre-training (Sect. 3.2), and segmentation steps (Sect. 3.3) in more detail. In Fig. 2, we present a high-level flowchart of our technique.

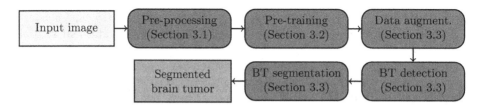

Fig. 2. A high-level flowchart of our brain tumor segmentation technique.

3.1   Data Pre-processing

Our pre-processing method involves two steps. In the first step, we perform volume-wise histogram equalization of each modality separately. We selected one volume from the training set (`BraTS19_CBICA_AAB_1`) as a template, which was characterized by a negligible number of high values (considered as outliers), a moderate tumor area, and lack of visible bias-related artifacts—an example process of histogram equalization is rendered in Fig. 3.

During training and inference, we apply the Z-score normalization of each modality (calculated volume-wise). For each patient, we take each sequence separately, and calculate the mean and standard deviation of the intensity values of all non-zero pixels (excluding the background) within this modality (denoted as $\bar{p}_M$ and σ_M, respectively). For each i-*th* non-zero pixel p_i, we subtract the mean and divide it by the standard deviation to obtain the normalized pixel p_i':

$$p_i' = \frac{p_i - \bar{p}_M}{\sigma_M}. \tag{1}$$

To reduce the memory requirements and accelerate the training and inference process, we crop volumes in the axial plane from (240×240) px to (192×192) px. The input tensor fed to our multi-modal U-Net is therefore of size $192 \times 192 \times 4$ (as we deal with four separate MRI modalities).

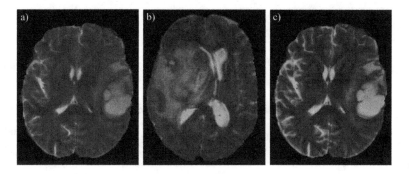

Fig. 3. Example histogram equalization result for a T2-FLAIR image: a) a source image, b) the template image, c) the result of the histogram equalization.

3.2 Unsupervised Pre-training

Erhan et al. [10] showed that unsupervised pre-training allows a model to extract useful features and learn the statistical distribution of the data. Since the U-Net architecture intrinsically resembles an encoder-decoder architecture, it requires small effort to add a pre-training phase to the training pipeline.

During the unsupervised pre-training phase, a model learns to compress the input data and then reconstruct the original image. By reducing the reconstruction loss, the weights of the model are iteratively tuned to better match the input data, which further speeds up the convergence during the fine-tuning phase. Moreover, Erhan et al. claim that pre-training allows the model to find better local minima, which improves the generalization performance [10]. Here, the mean square error acted as the reconstruction loss.

3.3 Segmentation of Brain Tumors Using Multi-modal U-Nets

Our models are based on a famous U-Net, with significant changes to the architecture (Fig. 4). Firstly, there are only three "down" convolutions instead of four, since we realized that such a high reduction of spatial dimensions can easily deteriorate the segmentation performance, especially in the case of smaller tumors. Secondly, instead of three plain convolutional layers, each level consists of three residual blocks [18] with three convolutional layers. Thirdly, in the original architecture, the number of filters was doubled at each down-block, starting from 64, whereas in our model it starts at 32 and goes to 48, and 64 at the very bottom. Finally, the sigmoid activation at the output layer was replaced by a softmax non-linearity for classification. In this work, we employ training-time data augmentation to improve the generalization abilities of our models.

Data Augmentation. To increase the robustness of the neural networks, we applied data augmentation during training [25,33]. We used a) horizontal flips, since they preserve the anatomical structure of the brain, b) random channel

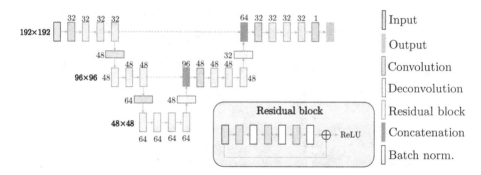

Fig. 4. Our U-Net-based architecture used for both detection and segmentation of brain tumors (note that the final layer of the architecture, one or three filters, is the only difference between the one-class classification and multi-class segmentation). We report the number of filters in each layer (each filter is 3×3), and the size of the feature map at each level of our architecture (this size is boldfaced).

intensity variations—the pixel values of each channel separately were multiplied by a value drawn from the uniform distribution of $[0.95, 1.05]$.

Brain Tumor Detection. The first U-Net (a detector) is trained on binary masked images, with all tumor classes merged into one. Such binary classification allows us to delineate the area of the tumor, making the final multi-class classification easier for the classifier.

Brain Tumor Classification. The second U-Net (a classifier) is trained on images with all but the tumor area put to zero. In this way, it can focus on the important task of classifying each pixel correctly, without any additional noise. The number of voxels belonging to each class (non-enhancing tumor core—NCR, ED, ET) is largely imbalanced; therefore, we introduce the weighting factors to accommodate for those differences and make the training process more effective. The weights were calculated for each batch separately, with average values of 4.6:1:2.2 for NCR:ED:ET, respectively.

4   Experimental Validation

4.1   Data

The Brain Tumor Segmentation (BraTS) dataset [2–5,29] encompasses MRI data of 335 patients with diagnosed gliomas: 259 high-grade glioblastomas (HGG), and 76 low-grade gliomas (LGG). Each study was manually annotated by one to four experienced readers. The data comes in four co-registered modalities: native pre-contrast (T1), post-contrast T1-weighted (T1Gd), T2-weighted (T2), and T2 Fluid Attenuated Inversion Recovery (FLAIR). All the pixels have one of four labels attached: healthy tissue, Gd-enhancing tumor (ET), peritumoral edema (ED), the necrotic and non-enhancing tumor core (NCR/NET).

The data was acquired with different clinical protocols, various scanners, and at 19 institutions, therefore the pixel intensity distribution may vary significantly across different studies. The studies were interpolated to the same shape (240 × 240× 155, hence 155 images of 240 × 240 size, with voxel size $1\,\mathrm{mm}^3$), and they were pre-processed (skull-stripping was applied). Overall, there are 335 patients in the training set T (259 HGG, 76 LGG), 125 patients in the validation set V, and 166 in the test set Ψ (the V and Ψ sets were provided without the ground-truth segmentation by the BraTS 2019 organizers).

4.2   Loss Function

Extremely imbalanced segmentations are quite common in medical image analysis, where the size of the target foreground region (here, tumorous tissue) is several orders of magnitude smaller than the background size. This represents a challenge in many cases because the foreground and the background terms have substantial differences in their values, which ultimately affects segmentation performance and training ability. Our purpose is to build a metric that will help mitigate the above-mentioned difficulties for imbalanced segmentations.

For this purpose, we employ the DICE score [30]:

$$\mathcal{L}_{\mathrm{DICE}}(A, B) = \frac{2 \cdot |A \cap B|}{|A| + |B|}, \tag{2}$$

where A and B are two segmentations, i.e., manual and automated. DICE ranges from zero to one (one is the perfect score), and we combine it with a boundary loss specially targeted towards highly imbalanced segmentation [22]:

$$\mathcal{L}_{\mathrm{B}}(\theta) = \mathrm{Dist}(\partial A, \partial B_\theta) = \int_\Omega \phi_A(q) b_\theta(q) dq, \tag{3}$$

which takes the form of a distance metric in the contour space (or region boundaries), with ∂A denoting a representation of the boundary of ground-truth region A (e.g., the set of points of A, which have a spatial neighbor in the background), and ∂B_θ denoting the boundary of the segmentation region defined by the network output. Here, $\phi_A : \Omega \to \mathbb{R}$ denotes the level set representation of boundary ∂A, and $b_\theta(q)$ are the softmax probability outputs of the network.

We employ the coefficient α as a hyper-parameter to balance the influence of every loss during training, according to a pre-set schedule, as precise boundary delineation becomes more important as training progresses:

$$\alpha \mathcal{L}_{\mathrm{DICE}}(A, B) + (1 - \alpha)\mathcal{L}_{\mathrm{B}}(\theta). \tag{4}$$

The boundary loss works if the area of the segmented object is already defined, meaning that it is problematic to use it predominantly from the very beginning of the training, when the predictions are still (more or less) random.

Therefore, similarly to [22], we start training with the loss function being pure DICE ($\alpha = 1.0$) for 8 epochs to make sure that the network can produce viable predictions, and then linearly decrease α at the end of each epoch. We also noticed that for the small values of α the network starts to diverge, hence the smallest α was selected to be equal to 0.33.

4.3   Experimental Setup

The deep neural network models were implemented using `Python3` with the `PyTorch` library. The experiments were run on a machine equipped with an Intel Xeon E5-1630 v4 (3.70 GHz) CPU with 64 GB RAM and two NVIDIA GeForce GTX 1080 Ti GPUs with 11 GB VRAM. The optimizer was SGD with the momentum of 0.9, and the initial learning rate of 0.1. The training ran for 24 epochs, and the learning rate was decreased to 0.01 after 18 epochs. The model's weights which were used to obtain the best score on the validation set were saved. To reduce over-fitting, we introduced the weight decay of 10^{-4}. The training time for one epoch is around 8 min.

The final prediction over the validation set was performed with an ensemble of 10 multi-modal U-Nets trained on different folds of the training set (we followed the 10-fold cross-validation setting over the training set). Additionally, we exploited test-time data augmentation—each frame in the volume was fed into the network in the original orientation and flipped horizontally, and the activation maps were averaged. Using an ensemble of 10 models (and averaging their outputs to elaborate the final prediction) with test-time augmentation was shown to improve the performance, and the inference took around 20 seconds per full volume (10 seconds without the test-time augmentation).

Table 1. Segmentation performance (DICE) over the validation set obtained using an ensemble of our multi-modal U-Nets. The scores are presented for whole tumor (WT), tumor core (TC), and enhancing tumor (ET) classes. The results were reported by the competition server (we used an ensemble of 10 U-Nets trained over different folds).

Dataset	Label	DICE	Sensitivity	Specificity
Training	ET	0.71082	0.80660	0.99823
	WT	0.89686	0.86828	0.99686
	TC	0.80150	0.78978	0.99770
Validation	ET	0.66336	0.76303	0.99816
	WT	0.89037	0.87780	0.99556
	TC	0.75105	0.71916	0.99774

4.4   Experimental Results

In Table 1, we gather the results (DICE) obtained over the training and validation sets. We report the average DICE for 10 non-overlapping folds of the

ET: 0.97, WT: 0.97, TC: 0.98   ET: 0.82, WT: 0.97, TC: 0.88   ET: 0.86, WT: 0.92, TC: 0.94

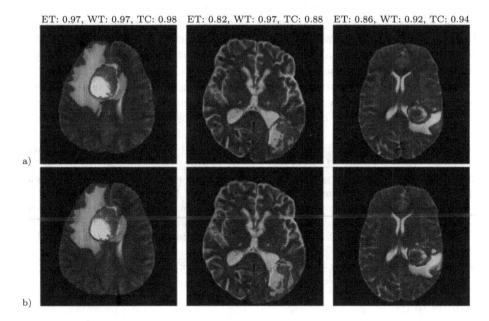

Fig. 5. Example a) the ground-truth segmentation, alongside our segmentation results b) overlaid onto T2 images. We report the DICE scores for ET (annotated in blue), WT, and TC (red). The peritumoral edema is highlighted in green.

Table 2. Segmentation performance: DICE and the 95th percentile of Hausdorff distance (H95) over the BraTS 2019 test set.

Measure	DICE (ET)	DICE (WT)	DICE (TC)	H95 (ET)	H95 (WT)	H95 (TC)
Mean	0.72204	0.85991	0.77112	5.59566	7.88359	9.42541
StdDev	0.26539	0.15694	0.27371	9.48298	13.37874	12.52158
Median	0.81329	0.90791	0.88134	2.23607	4	4.6365
25quantile	0.71983	0.85109	0.77579	1.41421	2.23607	2.23607
75quantile	0.87385	0.9387	0.93229	3.74166	6.7082	12.22697

training set T, and the final DICE over the validation set V. Note that the ground-truth data for V were not publicly-available during the challenge, therefore they could not be exploited to improve the models. In Fig. 5, we can see that our technique delivers accurate brain tumor delineation with only subtle differences between our segmentations and the ground truth. Finally, we present the results obtained over the unseen test set in Table 2.

5   Conclusion

In this paper, we presented multi-modal U-Nets with the adaptive loss function, pre-trained in an unsupervised manner, and applied them to the brain

tumor segmentation task. The results, obtained over the training and validation BraTS 2019 sets, revealed that pre-training step helps to achieve faster
convergence and high quality segmentations. Additionally, exploiting multiple
modalities allows us to obtain accurate delineation, and that our models can
be seamlessly trained in an end-to-end fashion without any additional overhead
of dealing with multiple modalities. The experiments showed that our method
offers very fast operation, even when coupled with test-time data augmentation.

Acknowledgments. This research was supported by the Silesian University of Technology (PRL: BKM-556/RAU2/2018, JN: 02/020/BKM19/0183,
02/020/RGH19/0185). We gratefully acknowledge the support of NVIDIA Corporation with the donation of the computing resources used for this research.

This paper is in memory of Dr. Grzegorz Nalepa, an extraordinary scientist, pediatric hematologist/oncologist, and a compassionate champion for kids at Riley Hospital for Children, Indianapolis, USA, who helped countless patients and their families
through some of the most challenging moments of their lives. JN thanks Dana K.
Mitchell for lots of inspiring discussions on (not only) brain MRI analysis.

References

1. Aljabar, P., Heckemann, R., Hammers, A., Hajnal, J., Rueckert, D.: Multi-atlas
   based segmentation of brain images: atlas selection and its effect on accuracy.
   NeuroImage **46**(3), 726–738 (2009)
2. Bakas, S., et al.: Advancing the cancer genome atlas glioma MRI collections with
   expert segmentation labels and radiomic features. Nat. Sci. Data **4**, 1–13 (2017).
   https://doi.org/10.1038/sdata.2017.117
3. Bakas, S., et al.: Segmentation labels and radiomic features for the pre-operative
   scans of the TCGA-GBM collection (2017). the Cancer Imaging Archive. https://
   doi.org/10.7937/K9/TCIA.2017.KLXWJJ1Q
4. Bakas, S., et al..: Segmentation labels and radiomic features for the pre-operative
   scans of the TCGA-LGG collection (2017). the Cancer Imaging Archive. https://
   doi.org/10.7937/K9/TCIA.2017.GJQ7R0EF
5. Bakas, S., et al.: Identifying the best machine learning algorithms for brain
   tumor segmentation, progression assessment, and overall survival prediction in the
   BRATS challenge. CoRR abs/1811.02629 (2018). http://arxiv.org/abs/1811.02629
6. Bauer, S., Seiler, C., Bardyn, T., Buechler, P., Reyes, M.: Atlas-based segmentation
   of brain tumor images using a Markov random field-based tumor growth model
   and non-rigid registration. In: Proceedings of IEEE EMBC, pp. 4080–4083 (2010).
   https://doi.org/10.1109/IEMBS.2010.5627302
7. Cabezas, M., Oliver, A., Lladó, X., Freixenet, J., Cuadra, M.B.: A review of atlas-
   based segmentation for magnetic resonance brain images. Comput. Methods Programs Biomed. **104**(3), e158–e177 (2011)
8. Chander, A., Chatterjee, A., Siarry, P.: A new social and momentum component
   adaptive PSO algorithm for image segmentation. Expert Syst. Appl. **38**(5), 4998–
   5004 (2011)
9. Dai, L., Li, T., Shu, H., Zhong, L., Shen, H., Zhu, H.: Automatic brain tumor
   segmentation with domain adaptation. In: Crimi, A., Bakas, S., Kuijf, H., Keyvan,

F., Reyes, M., van Walsum, T. (eds.) BrainLes 2018. LNCS, vol. 11384, pp. 380–392. Springer, Cham (2019). https://doi.org/10.1007/978-3-030-11726-9_34

10. Erhan, D., Bengio, Y., Courville, A., Manzagol, P.A., Vincent, P., Bengio, S.: Why does unsupervised pre-training help deep learning? J. Mach. Learn. Res. **11**, 625–660 (2010). http://dl.acm.org/citation.cfm?id=1756006.1756025

11. Fan, X., Yang, J., Zheng, Y., Cheng, L., Zhu, Y.: A novel unsupervised segmentation method for MR brain images based on fuzzy methods. In: Liu, Y., Jiang, T., Zhang, C. (eds.) CVBIA 2005. LNCS, vol. 3765, pp. 160–169. Springer, Heidelberg (2005). https://doi.org/10.1007/11569541_17

12. Fang, L., He, H.: Three pathways U-Net for brain tumor segmentation. In: Brain-lesion: Glioma, Multiple Sclerosis, Stroke and Traumatic Brain Injuries - 4th International Workshop, BrainLes 2018, Held in Conjunction with MICCAI 2018, Granada, Spain, Pre-Conference Proceedings, pp. 119–126 (2018)

13. Geremia, E., Clatz, O., Menze, B.H., Konukoglu, E., Criminisi, A., Ayache, N.: Spatial decision forests for MS lesion segmentation in multi-channel magnetic resonance images. NeuroImage **57**(2), 378–390 (2011)

14. Ghafoorian, M., et al.: Location sensitive deep convolutional neural networks for segmentation of white matter hyperintensities. CoRR abs/1610.04834 (2016). http://arxiv.org/abs/1610.04834

15. Ghafoorian, M., et al.: Transfer learning for domain adaptation in MRI: application in brain lesion segmentation. In: Descoteaux, M., Maier-Hein, L., Franz, A., Jannin, P., Collins, D.L., Duchesne, S. (eds.) MICCAI 2017. LNCS, vol. 10435, pp. 516–524. Springer, Cham (2017). https://doi.org/10.1007/978-3-319-66179-7_59

16. Gholipour, A., Kehtarnavaz, N., Briggs, R., Devous, M., Gopinath, K.: Brain functional localization: a survey of image registration techniques. IEEE Trans. Med. Imaging **26**(4), 427–451 (2007). https://doi.org/10.1109/TMI.2007.892508

17. Havaei, M., Dutil, F., Pal, C., Larochelle, H., Jodoin, P.-M.: A convolutional neural network approach to brain tumor segmentation. In: Crimi, A., Menze, B., Maier, O., Reyes, M., Handels, H. (eds.) BrainLes 2015. LNCS, vol. 9556, pp. 195–208. Springer, Cham (2016). https://doi.org/10.1007/978-3-319-30858-6_17

18. He, K., Zhang, X., Ren, S., Sun, J.: Deep residual learning for image recognition. In: 2016 IEEE Conference on Computer Vision and Pattern Recognition (CVPR), pp. 770–778, June 2016. https://doi.org/10.1109/CVPR.2016.90

19. Isensee, F., Kickingereder, P., Wick, W., Bendszus, M., Maier-Hein, K.H.: No new-net. In: Crimi, A., Bakas, S., Kuijf, H., Keyvan, F., Reyes, M., van Walsum, T. (eds.) BrainLes 2018. LNCS, vol. 11384, pp. 234–244. Springer, Cham (2019). https://doi.org/10.1007/978-3-030-11726-9_21

20. Ji, S., Wei, B., Yu, Z., Yang, G., Yin, Y.: A new multistage medical segmentation method based on superpixel and fuzzy clustering. Comput. Math. Methods Med. 747549:1–747549:13 (2014)

21. Kamnitsas, K., et al.: Ensembles of multiple models and architectures for robust brain tumour segmentation. In: Crimi, A., Bakas, S., Kuijf, H., Menze, B., Reyes, M. (eds.) BrainLes 2017. LNCS, vol. 10670, pp. 450–462. Springer, Cham (2018). https://doi.org/10.1007/978-3-319-75238-9_38

22. Kervadec, H., Bouchtiba, J., Desrosiers, C., Granger, E., Dolz, J., Ben Ayed, I.: Boundary loss for highly unbalanced segmentation. In: Cardoso, M.J., et al. (eds.) Proceedings of The 2nd International Conference on Medical Imaging with Deep Learning. Proceedings of Machine Learning Research, vol. 102, pp. 285–296. PMLR, London, 08–10 July 2019. http://proceedings.mlr.press/v102/kervadec19a.html

23. Korfiatis, P., Kline, T.L., Erickson, B.J.: Automated segmentation of hyperintense regions in FLAIR MRI using deep learning. Tomogr.: J. Imaging Res. **2**(4), 334–340 (2016). https://doi.org/10.18383/j.tom.2016.00166

24. Ladgham, A., Torkhani, G., Sakly, A., Mtibaa, A.: Modified support vector machines for MR brain images recognition. In: Proceedings of CoDIT, pp. 032–035 (2013). https://doi.org/10.1109/CoDIT.2013.6689515

25. Lorenzo, P.R., et al.: Segmenting brain tumors from FLAIR MRI using fully convolutional neural networks. Comput. Methods Programs Biomed. **176**, 135–148 (2019). https://doi.org/10.1016/j.cmpb.2019.05.006. http://www.sciencedirect.com/science/article/pii/S0169260718315955

26. Marcinkiewicz, M., Nalepa, J., Lorenzo, P.R., Dudzik, W., Mrukwa, G.: Segmenting brain tumors from MRI using cascaded multi-modal U-Nets. In: Crimi, A., Bakas, S., Kuijf, H., Keyvan, F., Reyes, M., van Walsum, T. (eds.) BrainLes 2018. LNCS, vol. 11384, pp. 13–24. Springer, Cham (2019). https://doi.org/10.1007/978-3-030-11726-9_2

27. McKinley, R., Meier, R., Wiest, R.: Ensembles of densely-connected CNNs with label-uncertainty for brain tumor segmentation. In: Crimi, A., Bakas, S., Kuijf, H., Keyvan, F., Reyes, M., van Walsum, T. (eds.) BrainLes 2018. LNCS, vol. 11384, pp. 456–465. Springer, Cham (2019). https://doi.org/10.1007/978-3-030-11726-9_40

28. Mei, P.A., de Carvalho Carneiro, C., Fraser, S.J., Min, L.L., Reis, F.: Analysis of neoplastic lesions in magnetic resonance imaging using self-organizing maps. J. Neurol. Sci. **359**(1–2), 78–83 (2015)

29. Menze, B.H., et al.: The multimodal brain tumor image segmentation benchmark (BRATS). IEEE Trans. Med. Imaging **34**(10), 1993–2024 (2015). https://doi.org/10.1109/TMI.2014.2377694

30. Milletari, F., Navab, N., Ahmadi, S.: V-net: fully convolutional neural networks for volumetric medical image segmentation. CoRR abs/1606.04797 (2016). http://arxiv.org/abs/1606.04797

31. Moeskops, P., Viergever, M.A., Mendrik, A.M., de Vries, L.S., Benders, M.J.N.L., Isgum, I.: Automatic segmentation of MR brain images with a convolutional neural network. IEEE Trans. Med. Imaging **35**(5), 1252–1261 (2016). https://doi.org/10.1109/TMI.2016.2548501

32. Myronenko, A.: 3D MRI brain tumor segmentation using autoencoder regularization. In: Crimi, A., Bakas, S., Kuijf, H., Keyvan, F., Reyes, M., van Walsum, T. (eds.) BrainLes 2018. LNCS, vol. 11384, pp. 311–320. Springer, Cham (2019). https://doi.org/10.1007/978-3-030-11726-9_28

33. Nalepa, J., et al.: Data augmentation via image registration. In: 2019 IEEE International Conference on Image Processing (ICIP), pp. 4250–4254, September 2019. https://doi.org/10.1109/ICIP.2019.8803423

34. Park, M.T.M., et al.: Derivation of high-resolution MRI atlases of the human cerebellum at 3T and segmentation using multiple automatically generated templates. NeuroImage **95**, 217–231 (2014)

35. Pinto, A., Pereira, S., Correia, H., Oliveira, J., Rasteiro, D.M.L.D., Silva, C.A.: Brain tumour segmentation based on extremely rand. forest with high-level features. In: Proceedings of IEEE EMBC, pp. 3037–3040 (2015). https://doi.org/10.1109/EMBC.2015.7319032

36. Pipitone, J., et al.: Multi-atlas segmentation of the whole hippocampus and subfields using multiple automatically generated templates. NeuroImage **101**, 494–512 (2014)

37. Rajendran, A., Dhanasekaran, R.: Fuzzy clustering and deformable model for tumor segmentation on MRI brain image: a combined approach. Procedia Eng. **30**, 327–333 (2012)
38. Ronneberger, O., Fischer, P., Brox, T.: U-Net: convolutional networks for biomedical image segmentation. CoRR abs/1505.04597 (2015)
39. Saha, S., Bandyopadhyay, S.: MRI brain image segmentation by fuzzy symmetry based genetic clustering technique. In: Proceedings of IEEE CEC, pp. 4417–4424 (2007)
40. Sembiring, R.W., Zain, J.M., Embong, A.: Dimension reduction of health data clustering. CoRR abs/1110.3569 (2011). http://arxiv.org/abs/1110.3569
41. Simi, V., Joseph, J.: Segmentation of glioblastoma multiforme from MR images - a comprehensive review. Egypt. J. Radiol. Nuclear Med. **46**(4), 1105–1110 (2015)
42. Soltaninejad, M., et al.: Automated brain tumour detection and segmentation using superpixel-based extremely randomized trees in FLAIR MRI. Int. J. Comput. Assist. Radiol. Surg. **12**(2), 183–203 (2017). https://doi.org/10.1007/s11548-016-1483-3
43. Taherdangkoo, M., Bagheri, M.H., Yazdi, M., Andriole, K.P.: An effective method for segmentation of MR brain images using the ant colony optimization algorithm. J. Digit. Imaging **26**(6), 1116–1123 (2013). https://doi.org/10.1007/s10278-013-9596-5
44. Verma, N., Cowperthwaite, M.C., Markey, M.K.: Superpixels in brain MR image analysis. In: Proc. IEEE EMBC. pp. 1077–1080 (2013). https://doi.org/10.1109/EMBC.2013.6609691
45. Wadhwa, A., Bhardwaj, A., Verma, V.S.: A review on brain tumor segmentation of MRI images. Magn. Reson. Imaging **61**, 247–259 (2019)
46. Wu, W., Chen, A.Y.C., Zhao, L., Corso, J.J.: Brain tumor detection and segmentation in a CRF framework with pixel-pairwise affinity and superpixel-level features. Int. J. Comput. Assist. Radiol. Surg. **9**(2), 241–253 (2014). https://doi.org/10.1007/s11548-013-0922-7
47. Zhao, X., Wu, Y., Song, G., Li, Z., Zhang, Y., Fan, Y.: A deep learning model integrating FCNNs and CRFs for brain tumor segmentation. CoRR abs/1702.04528 (2017)
48. Zhuge, Y., et al.: Brain tumor segmentation using holistically nested neural networks in MRI images. Med. Phys. **44**, 1–10 (2017). https://doi.org/10.1002/mp.12481
49. Zikic, D., et al.: Decision forests for tissue-specific segmentation of high-grade gliomas in multi-channel MR. In: Ayache, N., Delingette, H., Golland, P., Mori, K. (eds.) MICCAI 2012. LNCS, vol. 7512, pp. 369–376. Springer, Heidelberg (2012). https://doi.org/10.1007/978-3-642-33454-2_46

Multidimensional and Multiresolution Ensemble Networks for Brain Tumor Segmentation

Gowtham Krishnan Murugesan[1(✉)], Sahil Nalawade[1],
Chandan Ganesh[1], Ben Wagner[1], Fang F. Yu[1], Baowei Fei[3],
Ananth J. Madhuranthakam[1,2], and Joseph A. Maldjian[1,2]

[1] Department of Radiology, University of Texas Southwestern Medical Center,
Dallas, TX, USA
Gowtham.Murugesan@UTSouthwestern.edu
[2] Advanced Imaging Research Center,
University of Texas Southwestern Medical Center, Dallas, TX, USA
[3] Department of Bioengineering, University of Texas at Dallas,
Richardson, TX, USA

Abstract. In this work, we developed multiple 2D and 3D segmentation models with multiresolution input to segment brain tumor components and then ensembled them to obtain robust segmentation maps. Ensembling reduced overfitting and resulted in a more generalized model. Multiparametric MR images of 335 subjects from the BRATS 2019 challenge were used for training the models. Further, we tested a classical machine learning algorithm with features extracted from the segmentation maps to classify subject survival range. Preliminary results on the BRATS 2019 validation dataset demonstrated excellent performance with DICE scores of 0.898, 0.784, 0.779 for the whole tumor (WT), tumor core (TC), and enhancing tumor (ET), respectively and an accuracy of 34.5% for predicting survival. The Ensemble of multiresolution 2D networks achieved 88.75%, 83.28% and 79.34% dice for WT, TC, and ET respectively in a test dataset of 166 subjects.

Keywords: Residual Inception Dense Networks · Densenet-169 · Squeezenet · Survival prediction · Brain tumor segmentation

1 Introduction

Brain Tumors account for 85–90% of all primary CNS tumors. The most common primary brain tumors are gliomas, which are further classified into a high grade (HGG) and low grade gliomas (LGG) based on their histologic features. Magnetic Resonance Imaging (MRI) is a widely used modality in the diagnosis and clinical treatment of gliomas. Despite being a standard imaging modality for tumor delineation and treatment planning, brain tumor segmentation on MR images remains a

---

G. K. Murugesan and S. Nalawade—Equal Contribution.

A. Crimi and S. Bakas (Eds.): BrainLes 2019, LNCS 11993, pp. 148–157, 2020.
https://doi.org/10.1007/978-3-030-46643-5_14

challenging task due to the high variation in tumor shape, size, location, and particularly the subtle intensity changes relative to the surrounding normal brain tissue. Consequently, manual tumor contouring is performed, which is both time-consuming and subject to large inter- and intra-observer variability. Semi- or fully-automated brain tumor segmentation methods could circumvent this variability for better patient management (Zhuge et al. 2017). As a result, developing automated, semi-automated, and interactive segmentation methods for brain tumors has important clinical implications, but remains highly challenging. Efficient deep learning algorithms to segment brain tumors into their subcomponents may help in early clinical diagnosis, treatment planning, and follow-up of patients (Saouli et al. 2018).

The multimodal Brain Tumor Segmentation Benchmark (BRATS) dataset provided a comprehensive platform by outsourcing a unique brain tumor dataset with known ground truth segmentations performed manually by experts (Menze et al. 2014). Several advanced deep learning algorithms were developed on this unique platform provided by BRATS and benchmarked against standard datasets allowing comparisons between them. Convolutional Neural Networks (CNN)-based methods have shown advantages for learning the hierarchy of complex features and have performed the best in recent BRATS challenges. U-net (Ronneberger et al. 2015) based network architectures have been used for segmenting complex brain tumor structures. Pereira et al. developed a 2D CNN method with two CNN architectures for HGG and LGG separately and combined the outputs in the post-processing steps (Pereira et al. 2016). Havaei et al. developed a multi-resolution cascaded CNN architecture with two pathways, each of which takes different 2D patch sizes with four MR sequences as channels (Havaei et al. 2017). The BRATS 2018 top performer developed a 3D decoder encoder style CNN architecture with inter-level skip connections to segment the tumor (Myronenko 2019). In addition to the decoder part, a Variation Autoencoder (VAE) was included to add reconstruction loss to the model.

In this study, we propose to ensemble output from Multiresolution and Multidimensional models to obtain robust tumor segmentations. We utilized off-the-shelf model architectures (DensNET-169, SERESNEXT-101, and SENet-154) to perform segmentation using 2D inputs. We also implemented a 2D and 3D Residual Inception Densenet (RID) network to perform tumor segmentation with patch-based inputs (64×64 and $64 \times 64 \times 64$). The outputs from the model trained on different resolutions and dimensions were combined to eliminate false positives and post-processed using cluster analysis to obtain the final outputs.

2 Materials and Methods

2.1 Data and Preprocessing

The BRATS 2019 dataset included a total of 335 multi-institutional subjects (Menze et al. 2014; Bakas et al. 2017a; b and c), consisting of 259 HGGs and 76 LGGs. The standard preprocessing steps by the BRATS organizers on all MR images included co-registration to an anatomical template (Rohlfing et al. 2010), resampling to isotropic resolution ($1 \times 1 \times 1$ mm^3), and skull-stripping (Bakas et al. 2018). Additional preprocessing steps included N4 bias field correction (Tustison et al. 2014) for removing RF inhomogeneity and normalizing the multi-parametric MR images to zero mean and unit variance.

The purpose of the survival prediction task is to predict the overall survival of the patient based on the multiparametric pre-operative MR imaging features in combination with the segmented tumor masks. Survival prediction based on only imaging-based features (with age and resection status) is a difficult task. Additional information such as histopathology, genomic information, radiotracer based imaging, and other non-MR imaging features can be used to improve the overall survival prediction. Pooya et al. (Mobadersany et al. 2018) reported better accuracy by combining genomic information and histopathological images to form a genomic survival convolutional neural network architecture (GSCNN model). Several studies have reported predicting overall survival for cerebral gliomas using ^{11}C-acetate and ^{18}F-FDG PET/CT scans (Tsuchida et al. 2008; Yamamoto et al. 2008; Kim et al. 2018).

2.2 Network Architecture

We trained several models to segment tumor components. All network architectures used for the segmentation task, except Residual Inception dense Network, were imported using Segmentation models, a python package (Yakubovskiy 2019). The models selected for brain tumor segmentation had different backbones (DenseNet-169 (Huang et al. 2017), SERESNEXT-101 (Chen et al. 2018a) and SENet-154 (Hu et al. 2018)). The DenseNet architecture has shown promising results in medical data classification and image segmentation tasks (Islam and Zhang 2017; Chen et al. 2018b; Dolz et al. 2018). The DenseNet model has advantages in feature propagation from one dense block to the next and overcomes the problem of the vanishing gradient (Huang et al. 2017). The squeeze and excitation block was designed to improve the feature propagation by enhancing the interdependencies between features for the classification task. This helps in propagating more useful features to the next block and suppressing less informative features. This network architecture was the top performer at the ILSVC 2017 classification challenge. SENet-154 and SE-ResNeXt-101 have more parameters

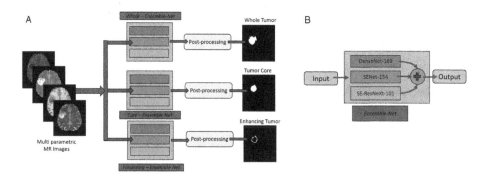

Fig. 1. A. Ensemble of segmentation models (DenseNET-169, SERESNEXT-101 and SENet-154). B. Ensemble methodology used to combine the outputs from Segmentation Models to produce output segmentation maps

and is computationally expensive but has shown good results on the ImageNet classification tasks (Hu, Shen, et al. 2018). Three of the proposed models were ensembled to obtain the final results. All of these models from the Segmentation Models package were trained with 2D axial slices of size 240 × 240 (Fig. 1).

The Residual Inception Dense Network (RID) was first proposed and developed by Khened et al. for cardiac segmentation. We incorporated our implementation of the RID network in Keras with a Tensorflow backend (Fig. 2). In the DenseNet architecture, the GPU memory footprint increases with the number of feature maps of larger spatial resolution. The skip connections from the down-sampling path to the up-sampling path use element-wise addition in this model, instead of the concatenation operation in DenseNet, to mitigate feature map explosion in the up-sampling path. For the skip connections, a projection operation was performed using Batch Normalization

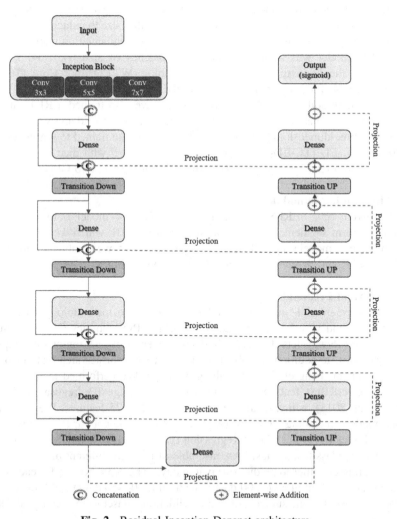

Fig. 2. Residual Inception Densnet architecture

(BN)-1 × 1-convolution-dropout to match the dimensions for element-wise addition (Fig. 3). These additions to the Densenet architecture help in reducing the parameters and the GPU memory footprint without affecting the quality of segmentation output. In addition to performing dimension reduction, the projection operation facilitates learning interactions of cross channel information (Lin et al. 2013) and faster convergence. Further, the initial layer of the RID networks includes parallel CNN branches similar to the inception module with multiple kernels of varying receptive fields. The inception module helps in capturing view-point dependent object variability and learning relations between image structures at multiple-scales.

2.2.1   Model Training and Ensemble Methodology

All models from the Segmentation models package were trained with full resolution axial slices of size 240 × 240 as input to segment the tumor subcomponents separately. The outputs of each component from the models were combined following post-processing steps that included removing clusters of smaller size to reduce false positives. Each tumor component was then combined to form the segmentation map (Fig. 1B).

The RID model was trained on 2D input patches of size 64 × 64. For each component of the brain tumor (e.g., Whole Tumor (WT), Tumor Core (TC), and Enhancing Tumor (ET)), we trained a separate RID model with axial as well as sagittal slices as input. In addition to the six RID models, we also trained a RID with axial slices as input with patch size of 64 × 64 to segment TC and Edema simultaneously (TC-ED). A three-dimensional RID network model was also trained to segment ET and a multiclass TC-ED (TC-ED-3D). All models were trained with dice loss and Adam optimizers with a learning rate of 0.001 using NVIDIA Tesla P40 GPU's.

2.2.2   Ensemble Methodology

The DenseNET-169, SERESNEXT-101, and SENet-154 model outputs were first combined to form segmentation maps, as shown in Fig. 1B, which we will refer to as the Segmentation model output. Then, for each component, we combined outputs from the RID models and Segmentation models, as shown in Fig. 4.

2.2.3   Survival Prediction

The tumor segmentation maps extracted from the above methodology was used to extract texture and wavelet based features using the PyRadiomics (Van Griethuysen et al. 2017) and Pywavelets (Lee et al. 2019) packages from each tumor subcomponent for each contrast. In addition, we also added volume and surface area features of each tumor component (Feng et al. 2019), along with age. We performed feature selection based on SelectKBest features using the sklearn package (Pedregosa et al. 2011; Buitinck et al. 2013), which resulted in a reduced set of 25 features. We trained four different models, including XGBoost (XGB), K-Nearest Neighbour (KNN), Extremely randomized trees (ET), and Linear Regression (LR) models (Chen and Guestrin 2016) for the survival classification task. An ensemble of the four different models was used to form a voting classifier to predict survival in days. These predictions for each subject were then separated into low (<300 days), medium (300–450 days) and long survivors (>450 days). Twenty-nine subjects from the validation dataset were used to validate the trained model.

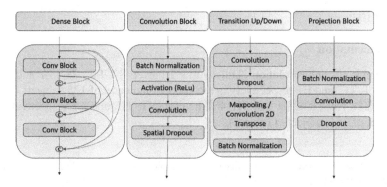

Fig. 3. Building blocks of residual inception network. From left to right, dense block, convolution block, transition block and projection block

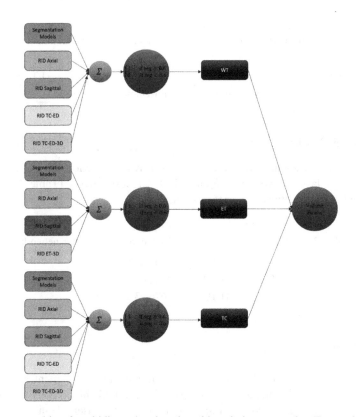

Fig. 4. An ensemble of multidimensional and multiresolution networks. Top to bottom, the ensemble for the Whole Tumor (WT), Tumor Core (TC), and Enhancing Tumor (ET), respectively.

3   Results

3.1   Segmentation

The Ensemble of multiresolution 2D networks achieved 89.79%, 78.43% and 77.97% dice for WT, TC, and ET respectively in the validation dataset of 125 subjects (Table 1, Fig. 5) and 88.75%, 83.28% and 79.34% dice for WT, TC, and ET respectively in the test dataset of 166 subjects (Table 2).

Table 1. Validation segmentation results for multiresolution 2D ensemble model and multidimensional multiresolution ensemble model

Models	WT	TC	ET
Multiresolution 2D Ensemble	0.892	0.776	0.783
Multidimensional and Multiresolution ensemble	0.898	0.78	0.784

Table 2. Testing segmentation results for the multidimensional and multiresolution ensemble model

Models	WT	TC	ET
Multidimensional and Multiresolution Ensemble	0.888	0.833	0.793

3.2   Survival Prediction

Accuracy and mean square error for overall survival prediction for the 29 subjects using a Voting Classifier were 51.7% and 117923.1, respectively (Table 3). In testing the proposed method achieved 41.1% accuracy.

Table 3. Validation survival results for the voting classifier network.

	Accuracy	Mean squared error
Validation	51.7%	117923.1
Testing	41.1%	446765.3

4   Discussion

We ensembled several models with multiresolution inputs to segment brain tumors. The RID network was parameter and memory efficient, and able to converge in as few as three epochs. This allowed us to train several models for ensemble in a short amount

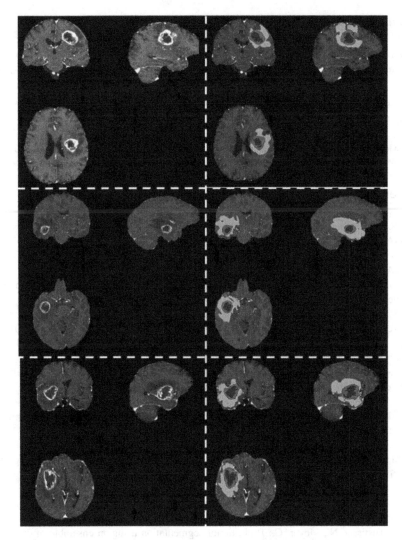

Fig. 5. Example tumor segmentation performance for 3 subjects shown in each row. (a) T1-post contrast (T1C), (b) Segmentation output, (c) Overlay of segmentation output on the T1-post contrast images. Colors: Blue = Non-enhancing tumor + Necrosis, Red = Enhancing Tumor, and Green = Edema (Color figure online)

of time. The proposed methodology of combining multidimensional models improved performance and achieved excellent segmentation results, as shown in Table 1. For survival prediction, we extracted numerous features based on texture, first-order statistics, and wavelets. Efficient model based feature selection allowed us to reduce the otherwise large feature set to 25 features per subject. We trained several classical machine learning models and then combined them to improve results on the validation dataset.

5 Conclusion

We demonstrated two-dimensional multiresolution ensemble network for automated brain tumor segmentation to generate robust segmentation of tumor subcomponents. We also predicted the overall survival based on the segmented mask using an xgboost model. These may assist in diagnosis, treatment planning and therapy response monitoring of brain tumor patients with more objective and reproducible measures.

Acknowledgement. This work was partly supported by the grant, NIH/NCI U01CA207091.

References

Bakas, S., et al.: Segmentation labels and radiomic features for the pre-operative scans of the TCGA-GBM collection. The Cancer Imaging Archive (2017a)

Bakas, S., et al.: Segmentation labels and radiomic features for the pre-operative scans of the TCGA-LGG collection. Cancer Imaging Archive **286** (2017b)

Bakas, S., et al.: Advancing the cancer genome atlas glioma MRI collections with expert segmentation labels and radiomic features. Sci. Data **4**, 170117 (2017c)

Bakas, S., et al.: Identifying the best machine learning algorithms for brain tumor segmentation, progression assessment, and overall survival prediction in the BRATS challenge (2018)

Buitinck, L., et al.: API design for machine learning software: experiences from the scikit-learn project (2013)

Chen, C.-F., Fan, Q., Mallinar, N., Sercu, T., Feris, R.: Big-little net: an efficient multi-scale feature representation for visual and speech recognition (2018)

Chen, L., Wu, Y., DSouza, A.M., Abidin, A.Z., Wismüller, A., Xu, C.: MRI tumor segmentation with densely connected 3D CNN. In: Medical Imaging 2018: Image Processing, International Society for Optics and Photonics (2018)

Chen, T., Guestrin, C.: Xgboost: a scalable tree boosting system. In: Proceedings of the 22nd ACM SIGKDD International Conference on Knowledge Discovery and Data Mining. ACM (2016)

Dolz, J., Gopinath, K., Yuan, J., Lombaert, H., Desrosiers, C., Ayed, I.B.: HyperDense-Net: a hyper-densely connected CNN for multi-modal image segmentation. IEEE Trans. Med. Imaging **38**(5), 1116–1126 (2018)

Feng, X., Tustison, N., Meyer, C.: Brain tumor segmentation using an ensemble of 3D U-Nets and overall survival prediction using radiomic features. In: Crimi, A., Bakas, S., Kuijf, H., Keyvan, F., Reyes, M., van Walsum, T. (eds.) BrainLes 2018. LNCS, vol. 11384, pp. 279–288. Springer, Cham (2019). https://doi.org/10.1007/978-3-030-11726-9_25

Havaei, M., et al.: Brain tumor segmentation with deep neural networks. Med. Image Anal. **35**, 18–31 (2017)

Hu, J., Shen, L., Sun, G.: Squeeze-and-excitation networks. In: Proceedings of the IEEE Conference on Computer Vision and Pattern Recognition (2018)

Huang, G., Liu, Z., Van Der Maaten, L., Weinberger, K.Q.: Densely connected convolutional networks. In: Proceedings of the IEEE Conference on Computer Vision and Pattern Recognition (2017)

Islam, J., Zhang, Y.J.: An ensemble of deep convolutional neural networks for Alzheimer's disease detection and classification (2017)

Kim, S., Kim, D., Kim, S.H., Park, M.-A., Chang, J.H., Yun, M.: The roles of 11 C-acetate PET/CT in predicting tumor differentiation and survival in patients with cerebral glioma. Eur. J. Nucl. Med. Mol. Imaging **45**(6), 1012–1020 (2018). https://doi.org/10.1007/s00259-018-3948-9

Lee, G., Gommers, R., Waselewski, F., Wohlfahrt, K., O'Leary, A.: PyWavelets: a python package for wavelet analysis. J. Open Source Softw. **4**(36), 1237 (2019)

Lin, M., Chen, Q., Yan, S.: Network in network. arXiv preprint arXiv:1312.4400 (2013)

Menze, B.H., et al.: The multimodal brain tumor image segmentation benchmark (BRATS). IEEE Trans. Med. Imaging **34**(10), 1993–2024 (2014)

Mobadersany, P., et al.: Predicting cancer outcomes from histology and genomics using convolutional networks. Proc. Natl. Acad. Sci. **115**(13), E2970–E2979 (2018)

Myronenko, A.: 3D MRI brain tumor segmentation using autoencoder regularization. In: Crimi, A., Bakas, S., Kuijf, H., Keyvan, F., Reyes, M., van Walsum, T. (eds.) BrainLes 2018. LNCS, vol. 11384, pp. 311–320. Springer, Cham (2019). https://doi.org/10.1007/978-3-030-11726-9_28

Pedregosa, F., et al.: Scikit-learn: machine learning in python. J. Mach. Learn. Res. **12**(Oct), 2825–2830 (2011)

Pereira, S., Pinto, A., Alves, V., Silva, C.A.: Brain tumor segmentation using convolutional neural networks in MRI images. IEEE Trans. Med. Imaging **35**(5), 1240–1251 (2016)

Rohlfing, T., Zahr, N.M., Sullivan, E.V., Pfefferbaum, A.: The SRI24 multichannel atlas of normal adult human brain structure. Human Brain Mapp. **31**(5), 798–819 (2010)

Ronneberger, O., Fischer, P., Brox, T.: U-Net: convolutional networks for biomedical image segmentation. In: Navab, N., Hornegger, J., Wells, W.M., Frangi, A.F. (eds.) MICCAI 2015. LNCS, vol. 9351, pp. 234–241. Springer, Cham (2015). https://doi.org/10.1007/978-3-319-24574-4_28

Saouli, R., Akil, M., Kachouri, R.: Fully automatic brain tumor segmentation using end-to-end incremental deep neural networks in MRI images. Comput. Methods Programs Biomed. **166**, 39–49 (2018)

Tsuchida, T., Takeuchi, H., Okazawa, H., Tsujikawa, T., Fujibayashi, Y.: Grading of brain glioma with 1-11C-acetate PET: comparison with 18F-FDG PET. Nucl. Med. Biol. **35**(2), 171–176 (2008)

Tustison, N.J., et al.: Large-scale evaluation of ANTs and FreeSurfer cortical thickness measurements. Neuroimage **99**, 166–179 (2014)

Van Griethuysen, J.J., et al.: Computational radiomics system to decode the radiographic phenotype. Cancer Res. **77**(21), e104–e107 (2017)

Yakubovskiy, P.: Segmentation models. GitHub repository (2019)

Yamamoto, Y., et al.: 11 C-acetate PET in the evaluation of brain glioma: comparison with 11 C-methionine and 18 F-FDG-PET. Mol. Imaging Biol. **10**(5), 281 (2008)

Zhuge, Y., et al.: Brain tumor segmentation using holistically nested neural networks in MRI images. Med. Phys. **44**(10), 5234–5243 (2017)

Hybrid Labels for Brain Tumor Segmentation

Parvez Ahmad[1(✉)], Saqib Qamar[2], Seyed Raein Hashemi[3], and Linlin Shen[2]

[1] National Engineering Research Center for Big Data Technology and System,
Services Computing Technology and System Lab,
Cluster and Grid Computing Lab, School of Computer Science and Technology,
Huazhong University of Science and Technology, Wuhan 430074, China
parvezamu@hust.edu.cn
[2] Computer Vision Institute, School of Computer Science and Software Engineering,
Shenzhen University, Shenzhen, China
{sqbqamar,llshen}@szu.edu.cn
[3] Brigham and Women's Hospital, 75 Francis Street, Boston, MA 02115, USA
hashemi.s@husky.neu.edu

Abstract. The accurate automatic segmentation of brain tumors enhances the probability of survival rate. Convolutional Neural Network (CNN) is a popular automatic approach for image evaluations. CNN provides excellent results against classical machine learning algorithms. In this paper, we present a unique approach to incorporate contexual information from multiple brain MRI labels. To address the problems of brain tumor segmentation, we implement combined strategies of residual-dense connections, multiple rates of an atrous convolutional layer on popular 3D U-Net architecture. To train and validate our proposed algorithm, we used BRATS 2019 different datasets. The results are promising on the different evaluation metrics.

Keywords: Deep learning · Convolutional neural networks · Residual-dense connections · Atrous rates · Brain tumor segmentation

1 Introduction

The growth of irregular cells in the central nervous system changes the normal working of human brains. The brain tumor is the synonym of the growth of irregular cells. According to the World Health Organization (WHO) [1], only cancer shares 1 million deaths in the year of 2018. Among them, high-grade glioblastoma (HGG) and low-grade glioblastoma (LGG) are very common. The categories of gliomas are essential and need to be studied with proper planning to improve the survival rate. In the development of an accurate algorithm, the variations among the shape, size and location of a tumor are obstacles. Feature learning has the great potential to handle this kind of problem. Convolutional neural network (CNN) is one kind of automatic feature learning approach. CNN learns features

© Springer Nature Switzerland AG 2020
A. Crimi and S. Bakas (Eds.): BrainLes 2019, LNCS 11993, pp. 158–166, 2020.
https://doi.org/10.1007/978-3-030-46643-5_15

layer by layer and gives an abstract representation of input data. The encoder-decoder architectures are prevalent for medical image segmentation. U-Net is such a kind of architecture. In which, the encoder part is responsible for reducing the sizes of input resolution using strided convolution or max-pooling while the decoder part recovers the resolution of input size. To avoid the possible loss information in the sub-sampling, the feature maps of the encoder are concatenated with the feature maps of the decoder. In deep architecture, the vanishing gradient problem occurs during training. To avoid it, we used residual connections [2]. These implementations are good with both natural and bio-medical image segmentations. A number of works have been introduced with Residual 3D U-Net for brain tumor segmentation [3], [4], [5]. To contribute more pixels for a current layer, the idea of dense atrous-spatial pyramid pooling (ASPP) was introduced [6]. In this paper, we present a variant form of 3D U-Net for brain tumor segmentation. Main contributions of our proposed work are as follows:

- We develop a variant form of 3D U-Net with the use of Residual and Dense connections for brain tumor segmentation.
- To learn multi-scale contextual information, we use residual-dense ASPP blocks on the segmentation layer at each level of the decoder.

2 Propose Model

Figure 1 shows our proposed architecture for brain tumor segmentation. Our proposed approach is based on 3D U-Net to process the patch sizes on the Brain MRIs dataset. To avoid the vanishing gradient problem in the deeper architecture, we implemented dense connections [2] and residual connections [7] in the proposed architecture. Recently, dense and residual connections are successfully implemented on different natural image segmentation as well as biomedical image segmentation with excellent results. We used both connections in each block of encoder-decoder sub-networks to reduces the size of parameters as well as to enhance the size of the receptive field to preserve more contextual information. We also used residual-dense ASPP [6] blocks on the segmentation layers at each level of the decoder to incorporate multi-scale contextual information. Our model has some similarities with the architectures of [5], [8] in case of residual connections and segmentation layers on different levels of decoder sub-network. While our proposed work utilised some unique strategies to improve segmentation outcomes. The residual connections help build the deeper models that are required in the complexity of brain tumors. However, three major drawbacks come with the utilization of residual connections:

- A large number of parameters are generated. Experimentally, we found that residual connection does not impact on the improvement of segmentation results.
- Generated feature maps of the previous layer are not used in the later layers. The combined features of the lower and higher layers help understand the complex tumors better.

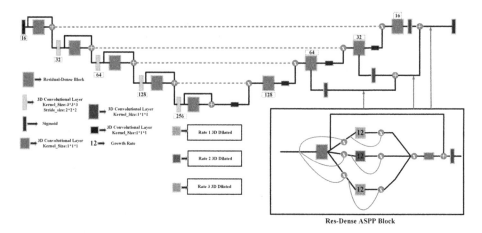

Fig. 1. Proposed model. Residual-Dense blocks with ASPP are implemented on different levels after segmentation layers.

- As discussed by [6], the sizes of the receptive field are small when atrous rates are implements with residual connections.

To overcome these limitations, we built architecture with a combination of dense connections and residual connection. Dense connections reduce the size of parameters and give the concept of the reusing feature maps from one layer to all subsequent layers. The residual-dense block of 3D U-Net contains two 3D convolution layer with a growth rate of 12. A 3D Convolution layer of kernel size $1 \times 1 \times 1$ is implemented to reduce the possible number of channels. This bottleneck layers also avoid the high memory requirement of a 3D conventional layer after the upsampling process. Our proposed work also justifies that the non-factorization of a 3D Convolution layer [4] can manage the high memory demands during the training. The total number of parameters of our proposed work is 3.5M with depth 5 and the initial number of channels is 16. The parameter size is quite lower than 8.5M in case of only residual connections. To capture multi-contexual information, Chen et al. [9] proposed a concept of ASPP for non-biomedical image segmentation. Multi-scaling improves the segmentation accuracy with the successful implementation of the works [10], [11], [12], [13], [14], [15]. Residual-Dense ASPP [6] blocks can be implemented at any part of a 3D U-Net. In our work, we developed a unique architecture to learn more useful contexual information from the labels. Initially, an intermediate layer is implemented after the segmentation layer to increase the number of channels, followed by a residual-dense ASPP block to learning the multi-scale information. Then another segmentation layer is used to receive output on labels and then combine each output on the above level of decoders to generate the outcomes.

3 Implementation Details

3.1 Dataset

To train and evaluate our model, we used the Brats 2019 training and validation datasets [16–20]. The organizers increased the number of patients to 335 in this year, in which 259 subjects are related to high-grade glioblastoma (HGG) and 76 to low-grade glioblastoma (LGG). Four different types of multimodal brain volumes related to each patient: T1, T2, Flair, and T1ce. Three types of labels are: 4 represents ET, 2 represents ED while 1 for NCR/NET [16–20]. The organizers performed necessary pre-processing steps for simplicity. The validation and testing datasets consist of 125 and 166 patients respectively without the ground truth.

3.2 Preprocessing and Training

A bias field is a common problem of MRI modalities. To address this problem, we used an N4ITK bias field correction algorithm on all MRI modalities. Moreover, we used the normalization process for all the MRI scans with 0 mean and 1 standard deviation. The entire training dataset is divided into five different folds to give an equal chance of each patient in training. Moreover, for each training process, 267 subjects are selected for training and 68 for the validation sets. Previously, different patching sampling strategies were proposed [21], we choose random sampling to extract the patches of size 128. In patch creation, we have to be more careful to include all the labels in each patch. The nature of 3D U-Net encouraged us to choose as our baseline model. All the modalities are used in the patch extraction process. Different regularization techniques such as weight decay of 10^{-5}, dropout layers, and multiple augmentations are used to avoid the common overfitting problem. In the entire training of the network, we used 7×10^{-5} as an initial learning rate. The loss function, as well as deep architecture, is also important to solve the problem of segmentation. Recently, several works have been implemented with weighted loss functions and given excellent results as compared to other state-of-the-art methods. However, the weight factor is an extra hyper-parameter in the tuning of model training. We used the loss function [22], which is free from any weight scaling.

4 Results

4.1 Quantitative Analysis

Adam optimizer used in the training process with batch size 1. To calculate the results from five different training as well as five different validation folds, we used the same strategy that is implemented by [5]. The combined folds improve the results for brain tumor segmentation. To analyse the capability of our proposed architecture, we split the results into three categories which are based on the different evaluation metrics. In the first category, we have only used both training and validation datasets of HGG for the evaluation of brain tumor results. In the second category, we add few LGG patients on whole HGG patients of datasets. In the final category, we implemented our approach on both LGG and HGG training and the validation datasets. Tables 1, 2 and 3 shows the training and validation results of each category. Finally, the test results are abbreviated in Table 4. All the effects of three types of datasets are evaluated using CBICAs Image Processing Portal of Multimodal Brain Tumor Segmentation Challenge 2019.

Table 1. Dice and Hausdorff95 Scores of Brats 2019 Training and Validation Datasets. Only HGG patients are used in the evaluation.

	Dice scores			Hausdorff95		
	Enhancing	Whole	Core	Enhancing	Whole	Core
Training	0.78393	0.87142	0.84556	4.60341	7.31335	6.22995
Validation	0.78998	0.86800	0.83537	3.86609	12.33679	21.64817

Table 2. Dice and Hausdorff95 Scores of Brats 2019 Training and Validation Datasets. Combined (Added few LGG cases with the entire HGG cases).

	Dice scores			Hausdorff95		
	Enhancing	Whole	Core	Enhancing	Whole	Core
Training	0.75108	0.86807	0.80464	6.13611	7.53247	7.41123
Validation	0.77597	0.86092	0.79611	5.73398	7.89983	9.33096

Table 3. Dice and Hausdorff95 Scores of Brats 2019 Training and Validation Datasets. Both HGG and LGG cases are evaluated.

	Dice scores			Hausdorff95		
	Enhancing	Whole	Core	Enhancing	WholeCore	
Training	0.65937	0.85675	0.76591	8.86077	8.89822	9.89342
Validation	0.62301	0.85184	0.75762	8.46826	9.00831	10.67437

4.2 Qualitative Analysis

For the qualitative analysis of our proposed architecture, we select a few cases in the pictorial form. Figure 2 shows the segmentation results for subject BraTS19_CBICA_AAB_1. Our proposed model segments the intricate details of the different types of brain tumors, as shown in Fig. 2. However, In some cases,

Table 4. Dice and Hausdorff95 scores of Brats 2019 Test Dataset.

	Dice			Hausdorff95		
	Enhancing	Whole	Core	Enhancing	Whole	Core
Mean	0.64928	0.80744	0.70716	8.58525	11.57987	11.0498

Ground Truth (T1ce)   Segmented (T1ce)

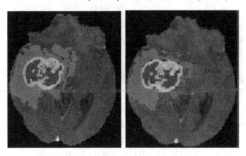

Fig. 2. Segmentation results of different tumor classes. Tumor core visualize in red, Whole tumor in green, and Enhancing tumor in yellow. (Color figure online)

Ground Truth (T1ce)   Segmented (T1ce)

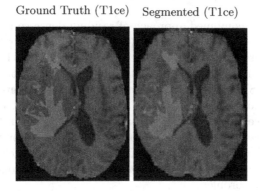

Fig. 3. Segmentation results of different tumor classes. Tumor core visualize in red, Whole tumor in green, and Enhancing tumor in yellow. Images shows the zero values for tumor core and enhancing tumor labels. This is the visualization of subject BraTS19_CBICA_AQA_1. (Color figure online)

the presence of zero values in the enhancing tumor and the tumor core labels in the training dataset hinder the development of accurate automatic segmentation algorithms. To confirm the above statement, Fig. 3 and Fig. 4 visualizes some subjects from the training dataset.

Ground Truth (T1ce)   Segmented (T1ce)

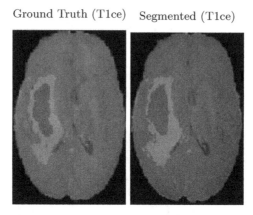

Fig. 4. Segmentation results of different tumor classes. Red color represents Tumor core, Whole tumor in green, and Enhancing tumor in yellow. Zero values represent for label enhancing tumor in both ground truth as well as in prediction. This is the visualisation of a subject BraTS19_TCIA09_462_1. (Color figure online)

5   Discussion and Conclusion

We have proposed a combined approach of residual-dense connections, multiple rates of the atrous convolutional layer (ASPP) on a popular 3D U-Net architecture. ASPP approach have given excellent results on both biomedical [23] as well as non-biomedical image segmentations [9]. In the proposed architecture, we implemented ASPP blocks with the combined residual-dense connections to incorporate more contexual information for Brain MRI data. Furthermore, the dense approach gives the advantage of parameter reduction in the development of a model. However, residual connections have a huge amount of parameters but no impact on the results. We applied a combined approach to gain benefits residual and dense connections. We also have utilized the proper techniques to avoid the overfitting problem.

Our proposed model has strong capability for delivering excellent results on the training and validation datasets of HGG patients. Furthermore, this approach also offers handy effects on the combined whole HGG and several LGG patients of BraTS 2019 datasets. In the meantime, the result of our proposed model is very promising for the entire training and validation datasets. However, the final scores also exhibit the presence of zero value of the enhancing tumor in the training dataset. In future, we will perform some post-processing steps by using different architectures such as R-CNN [24] and GAN [25] to improve the final scores.

Acknowledgment. This work is supported by the National Natural Science Foundation of China under Grant No.61672250 and the Hubei Provincial Development and Reform Commission Project in China.

References

1. World Health Organisation: Cancer (2019). https://www.who.int/cancer/en/. Accessed 20 Nov 2019
2. Huang, G., Liu, Z., Weinberger, K.Q.: Densely connected convolutional networks. CoRR abs/1608.0 (2016). http://arxiv.org/abs/1608.06993
3. Kermi, A., Mahmoudi, I., Khadir, M.T.: Deep convolutional neural networks using U-Net for automatic brain tumor segmentation in multimodal MRI volumes. In: Crimi, A., Bakas, S., Kuijf, H., Keyvan, F., Reyes, M., van Walsum, T. (eds.) BrainLes 2018. LNCS, vol. 11384, pp. 37–48. Springer, Cham (2019). https://doi. org/10.1007/978-3-030-11726-9_4
4. Chen, W., Liu, B., Peng, S., Sun, J., Qiao, X.: S3D-UNet: separable 3D U-Net for brain tumor segmentation. In: Crimi, A., Bakas, S., Kuijf, H., Keyvan, F., Reyes, M., van Walsum, T. (eds.) BrainLes 2018. LNCS, vol. 11384, pp. 358–368. Springer, Cham (2019). https://doi.org/10.1007/978-3-030-11726-9_32
5. Isensee, F., Kickingereder, P., Wick, W., Bendszus, M., Maier-Hein, K.H.: Brain tumor segmentation and radiomics survival prediction: contribution to the BRATS 2017 challenge. CoRR abs/1802.1 (2018). http://arxiv.org/abs/1802.10508
6. Yang, M., Yu, K., Zhang, C., Li, Z., Yang, K.: DenseASPP for semantic segmentation in street scenes. In: 2018 IEEE/CVF Conference on Computer Vision and Pattern Recognition, pp. 3684–3692 (2018)
7. He, K., Zhang, X., Ren, S., Sun, J.: Deep residual learning for image recognition. CoRR abs/1512.0 (2015). http://arxiv.org/abs/1512.03385
8. Kayalibay, B., Jensen, G., van der Smagt, P.: CNN-based segmentation of medical imaging data. CoRR abs/1701.0 (2017). http://arxiv.org/abs/1701.03056
9. Chen, L.C., Papandreou, G., Schroff, F., Adam, H.: Rethinking atrous convolution for semantic image segmentation. CoRR abs/1706.0 (2017). http://arxiv.org/abs/ 1706.05587
10. Ahmad, P., et al.: 3D dense dilated hierarchical architecture for brain tumor segmentation. In: Proceedings of the 2019 4th International Conference on Big Data and Computing, ICBDC 2019, pp. 304–307. ACM, New York (2019). http://doi. acm.org/10.1145/3335484.3335516
11. Chen, L., Wu, Y., DSouza, A.M., Abidin, A.Z., Wismüller, A., Xu, C.: MRI tumor segmentation with densely connected 3D CNN. In: Medical Imaging: Image Processing (2018)
12. Dolz, J., Gopinath, K., Yuan, J., Lombaert, H., Desrosiers, C., Ayed, I.B.: HyperDense-Net: a hyper-densely connected CNN for multi-modal image segmentation. CoRR abs/1804.0 (2018). http://arxiv.org/abs/1804.02967
13. Kamnitsas, K., et al.: Efficient multi-scale 3D CNN with fully connected CRF for accurate brain lesion segmentation. CoRR abs/1603.0 (2016). http://arxiv.org/ abs/1603.05959
14. Qamar, S., Jin, H., Zheng, R., Ahmad, P.: 3D hyper-dense connected convolutional neural network for brain tumor segmentation. In: 2018 14th International Conference on Semantics, Knowledge and Grids (SKG), pp. 123–130 (2018)
15. Qamar, S., Jin, H., Zheng, R., Ahmad, P.: Multi stream 3D hyper-densely connected network for multi modality isointense infant brain MRI segmentation. Multimed. Tools Appl. **78**, 25807–25828 (2019). https://doi.org/10.1007/s11042-019-07829-1
16. Bakas, S., et al.: Segmentation labels and radiomic features for the pre-operative scans of the TCGA-GBM collection. The Cancer Imaging Archive (2017). https:// doi.org/10.7937/K9/TCIA.2017.KLXWJJ1Q

17. Bakas, S., et al.: Segmentation labels and radiomic features for the pre-operative scans of the TCGA-LGG collection. The Cancer Imaging Archive (2017). https://doi.org/10.7937/K9/TCIA.2017.GJQ7R0EF

18. Bakas, S., et al.: Advancing the cancer genome atlas glioma MRI collections with expert segmentation labels and radiomic features. Sci. Data **4**, 170117 (2017). https://doi.org/10.1038/sdata.2017.117

19. Bakas, S., et al.: Identifying the best machine learning algorithms for brain tumor segmentation, progression assessment, and overall survival prediction in the BRATS challenge. CoRR abs/1811.0 (2018). http://arxiv.org/abs/1811.02629

20. Menze, B.H., et al.: The multimodal brain tumor image segmentation benchmark (BRATS). IEEE Trans. Med. Imaging (10), 1993–2024. https://doi.org/10.1109/TMI.2014.2377694

21. Feng, X., Tustison, N.J., Meyer, C.H.: Brain tumor segmentation using an ensemble of 3D U-Nets and overall survival prediction using radiomic features. CoRR abs/1812.0 (2018). http://arxiv.org/abs/1812.01049

22. Milletari, F., Navab, N., Ahmadi, S.A.: V-Net: fully convolutional neural networks for volumetric medical image segmentation. CoRR abs/1606.0 (2016). http://arxiv.org/abs/1606.04797

23. Sarker, M.M.K., et al..: SLSDeep: skin lesion segmentation based on dilated residual and pyramid pooling networks. CoRR abs/1805.1 (2018). http://arxiv.org/abs/1805.10241

24. Kopelowitz, E., Englehard, G.: Lung nodules detection and segmentation using 3D mask-RCNN (2019)

25. Mondal, A.K., Dolz, J., Desrosiers, C.: Few-shot 3D multi-modal medical image segmentation using generative adversarial learning. CoRR abs/1810.12241 (2018). http://arxiv.org/abs/1810.12241

Two Stages CNN-Based Segmentation of Gliomas, Uncertainty Quantification and Prediction of Overall Patient Survival

Thibault Buatois, Élodie Puybareau[✉], Guillaume Tochon,
and Joseph Chazalon

EPITA Research and Development Laboratory (LRDE), Le Kremlin-Bicêtre, France
{thibault.buatois,elodie.puybareau,guillaume.tochon,
joseph.chazalon}@lrde.epita.fr

Abstract. This paper proposes, in the context of brain tumor study, a fast automatic method that segments tumors and predicts patient overall survival. The segmentation stage is implemented using two fully convolutional networks based on VGG-16, pre-trained on ImageNet for natural image classification, and fine tuned with the training dataset of the MICCAI 2019 BraTS Challenge. The first network yields to a binary segmentation (background vs lesion) and the second one focuses on the enhancing and non-enhancing tumor classes. The final multiclass segmentation is a fusion of the results of these two networks. The prediction stage is implemented using kernel principal component analysis and random forest classifiers. It only requires a predicted segmentation of the tumor and a homemade atlas. Its simplicity allows to train it with very few examples and it can be used after any segmentation process.

Keywords: Glioma · Tumor segmentation · Fully convolutional network · Random forest · Survival prediction

1 Introduction

1.1 Motivation

Gliomas are the most common brain tumors in adults, growing from glial cells and invading the surrounding tissues [10]. Two classes of tumors are observed. The patients with the more aggressive ones, classified as high-grade gliomas (HGG), have a median overall survival of two years or less and imply immediate treatment [13,16]. The less aggressive ones, the low-grade gliomas (LGG), allow an overall survival of several years, with no need of immediate treatment. Multimodal magnetic resonance imaging (MRI) helps pratitioners to evaluate the degree of the disease, its evolution and the response to treatment. Images are analyzed based on qualitative or quantitative measures of the lesion [8,22]. Developing automated brain tumor segmentation techniques that are able to analyze these tumors is challenging, because of the highly heterogeneous appearance and

© Springer Nature Switzerland AG 2020
A. Crimi and S. Bakas (Eds.): BrainLes 2019, LNCS 11993, pp. 167–178, 2020.
https://doi.org/10.1007/978-3-030-46643-5_16

shapes of these lesions. Manual segmentations by experts can also be a challenging task, as they show significant variations in some cases. Despite the relevance of glioma segmentation, this segmentation is challenging due to the high heterogeneity of tumors. The development of an algorithm that can perform fully automatic glioma segmentation and overall prediction of survival would be an important improvement for patients and practitioners. During the past 20 years, different algorithms for segmentation of tumor structures has been developed and reviewed [1,6,7]. However, a fair comparison of algorithms implies a benchmark based on the same dataset, as it has been proposed during MICCAI BraTS Challenges [5,15].

1.2   Context

The work proposed in this article has been done in the context of the MICCAI 2019 Multimodal Brain Tumor Segmentation Challenge (BraTS)[1]. The overall goal of this challenge is to establish a fair comparison between state-of-the-art methods, and to release a large annotated dataset. The objective of the work conducted within the scope of BraTS challenge is three-fold:

Task 1 providing a fully automated pipeline for the segmentation of the glioma from multimodal MRI scans without any manual assistance,

Task 2 predicting the patient overall survival from pre-operative scans,

Task 3 estimating the uncertainty in segmentation results provided within the scope of Task 1. Note that, unlike the two previously mentioned tasks which were already established in the former BraTS challenges, Task 3 is exclusive to the 2019 BraTS challenge.

We received data of 335 patients, with associated masks to develop our method. The data, avaliable online, have been annotated and preprocessed [2–4]. The volumes given are T1, T1ce, T2 and FLAIR. Our method is then evaluated on new volumes:

– a validation set composed of 125 patients, released by the organizers without the manual segmented masks (these masks will not be released), to obtain preliminary results.
– a test set comprising 166 patients, used for the challenge evaluation. In that case, our team was asked to process those data and send the corresponding results in the expected formats within 48 h after receiving the data.

1.3   Related Works

In the framework of BraTS 2018 challenge, we carried out the tumor segmentation task thanks to a fully convolutional network (FCN) approach [17]. The used network was VGG (Visual Geometry Group) [20], pre-trained on the ImageNet dataset and fine-tuned to the BraTS challenge dataset thanks to transfer learning strategy. As VGG expects 2D color images as input images (thus,

---

[1] http://braintumorsegmentation.org/.

2D images with 3 RGB channels), we made use of the "pseudo-3D" strategy, originating from [23]. Initially applied to the segmentation of 3D brain MR volumes, the pseudo-3D idea consists in creating a series of RGB images of the 3D input volume by selecting the $(n-1)^{\text{th}}, n^{\text{th}}$ and $(n+1)^{\text{th}}$ slices (with n sliding through the 3D volume) as the R, G and B channels, respectively. This pseudo-3D idea proved to be also successful for left atrium segmentation [18]. Note that in [17], this pseudo-3D idea was actually used in a multimodality fashion since the $(n-1)^{\text{th}}$ and $(n+1)^{\text{th}}$ slices were extracted from the T1ce modality while the nth slice was selected in the T2 modality.

However, this method yielded poor results for the tumor segmentation task of BraTS 2018 challenge. Thus, we decided to improve our approach, still using VGG and 3D-like images, but with a two-stage segmentation scheme, as detailed in the following Sect. 2.1. Based on those poor quality tumor segmentations, we nevertheless reached the second place for the survival prediction task of BraTS 2018 challenge [17]. We also attempted to improve our survival prediction algorithm, as exposed in Sect. 2.2.

2   Proposed Methodology

2.1   Tumor Segmentation

Overall FCN Architecture. As for the BraTS 2018 Challenge [17], our FCN architecture for the segmentation task relies on the 16-layer VGG network [20], which was pre-trained on ImageNet for image classification purposes [11]. We keep only the first 4 convolutional stages, and we discard the fully connected layers at the end of VGG network. Each stage is composed of convolutional layers, followed by Rectified Linear Unit (ReLU) layers and a max-pooling layer, corresponding to four fine-to-coarse feature maps. Inspired by the work in [12,14], we add specialized convolutional layers (with a 3×3 kernel size) with K (e.g. $K = 16$) feature maps after the last convolutional layer of each stage. All these specialized layers are then rescaled to the original image size, and concatenated together. We add a last convolutional layer with kernel size 1×1 at the end. This last step combines linearly the fine-to-coarse feature maps in the concatenated specialized layers, and provide the final segmentation result. The parameters of the network are the same than in [17].

This FCN architecture is depicted in its generic form in Fig. 1, where it accepts as input a RGB image and outputs its corresponding segmentation map.

Pre-processing. Let n, m be respectively the minimum non-null and maximum gray-level value of an input 3D volume. For each patient, we first requantize all voxel values using a linear function so that the gray-level range $[n, m]$ is mapped to $[-127, 127]$.

For our application, the question amounts to how to prepare appropriate inputs (RGB input images) given that a brain MR image is a 3D volume.

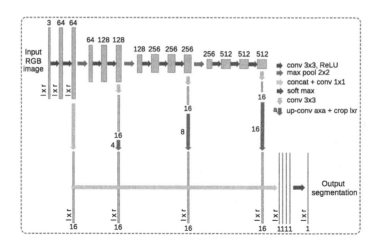

Fig. 1. Proposed FCN generic architecture, based on the pre-trained VGG network.

To that aim, we propose to stack 2D slices of different modalities according to the segmentation classes studied.

Precisely, to form an input artificial color image for the pre-trained network to segment the ith slice, we defined the input image this way:

– in the channel green, put the slice i of modality 1
– in the channel red, put the slice $n - x$ of modality 2
– in the channel blue, put the slice $n + x$ of the modality 3

The x parameter, namely the "offset", can bring 3D information for 2D segmentation while the choice of modalities (modality 1, 2 and 3 can be the same) can bring information from one to three modalities. For the challenge, we use the slice i of one volume modality in the green channel, the slice i of an other modality for the blue and red channels, and the offset x is set to 0. These parameters (modalities and x) have been selected after testing all the combinations.

Segmentation. During our experiments, we noticed that our network confuses the background with the edema, and the necrotic and non-enhancing tumor core with the enhancing tumor. Hence, we decided to performed the segmentation in two steps: the first step is a binary segmentation of the whole tumor (vs. background), and the second is a multi-label segmentation of two parts of the tumor (necrotic and non-enhancing tumor core vs. enhancing tumor). The final segmentation resulted of the fusion of these two results, as depicted in Fig. 2.

As the different modalities enhance different structures, we decided to take the advantage of each modality for a specific task. For the segmentation of the whole tumor vs. background, we put the T2 modality in the green channel of our RGB input image and the FLAIR in the blue and red channels. For the segmentation of necrotic and non-enhancing tumor core vs. enhancing tumor,

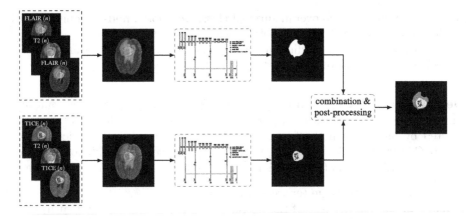

Fig. 2. Workflow of our procedure for the segmentation task.

we put the T2 modality in the green channel of our RGB input image and the T1ce in the blue and red channels.

Post-processing. The output of our process for one slice during the inference phase is two 2D segmented slice: one for the tumor vs. background, and one for the inner parts of the tumor. We first fused the two segmentation results: we kept the segmentation of the inner parts of the tumor as classes *necrotic and non-enhancing tumor core* and *enhancing tumor*. The last class, the *oedema* is the non-zero part of the whole tumor that was not detected as an other class in the tumor part segmentations.

After treating all the slices of the volume, the segmented slices are stacked to recover a 3D volume with the same shape as the initial volume, and containing only the segmented lesions. A spatial regularization is processed to ensure 3D cohesion, and we remove the 3D connected components that have as label only *oedema*.

2.2 Patient Survival Prediction

The second task of the MICCAI 2019 BraTS challenge was concerned with the prediction of patient overall survival from pre-operative scans (only for subjects with gross total resection (GTR) status). As precognized by the evaluation framework, the classification procedure has been conducted by labeling subjects into three classes: short-survivors (less than 10 months), mid-survivors (between 10 and 15 months) and long-survivors (greater than 15 months). For post-challenge analyses, prediction results were also compared in terms of mean and median square error of survival time predictions, expressed in days. For that reason, our proposed patient survival prediction algorithm was organized in two steps:

1 We first predicted the overall survival class, *i.e.* short-, mid- or long-survival (hereafter denoted by class/label 1, 2 and 3, respectively).
2 We then adjusted our prediction within the predicted class by means of linear regression, in order to express the survival time in days.

Definition and Extraction of Relevant Features. Extracting relevant features is critical for classification purposes. Here, we re-used the features implemented by our team in the framework of the patient survival prediction task of MICCAI 2018 BraTS challenge, which ranked tie second [17]. Those features were chosen after in-depth discussions with a practicioner and are the following:

feature 1: the patient age (expressed in years).
feature 2: the relative size of the necrosis (labeled 1 in the groundtruth) class with respect to the brain size.
feature 3: the relative size of the edema class (labeled 2 in the groundtruth) with respect to the brain size.
feature 4: the relative size of the active tumor class (labeled 4 in the groundtruth) with respect to the brain size.
feature 5: the normalized coordinates of the binarized enhanced tumor (thus only considering necrosis and active tumor classes).
feature 6: the normalized coordinates of the region that is the most affected by necrosis, in a home made brain atlas.

For the training stage, features 2, 3 and 4 were computed thanks to the patient ground truth map for each patient. As this information was unknown during the test stage, the segmented volumes predicted by our Deep FCN architecture are used instead. In any case, these size features were expressed relatively to the total brain size (computed as the number of voxels in the T2 modality whose intensity is greater than 0).

In addition, we also re-used the home-made brain atlas that we also developed for the 2018 BraTS challenge. This atlas was divided into 10 crudely designed regions accounting for the frontal, parietal, temporal and occipital lobes and the cerebellum for each hemisphere (see [17] for more details regarding this atlas and how it was adjusted to each patient brain size). Feature 6 was defined as the coordinates of the centroid of the region within the altas that was the most affected by the necrosis class (*i.e.*, the region that had the most voxels labeled as necrosis with respect to its own size). Note that this feature, as well as feature 5, were then normalized relatively to the brain bounding box. This led to a feature vector with 10 components per patient (since both centroids coordinates were 3-dimensionals).

Training Phase. For the training phase, we modified our previous work [17] in the following way: while we maintained the final learning stage through random forest (RF) classifiers [21], we replaced the principal component analysis (PCA) transformation, acting as preprocessing step for the learning stage, by its kernel counterpart (kPCA) [19]. The rationale is that we hoped to increase the RFs

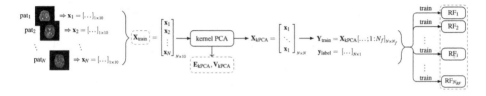

Fig. 3. Workflow of the proposed class-based training procedure. The information stored after the training phase (necessary for the test phase) is written in red or encircled in dashed red. (Color figure online)

performances in terms of classification/prediction as the input features are highly non-linear in terms of survival labels.

More specifically, the training stage of our prediction algorithm was as follows:

1. The feature vector $\mathbf{x}_i \in \mathbb{R}^{10}$ of each of the N patients in the training set is extracted as described in the previous Sect. 2.2. All those feature vectors are then stacked in a $N \times 10$ feature matrix $\mathbf{X}_{\text{train}}$
2. A kPCA is performed on $\mathbf{X}_{\text{train}}$, yielding the $N \times N$ matrix $\mathbf{X}_{\text{kPCA}}$. This matrix is obtained through the computation, normalization and diagonalization of the so-called *kernel matrix* which represents the dot product between the N features vectors when mapped in the feature space through a kernel function (here defined as a polynomial kernel with degree $d = 3$).
4. The $N \times N_f$ matrix $\mathbf{Y}_{\text{train}}$ is defined from $\mathbf{X}_{\text{train}}$ by retaining the first N_f columns (corresponding to the leading N_f features in the feature space, here set to $N_f = 10$). N_{RF} RF classifiers [21] are finally trained on all rows of $\mathbf{Y}_{\text{train}}$ to learn to predict the survival class of each training patient using the true label vector $\mathbf{y}_{\text{label}}$ as target values. The used RF parameters (number of decision trees per RF, splitting criterion, total number of RFs N_{RF}) are defined as in [17].
5. Three linear regressors (one per survival class) are finally trained using the patient age and its whole tumor size (relatively to its brain size) as explanatory variables and its true survival time (expressed in days) as measured variable.

Steps 1. to 4. are depicted by the workflow in Fig. 3. In addition to the three linear regressors, we also store (for the test phase) the training feature matrix $\mathbf{X}_{\text{train}}$, the eigenvector matrix V_{kPCA} and eigenvalues E_{kPCA} of the kernel matrix, and the number of retained features N_f after kPCA.

Test Phase. The test phase has been conducted in a similar fashion as the training phase. Given some input test patient, its overall survival class was first predicted, before being refined and expressed in terms of number of days. More specifically:

Fig. 4. Workflow of the proposed test procedure.

1. The features vector $\mathbf{x}_{\text{test}}$ of the test patient was retrieved as described previously.
2. This feature vector is then projected onto the principal axes learnt by the kPCA during the training phase. For that purpose, a new kernel matrix is computed and centered (hence the need for $\mathbf{X}_{\text{train}}$) before proper projection (through $\mathbf{V}_{\text{kPCA}}$) and scaling (with $\mathbf{E}_{\text{kPCA}}$).
3. This results in the projected vector $\mathbf{x}_{\text{kPCA}} \in \mathbb{R}^N$ from which the first N_f features are retained, yielding the test vector $\mathbf{y}_{\text{test}}$. This vector is then fed to the N_{RF} RF classifiers, leading to N_{RF} independent class label predictions. The final label prediction y_{pred} (1, 2 and 3 for short-, mid- and long-survivors, respectively) is eventually obtained by majority voting.
4. Once the survival class has been established, the final patient survival rate is predicted by means of the appropriate learnt linear regressor.

Steps 1. to 3. are illustrated by the worflow in Fig. 4.

2.3   Quantification of Uncertainty in Segmentation

The third task of MICCAI 2019 BraTS challenge focused on the estimation of the uncertainty in segmentation results provided within the scope of Task 1. For that purpose, we focused on the study of a lightweight technique which estimate uncertainty by considering the instability at the spatial boundary between two regions predicted to belong to different classes. We believed that such approach can be complementary to approaches based on the stability of the prediction under perturbations like Monte Carlo Dropout [9] which tend to be computationally demanding.

The resulting indicator assigns a maximal uncertainty (100) at the boundary between two regions, and linearly decreases this uncertainty to the minimal value (0) at a given distance from the boundary. This distance defines the (half) width of an "uncertainty border" between two regions. It is calibrated independently for each class, and was estimated according to the 95th percentile of the Hausdorff distance metric reported in Table 1 for our segmentation method for this particular class. In practice, we used a half-width of 9 voxels for the whole tumor (WT), a half-width of 12 voxels for the tumor core (TC) and a half-width of 7 voxels for the enhancing tumor (ET).

To compute this indicator, we first compute the Boundary Distance Transform $\text{BDT} = max(\text{DT}(\mathcal{R}), \text{DT}(\overline{\mathcal{R}}))$ using the Distance Transform DT to the given sub-region $\mathcal{R}$ and its complement $\overline{\mathcal{R}}$. Then, we inverted, shift and clip the BDT such that the map is maximal on the boundary and have 0 values at a distance greater or equal to the half-width of the border. We finally scale the resulting map so its values are comprised between 0 (far from the boundary) and 100 (on the boundary). The resulting uncertainty map for a given class exhibits a triangular activation shape on the direction perpendicular to the boundary of the objects detected by the segmentation stage.

3   Experiments and Results

Task 1: Tumor Segmentation. Table 1 presents the results obtained for the segmentation task by our proposed method. Overall, we perform better in terms of Dice coefficient on the whole tumor (Dice_WT column) than on the enhanced tumor (Dice_ET, corresponding to the active tumor class) and the tumor core (Dice_TC, corresponding to the necrosis class). Visual inspection on the segmentation infered on the training data set indicated that our method sometimes indeed tends to confuse those two classes together. This is confirmed by the fact that the median Dice is significantly higher than the mean Dice on the whole test data set, implying that eventhough it performs well on most cases, our method seems to really fail in the segmentation of necrosis and enhanced tumor classes on a few cases.

Regarding the Haussdorf distance metric, our method however performs better on the enhanced tumor and tumor core. We believe that this is due to the regularization step performed on the segmented infered by the used FCN architecture, which smoothes the external contours of the edema class, thus impacting the shape of the whole tumor but not the enhanced tumor and tumor core.

Table 1. Dice and Haussdorf distance metrics for the proposed segmentation method on the test data set.

Metric	Dice_ET	Dice_WT	Dice_TC	HD95_ET	HD95_WT	HD95_TC
Mean	0.750	0.854	0.800	3.084	7.047	5.961
StdDev	0.230	0.129	0.247	4.320	7.609	8.847
Median	0.821	0.897	0.891	2	5.099	3.606
25quantile	0.733	0.838	0.821	1.414	3.742	2
75quantile	0.877	0.919	0.936	3.162	7.071	6.224

Task 2: Survival Prediction. Table 2 presents the various classification performance metrics, namely the class-based accuracy, the mean, median and standard deviation square errors and Spearman R coefficient for survival predictions expressed in days, for the proposed prediction algorithm for the validation data

Table 2. Classification metrics of the proposed survival prediction method for the validation (with comparison with [17]) and test data sets.

	Data set	Accuracy	MSE	medianSE	stdSE	SpearmanR
Previous work [17]	Validation	0.379	131549	72900	169116	0.235
Proposed	Validation	0.517	127727	40645	191729	0.429
	Test	0.523	428641	59539	1162454	0.36

set and the test data set. For comparison purposes, results obtained by our previous work [17] applied on the validation data set are also presented. The validation and test data sets are comprised of $N = 27$ and $N = 107$ patients, respectively.

As it can be seen on Table 2, replacing the conventional PCA (as done in [17]) by the kPCA does indeed improve the class-based classification accuracy since it increases from 0.379 to 0.517 on the validation data set. This accuracy also remains stable when going from the validation data set to the test data set (slightly increasing from 0.517 to 0.523). This notably validates the capacity of kPCA to RFs performances in terms of class-wise prediction with respect to classical PCA. Metrics devoted to the evaluation of the survival prediction in days (namely the mean and median square errors) however do not allow to conclude with respect to the soundness of the final linear regression step for the validation data set when comparing [17] with the currently proposed algorithm.

Task 3: Uncertainty Quantification. As summarized by Table 3, our uncertainty estimation method produces encouraging results for the whole tumor (WT) class as it permits to efficiently filter false positives and increase the overall confidence from an original DICE score of 88.0%, as show by the *DICE AUC* metric. This, however, comes with the price of a large filtering of true positives, as shown by the *FTP Ratio AUC* metric. Regarding the tumor core (TC) and enhancing tumor (ET) classes, the results are less promising as the DICE score degrades from original DICE scores of 74.6% (TC) and 70.1% (ET), mainly because of an over-aggressive filtering. Those results let us believe that this uncertainty estimation method is better suited for cases were the underlying segmentation method already performs quite well. Because of its simplicity and its fast computation, it may be a natural baseline for more complex methods to be compared against.

Table 3. Performance reported by the automated evaluation platform for uncertainty estimation for our method on the validation data set.

Class	WT	TC	ET
DICE AUC (%) ↑	89.7	72.5	61.7
FTP Ratio AUC (%) ↓	48.3	71.6	70.6

4 Conclusion

In this article, we present the work submitted for the MICCAI Challenge BraTS 2019, for the segmentation, prediction and uncertaincy tasks.

- The segmentation procedure is performed using two VGG-based segmentation networks;
- The prediction procedure relies on feature extraction and random forests;
- The uncertainty procedure uses the results of the segmentation and gives a confidence score to each pixel based on its distance to the background.

The strength of this method is its modularity and its simplicity. It is easy to implement and fast. From the obtained segmentation result, we propose a simple method to predict the patient overall survival, based on Random Forests, based on the same procedure than during the BraTS 2018 Challenge where we reached the 2nd place. This method only needs as input a segmentation, a brain atlas and a brain volume for atlas registration. It means that our method is robust to the different acquisitions, does not need a special modality or setting, yielding to a method robust to inter-base variations.

References

1. Angelini, E.D., Clatz, O., Mandonnet, E., Konukoglu, E., Capelle, L., Duffau, H.: Glioma dynamics and computational models: a review of segmentation, registration, and in silico growth algorithms and their clinical applications. Curr. Med. Imaging Rev. **3**(4), 262–276 (2007)
2. Bakas, S., Akbari, H., Sotiras, A., Bilello, M., Rozycki, M., Kirby, J., et al.: Segmentation labels and radiomic features for the pre-operative scans of the TCGA-GBM collection. Cancer Imaging Arch. (2017). https://doi.org/10.7937/K9/TCIA.2017.KLXWJJ1Q
3. Bakas, S., Akbari, H., Sotiras, A., Bilello, M., Rozycki, M., Kirby, J., et al.: Segmentation labels and radiomic features for the pre-operative scans of the TCGA-LGG collection. Cancer Imaging Arch. (2017). https://doi.org/10.7937/K9/TCIA.2017.GJQ7R0EF
4. Bakas, S., et al.: Advancing the cancer genome atlas glioma MRI collections with expert segmentation labels and radiomic features. Sci. Data **4**, 170117 (2017)
5. Bakas, S., et al.: Identifying the best machine learning algorithms for brain tumor segmentation, progression assessment, and overall survival prediction in the BRATS challenge. arXiv preprint arXiv:1811.02629 (2018)
6. Bauer, S., Wiest, R., Nolte, L.P., Reyes, M.: A survey of MRI-based medical image analysis for brain tumor studies. Phys. Med. Biol. **58**(13), R97 (2013)
7. Bonnín Rosselló, C.: Brain lesion segmentation using Convolutional Neuronal Networks. B.S. thesis, Universitat Politècnica de Catalunya (2018)
8. Eisenhauer, E.A., et al.: New response evaluation criteria in solid tumours: revised RECIST guideline (version 1.1). Eur. J. Cancer **45**(2), 228–247 (2009)
9. Gal, Y., Ghahramani, Z.: Dropout as a bayesian approximation: representing model uncertainty in deep learning. In: Proceedings of the 33rd International Conference on Machine Learning (ICML 2016), pp. 1050–1059, June 2015

10. Holland, E.C.: Progenitor cells and glioma formation. Curr. Opin. Neurol. **14**(6), 683–688 (2001)
11. Krizhevsky, A., Sutskever, I., Hinton, G.E.: ImageNet classification with deep convolutional neural networks. In: Advances in Neural Information Processing Systems, pp. 1097–1105 (2012)
12. Long, J., Shelhamer, E., Darrell, T.: Fully convolutional networks for semantic segmentation. In: Proceedings of IEEE International Conference on Computer Vision and Pattern Recognition, pp. 3431–3440 (2015)
13. Louis, D.N., et al.: The 2007 who classification of tumours of the central nervous system. Acta Neuropathol. **114**(2), 97–109 (2007)
14. Maninis, K.-K., Pont-Tuset, J., Arbeláez, P., Van Gool, L.: Deep retinal image understanding. In: Ourselin, S., Joskowicz, L., Sabuncu, M.R., Unal, G., Wells, W. (eds.) MICCAI 2016. LNCS, vol. 9901, pp. 140–148. Springer, Cham (2016). https://doi.org/10.1007/978-3-319-46723-8_17
15. Menze, B.H., et al.: The multimodal brain tumor image segmentation benchmark (BRATS). IEEE Trans. Med. Imaging **34**(10), 1993 (2015)
16. Ohgaki, H., Kleihues, P.: Population-based studies on incidence, survival rates, and genetic alterations in astrocytic and oligodendroglial gliomas. J. Neuropathol. Exp. Neurol. **64**(6), 479–489 (2005)
17. Puybareau, E., Tochon, G., Chazalon, J., Fabrizio, J.: Segmentation of gliomas and prediction of patient overall survival: a simple and fast procedure. In: Crimi, A., Bakas, S., Kuijf, H., Keyvan, F., Reyes, M., van Walsum, T. (eds.) BrainLes 2018. LNCS, vol. 11384, pp. 199–209. Springer, Cham (2019). https://doi.org/10.1007/978-3-030-11726-9_18
18. Puybareau, É., et al.: Left atrial segmentation in a few seconds using fully convolutional network and transfer learning. In: Pop, M., Sermesant, M., Zhao, J., Li, S., McLeod, K., Young, A., Rhode, K., Mansi, T. (eds.) STACOM 2018. LNCS, vol. 11395, pp. 339–347. Springer, Cham (2019). https://doi.org/10.1007/978-3-030-12029-0_37
19. Schölkopf, B., Smola, A., Müller, K.-R.: Kernel principal component analysis. In: Gerstner, W., Germond, A., Hasler, M., Nicoud, J.-D. (eds.) ICANN 1997. LNCS, vol. 1327, pp. 583–588. Springer, Heidelberg (1997). https://doi.org/10.1007/BFb0020217
20. Simonyan, K., Zisserman, A.: Very deep convolutional networks for large-scale image recognition. CoRR abs/1409.1556 (2014)
21. Svetnik, V., Liaw, A., Tong, C., Culberson, J.C., Sheridan, R.P., Feuston, B.P.: Random forest: a classification and regression tool for compound classification and QSAR modeling. J. Chem. Inf. Comput. Sci. **43**(6), 1947–1958 (2003)
22. Wen, P.Y., et al.: Updated response assessment criteria for high-grade gliomas: response assessment in neuro-oncology working group. J. Clin. Oncol. **28**(11), 1963–1972 (2010)
23. Xu, Y., Géraud, T., Bloch, I.: From neonatal to adult brain MR image segmentation in a few seconds using 3D-like fully convolutional network and transfer learning. In: Proceedings of the 23rd IEEE International Conference on Image Processing (ICIP), pp. 4417–4421, Beijing, China, September 2017

Detection and Segmentation of Brain Tumors from MRI Using U-Nets

Krzysztof Kotowski[1], Jakub Nalepa[1,2(✉)], and Wojciech Dudzik[1]

[1] Future Processing, Gliwice, Poland
{kkotowski,jnalepa,wdudzik}@future-processing.com
[2] Silesian University of Technology, Gliwice, Poland

Abstract. In this paper, we exploit a cascaded U-Net architecture to perform detection and segmentation of brain tumors (low- and high-grade gliomas) from magnetic resonance scans. First, we detect tumors in a binary-classification setting, and they later undergo multi-class segmentation. The total processing time of a single input volume amounts to around 15 s using a single GPU. The preliminary experiments over the BraTS'19 validation set revealed that our approach delivers high-quality tumor delineation and offers instant segmentation.

Keywords: Brain tumor · Segmentation · Deep learning · CNN

1 Introduction

Automated brain tumor detection and segmentation are the vital research topics in the field of medical image analysis due to their wide clinical applicability. The state-of-the-art techniques for these tasks are often split into the *atlas-based, unsupervised, supervised,* and *hybrid* approaches (Fig. 1). In the *atlas-based* algorithms, manually segmented images (*atlases*) are used to segment unseen scans [26]. These atlases model the anatomical variability of the brain tissue [24]. Atlas images are extrapolated to the incoming frames by warping and applying other, very often non-rigid registration techniques. Hence, the effectiveness of such algorithms strongly depends on the quality of registration. An important shortcoming of such approaches is the necessity of creating representative annotated reference sets, which is a cumbersome and very user-dependent task in practice. The atlases that are not representative or incorrect may easily jeopardize the entire segmentation process of the atlas-based techniques [1,6].

In the *unsupervised* techniques, we seek for hidden structures within unlabeled data [7,8,18,34]. Such algorithms, which do not require annotated training sets, include various clustering techniques [13,29,35], Gaussian modeling [32], the approaches based on non-negative matrix factorization, and many more [30]. On the other hand, the *supervised* techniques benefit from manually-annotated training sets which are fed into the training procedure of a supervised learner.

We applied the *sequence-determines-credit* approach for the sequence of authors.

© Springer Nature Switzerland AG 2020
A. Crimi and S. Bakas (Eds.): BrainLes 2019, LNCS 11993, pp. 179–190, 2020.
https://doi.org/10.1007/978-3-030-46643-5_17

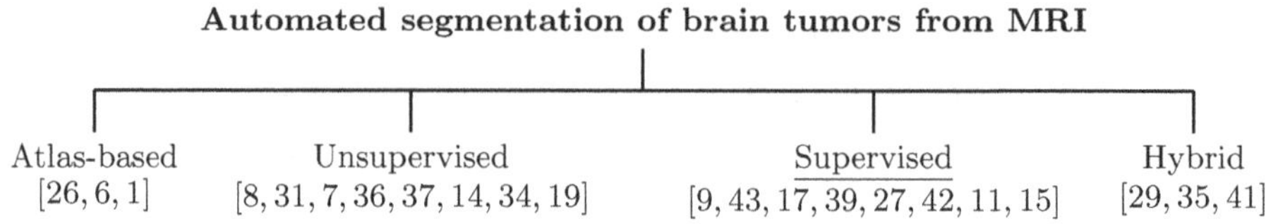

Automated segmentation of brain tumors from MRI

Atlas-based	Unsupervised	Supervised	Hybrid
[26, 6, 1]	[8, 31, 7, 36, 37, 14, 34, 19]	[9, 43, 17, 39, 27, 42, 11, 15]	[29, 35, 41]

Fig. 1. Automated delineation of brain tumors from MRI—a taxonomy [17]. In this paper, we focused on a supervised approached (underlined in this taxonomy), specifically a deep learning-powered segmentation technique which utilizes U-Nets.

Such algorithms include decision forests [9,41], conditional random fields [37], support vector machines [16], extremely randomized trees [25], and more [38].

Deep learning has established the state of the art in a variety of pattern-recognition and computer-vision tasks, and brain tumor detection and segmentation are not the exceptions here [5,11,15,20,21]. Holistically nested neural nets for MRI were introduced in [40]. White matter was segmented in [10]. The winning BraTS'17 algorithm used deep neural nets ensembles for better regularization [14]. The recent methods applied over the BraTS data showed that the architectural novelties can bring some improvements into the segmentation process, but it is the data pre-/post-processing, data augmentation, and appropriate regularization what can substantially enhance the abilities of (even basic) ensembles of deep models [12,21]. Interestingly, Isensee et al. [12] showed that U-Nets can outperform much more sophisticated and deeper networks.

Finally, the *hybrid* techniques combine various methods belonging to different categories [27,33,39]. They are often tailored to detect a specific lesion type. Superpixel processing was coupled with support vector machines and extremely randomized trees in [33]. Although the results appeared promising, the authors did not report the computation time nor cost of their method, and did not provide any insights into the classifier parameters. Tuning such kernel-based algorithms is very difficult and expensive in practice [23], and improperly selected parameters can easily deteriorate the classifier abilities [22]. In [27], the authors pointed out that deformable models are extensively used for brain tumor segmentation. Such methods suffer from poor convergence to lesion boundaries.

In this paper, we tackle the problem of brain tumor detection and segmentation in a two-step process, inspired by our previous work [17] (Sect. 3). First, a single- or multi- modal U-Net is applied to perform one-class classification (hence, assigning a tumorous or healthy label to each pixel), in order to determine the voxels of interest within an input magnetic resonance images (MRIs). Here, we show that our U-Nets can effectively detect brain tumors using just a single modality (T2 Fluid Attenuated Inversion Recovery, T2-FLAIR), which can greatly speed up the inference. Although multi-modal processing is pivotal for segmenting brain lesions into subregions, i.e., edema, non-enhancing solid core, necrotic core, and enhancing core, the whole area of brain tumors (without subdividing it into subregions) can be delineated from T2-FLAIR MRI that evaluates tissue architecture [36]. This approach, although giving slightly lower segmentation scores, is much easier to implement, train, and deploy, and infers in shorter time, especially due to the lack of the sequence co-registration step required in multi-modal techniques. Once a brain tumor is detected, we perform multi-class classification of its voxels (three classes of interest include the enhancing tumor, whole tumor, and tumor core) using a multi-modal U-Net (over the T2-FLAIR, T2-weighted, and post-contrast T1-weighted modalities). The experimental evidence (Sect. 4) elaborated over the newest release of the Brain Tumor Segmentation dataset (Sect. 2) shows that our approach delivers high-quality tumor segmentation and offers instant operation.

2   Data

The newest release of the Brain Tumor Segmentation (BraTS'19) dataset [2–5,19] includes MRI data of 335 patients with diagnosed gliomas—259 high-grade glioblastomas (HGG), and 76 low-grade gliomas (LGG). Each study was manually annotated by one to four experienced and trained readers. The data comes in four co-registered modalities: native pre-contrast (T1), post-contrast T1-weighted (T1Gd), T2-weighted (T2), and T2 Fluid Attenuated Inversion Recovery (T2-FLAIR). All pixels are labeled, and the following classes are considered: healthy tissue, Gd-enhancing tumor (ET), peritumoral edema (ED), the necrotic and non-enhancing tumor core (NCR/NET).

The data was acquired with different clinical protocols, various scanners, and at 19 institutions, hence the pixel intensity distribution may vary significantly. The studies were interpolated to the same shape ($240 \times 240 \times 155$, therefore there are 155 images of 240×240 size, with voxel size of $1\ mm^3$), and they were pre-processed (skull-stripping was applied). Overall, there are 335 patients in the training set T (259 HGG, 76 LGG), 125 patients in the validation set V (without ground-truth data provided by the BraTS'19 organizers; see example validation images rendered in Fig. 2).

3   Methods

In this work, we perform brain tumor detection and segmentation using U-Nets [28]—in the first step, we detect tumorous pixels within each image of an input MRI scan (using either a single-modality U-Net over T2-FLAIR, or three different modalities, T2-FLAIR, T2, and T1c, co-registered to the same anatomical template). Then, the multi-modal U-Net (T2-FLAIR, T2, and T1c) is exploited for multi-class classification, in which a class label is assigned to each tumorous pixel.

In Fig. 3, we render our deep-network architecture. Note that it is almost the same for both detection and segmentation; the only difference is that in the latter procedure we utilize three modalities stored as separate channels of an input image—the activations produced for each modality are merged (joined) before entering the expanding path, as presented in [17]. Since different modalities are processed separately during the multi-class segmentation, we effectively deal

| T1 | T1Gd | T2 | T2-FLAIR |

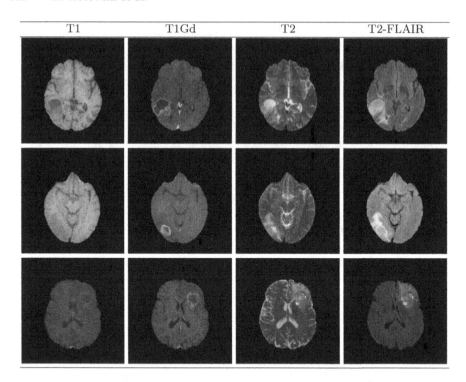

Fig. 2. Example images (all modalities: T1, T1Gd, T2, and T2-FLAIR) of three cases (each row presents a separate patient) from the BraTS 2019 validation set.

with three contracting paths within the segmentation U-Net. As proposed in our previous work [17], the number of filters is kept constant (48 filters of the size 3 × 3 each, with stride 1) at each level (except the very bottom part of the network, where we concatenate and merge the paths, in which the number of filters is doubled). In the expanding path, each upsampling block doubles the size of an activation map, which is followed by two convolutions (48 filters of the size 3 × 3 each, with stride 1). At the very last layer, we have a 1 × 1 convolution with 1 filter in the detection or 3 filters in the multi-class classification.

3.1   Data Standardization and Data Augmentation

The data was acquired with different clinical protocols, various scanners, and at different institutions, therefore the pixel intensity distribution across the scans may vary. To reduce the impact of varying data characteristics, we perform pre-processing of the input volumes which includes standardization of each modality (in this process, we exclude the first-quartile pixels according to their intensity). Specifically, we employ the Z-score normalization:

$$z = \frac{p_i - \mu}{\sigma}, \tag{1}$$

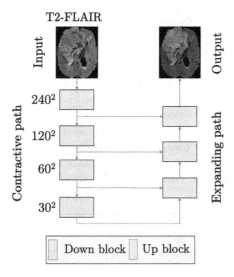

Fig. 3. The exploited U-Net architecture. At each level (each set of down blocks) the output is concatenated and sent to a corresponding up block. At the very bottom of this architecture, there is an additional merging block in the segmentation U-Net, where all the features are merged before entering the expanding path (for details, see [17]). The output layer is a 1×1 convolution with one filter for the first stage, and three filters for the second stage.

where p_i is the *i-th* non-zero pixel in the input volume, μ denotes the average of all pixels, and σ is their standard deviation.

To increase the representativeness of the training sets, we perform random (with a probability 0.5) horizontal flipping of input images.

3.2 Post-processing

After the detection process, the U-Net response is binarized using a threshold of 0.5. This binary mask is then post-processed using the 3D connected components analysis (the largest connected component is further analyzed), and the set of binary masks of a tumor is an input to the segmentation U-Net. It produces a $240 \times 240 \times 3$ sigmoid activation map, where the last dimension represents the number of classes. The activation is finally passed through a softmax operation, which performs the classification.

3.3 Ensemble of U-Nets

In this work, the BraTS'19 training dataset was divided into four folds with the size of 70, 74, 73, 64 patients, respectively. Additionally, we extracted one separate test set of 50 patients which was kept aside during the training process, and was used for validating the trained models over the unseen data. All folds and the test set were stratified and contain a similar ratio of patients with HGG/LGG, different origins, and different tumor sizes (here, we divided all tumors into "small", "regular", and "big" ones, where the volume of the small/big tumors falls into the first/fourth quartile of the volume distribution in the corresponding training/validation set). Therefore, we ensure that both training and validation sets within the separate folds are as representative as possible, and encompass tumors of various characteristics.

Once four models are trained, we form an ensemble and average the responses of base learners. As already mentioned, we apply the threshold of 0.5 to binarize the output of an ensemble and assign a single label during the detection process (healthy/tumorous). Similar thresholding is executed during the segmentation.

4   Results

4.1   Experimental Setup

The DNN models were implemented using `Python3` with the `Keras` library over CUDA 9.0 and CuDNN 5.1. The experiments were run on a machine equipped with an Intel i7–6850K (15 MB Cache, 3.80 GHz) CPU with 32 GB RAM and NVIDIA GTX Titan X GPU with 12 GB VRAM.

4.2   Training and Inference

The metric for training was the DICE score for both stages. The optimizer was Nadam (Adam with Nesterov momentum) with the initial learning rate of 10^{-4}, and the optimizer parameters were $\beta_1 = 0.9$, $\beta_2 = 0.999$. The training ran until DICE score over the validation set did not increase by at least 0.001 in 7 epochs. The training times for one epoch in both the detection and classification U-Net are around 10 min. The training converges in around 30 epochs (the complete training of a full ensemble encompassing four U-Nets takes approximately 28 hours). Both networks are relatively small, which directly translates to the low computational requirements during inference; one complete test volume (belonging to the validation set provided by the BraTS'19 organizers) is analyzed within around 15 seconds using four-models ensemble. Additionally to boost the results, the set of test-time augmentations was applied including horizontal flip, rotations by -10 and 10 degrees, and rescaling by 0.97 and 1.03 ratio.

4.3   The Preliminary Results

In our preliminary experimental study, the U-Net detector was trained in two settings: exploiting (a) T2-FLAIR only, and in (b) the multi-modal scenario with three modalities (T2-FLAIR, T2, and T1c). Note that the segmentation U-Net is always trained in a multi-modal fashion (T2-FLAIR, T2, and T1c). For (a), we performed four-fold cross-validation over the training BraTS'19 set, and obtained 0.8530 DICE on average, and 0.9231 over the entire training set.

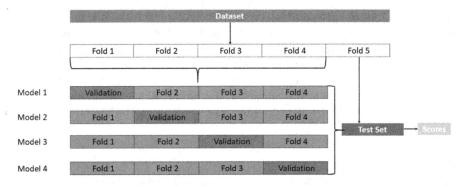

Fig. 4. The process of creating an ensemble of U-Nets includes dividing the BraTS'19 training set into four non-overlapping folds, where each training and validation set within each fold is stratified (to ensure the representativeness of these sets with respect to the tumor characteristics). A single test set is kept aside and used for the final validation of the trained ensemble.

We also kept our internal test set aside during the training process (as shown in Fig. 4), and the DICE score for this set amounted to 0.8768.

On the other hand, in the (b) setting, we executed five-fold cross-validation (with 0.8876 average DICE, and 0.9242 DICE for the entire training set; note that the results for (a) and (b) are statistically the same at $p < 0.05$, according to the Wilcoxon test), and we retrieved 0.8453 DICE over the BraTS'19 validation set (the result returned by the evaluation server; sensitivity was 0.82688, and specificity amounted to 0.99426; our team is Future_Healthcare). In Tables 1–2, we report the segmentation results returned by the competition server for both (a) T2-FLAIR only, and (b) the multi-modal settings. The preliminary results

Table 1. Segmentation performance (DICE, sensitivity, and specificity) over the BraTS'19 validation set obtained using our method (returned by the validation server). The scores are presented for the whole tumor (WT), tumor core (TC), and enhancing tumor (ET). We report the results obtained with the detection U-Net trained on (a) T2-FLAIR, and (b) T2-FLAIR, T2, and T1c (the segmentation U-Net is always trained with T2-FLAIR, T2, and T1c).

Var	Meas	Dice_ET	Dice_WT	Dice_TC	Sens_ET	Sens_WT	Sens_TC	Spec_ET	Spec_WT	Spec_TC
(a)	Mean	0.615	0.781	0.670	0.650	0.748	0.645	0.990	0.986	0.989
	StdDev	0.330	0.215	0.289	0.316	0.239	0.314	0.089	0.089	0.089
	Median	0.776	0.869	0.780	0.778	0.834	0.758	0.999	0.997	0.999
	25q	0.370	0.744	0.505	0.463	0.685	0.407	0.998	0.992	0.997
	75q	0.867	0.914	0.894	0.871	0.918	0.899	1.000	0.999	1.000
(b)	Mean	0.684	0.838	0.735	0.738	0.813	0.732	0.998	0.995	0.996
	StdDev	0.312	0.174	0.272	0.286	0.207	0.288	0.003	0.006	0.008
	Median	0.829	0.896	0.853	0.853	0.903	0.846	0.999	0.997	0.998
	25q	0.611	0.832	0.684	0.687	0.756	0.677	0.997	0.994	0.996
	75q	0.887	0.935	0.913	0.921	0.938	0.927	1.000	0.999	0.999

show that our technique provides accurate detection of tumors from the unseen MRI data, and offers instant operation. The Hausdorff distance values gathered in Table 2 (being the maximum distance of all points from the segmented lesion to the corresponding nearest point of the ground-truth segmentation [31]) significantly dropped in the case of multi-modal tumor segmentation, hence the maximum segmentation error quantified by this metric was notably reduced. Finally, we present the results obtained over the unseen BraTS 2019 test set in Table 3.

Table 2. Segmentation performance (Hausdorff distance) over the BraTS'19 validation set obtained using our method (returned by the validation server). The scores are presented for the whole tumor (WT), tumor core (TC), and enhancing tumor (ET). We report the results obtained with the detection U-Net trained on (a) T2-FLAIR, and (b) T2-FLAIR, T2, and T1c (the segmentation U-Net is always trained with T2-FLAIR, T2, and T1c).

Var	Meas	Haus_ET	Haus_WT	Haus_TC
(a)	Mean	45.568	22.735	22.826
	StdDev	108.778	40.321	58.422
	Median	4.123	7.078	9.000
	25q	2.000	3.742	4.123
	75q	17.000	26.926	16.882
(b)	Mean	10.122	12.185	14.496
	StdDev	17.891	17.390	18.597
	Median	2.4490	5.0990	6.6180
	25q	1.7320	3.0000	2.8280
	75q	6.5160	10.227	18.114

Table 3. Segmentation performance: DICE and the 95th percentile of Hausdorff distance (H95) over the BraTS 2019 test set.

Measure	DICE (ET)	DICE (WT)	DICE (TC)	H95 (ET)	H95 (WT)	H95 (TC)
Mean	0.68115	0.77412	0.72463	11.33429	22.16226	16.46728
StdDev	0.27797	0.21855	0.29184	20.55273	26.82085	21.97891
Median	0.78563	0.86684	0.84105	3	6.04138	7
25quantile	0.62971	0.71212	0.68159	1.41421	3	3.31662
75quantile	0.86193	0.91875	0.91849	6.94562	44.04073	17.78817

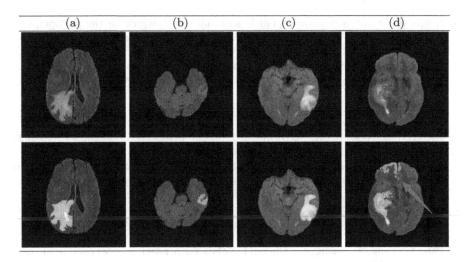

Fig. 5. Example results of the detection phase overlaid (in green) onto the T2-FLAIR sequences. In (a)–(c), we can appreciate fairly accurate delineation of the whole-tumor area, whereas in (d), there are clearly false positives (annotated with a blush arrow). (Color figure online)

5   Conclusions

In this paper, we presented our approach to the BraTS'19 segmentation challenge. We exploited the ensembles of U-Nets for detection and segmentation of brain tumors (performed as separate steps within our processing pipeline). The overall training time amounted to around 28 hours using a single GPU, while the average inference time for a whole volume was 15 seconds with an ensemble of four models applied in both detection and segmentation. The experimental study revealed that our techniques allow for obtaining fairly accurate segmentation (see examples in Fig. 5), and infer very fast. Finally, we showed that the brain tumors may be effectively segmented from a single modality (T2-FLAIR).

Acknowledgments. This research was supported by the National Centre for Research and Development (POIR.01.02.00-00-0030/15). JN was supported by the Silesian University of Technology funds (02/020/BKM19/0183).

References

1. Aljabar, P., Heckemann, R., Hammers, A., Hajnal, J., Rueckert, D.: Multi-atlas based segmentation of brain images: atlas selection and its effect on accuracy. NeuroImage **46**(3), 726–738 (2009)
2. Bakas, S., et al.: Advancing the cancer genome atlas glioma MRI collections with expert segmentation labels and radiomic features. Nat. Sci. Data **4**, 1–13 (2017). https://doi.org/10.1038/sdata.2017.117

3. Bakas, S., et al.: Segmentation labels and radiomic features for the pre-operative scans of the TCGA-GBM collection. Cancer Imaging Arch. (2017). https://doi.org/10.7937/K9/TCIA.2017.KLXWJJ1Q

4. Bakas, S., et al.: Segmentation labels and radiomic features for the pre-operative scans of the TCGA-LGG collection. Cancer Imaging Arch. (2017). https://doi.org/10.7937/K9/TCIA.2017.GJQ7R0EF

5. Bakas, S., et al.: Identifying the best machine learning algorithms for brain tumor segmentation, progression assessment, and overall survival prediction in the BRATS challenge (2018). CoRR abs/1811.02629. http://arxiv.org/abs/1811.02629

6. Bauer, S., Seiler, C., Bardyn, T., Buechler, P., Reyes, M.: Atlas-based segmentation of brain tumor images using a Markov random field-based tumor growth model and non-rigid registration. In: Proceedings of the IEEE EMBC, pp. 4080–4083 (2010). https://doi.org/10.1109/IEMBS.2010.5627302

7. Chander, A., Chatterjee, A., Siarry, P.: A new social and momentum component adaptive PSO algorithm for image segmentation. Exp. Syst. Appl. **38**(5), 4998–5004 (2011)

8. Fan, X., Yang, J., Zheng, Y., Cheng, L., Zhu, Y.: A novel unsupervised segmentation method for MR brain images based on fuzzy methods. In: Liu, Y., Jiang, T., Zhang, C. (eds.) CVBIA 2005. LNCS, vol. 3765, pp. 160–169. Springer, Heidelberg (2005). https://doi.org/10.1007/11569541_17

9. Geremia, E., Clatz, O., Menze, B.H., Konukoglu, E., Criminisi, A., Ayache, N.: Spatial decision forests for MS lesion segmentation in multi-channel magnetic resonance images. NeuroImage **57**(2), 378–390 (2011)

10. Ghafoorian, M., et al.: Location sensitive deep convolutional neural networks for segmentation of white matter hyperintensities (2016). CoRR abs/1610.04834

11. Ghafoorian, M., et al.: Transfer learning for domain adaptation in MRI: application in brain lesion segmentation. In: Descoteaux, M., Maier-Hein, L., Franz, A., Jannin, P., Collins, D.L., Duchesne, S. (eds.) MICCAI 2017. LNCS, vol. 10435, pp. 516–524. Springer, Cham (2017). https://doi.org/10.1007/978-3-319-66179-7_59

12. Isensee, F., Kickingereder, P., Wick, W., Bendszus, M., Maier-Hein, K.H.: No new-net. In: Crimi, A., Bakas, S., Kuijf, H., Keyvan, F., Reyes, M., van Walsum, T. (eds.) BrainLes 2018. LNCS, vol. 11384, pp. 234–244. Springer, Cham (2019). https://doi.org/10.1007/978-3-030-11726-9_21

13. Ji, S., Wei, B., Yu, Z., Yang, G., Yin, Y.: A new multistage medical segmentation method based on superpixel and fuzzy clustering. Comp. Math. Meth. Med. **2014**, 747549:1–747549:13 (2014)

14. Kamnitsas, K., et al.: Ensembles of multiple models and architectures for robust brain tumour segmentation. In: Crimi, A., Bakas, S., Kuijf, H., Menze, B., Reyes, M. (eds.) BrainLes 2017. LNCS, vol. 10670, pp. 450–462. Springer, Cham (2018). https://doi.org/10.1007/978-3-319-75238-9_38

15. Korfiatis, P., Kline, T.L., Erickson, B.J.: Automated segmentation of hyperintense regions in FLAIR MRI using deep learning. Tomography J. Imaging Res. **2**(4), 334–340 (2016). https://doi.org/10.18383/j.tom.2016.00166

16. Ladgham, A., Torkhani, G., Sakly, A., Mtibaa, A.: Modified support vector machines for MR brain images recognition. In: Proceedings of CoDIT, pp. 032–035 (2013). https://doi.org/10.1109/CoDIT.2013.6689515

17. Marcinkiewicz, M., Nalepa, J., Lorenzo, P.R., Dudzik, W., Mrukwa, G.: Segmenting brain tumors from MRI Using cascaded multi-modal U-Nets. In: Crimi, A., Bakas, S., Kuijf, H., Keyvan, F., Reyes, M., van Walsum, T. (eds.) BrainLes 2018. LNCS, vol. 11384, pp. 13–24. Springer, Cham (2019). https://doi.org/10.1007/978-3-030-11726-9_2

18. Mei, P.A., de Carvalho Carneiro, C., Fraser, S.J., Min, L.L., Reis, F.: Analysis of neoplastic lesions in magnetic resonance imaging using self-organizing maps. J. Neurol. Sci. **359**(1–2), 78–83 (2015)
19. Menze, et al.: The multimodal brain tumor image segmentation benchmark (brats). IEEE Trans. Med. Imag. **34**(10), 1993–2024 (2015). https://doi.org/10.1109/TMI.2014.2377694
20. Moeskops, P., Viergever, M.A., Mendrik, A.M., de Vries, L.S., Benders, M.J.N.L., Isgum, I.: Automatic segmentation of MR brain images with a convolutional neural network. IEEE Trans, Med. Imaging **35**(5), 1252–1261 (2016). https://doi.org/10.1109/TMI.2016.2548501
21. Myronenko, A.: 3D MRI brain tumor segmentation using autoencoder regularization. In: Crimi, A., Bakas, S., Kuijf, H., Keyvan, F., Reyes, M., van Walsum, T. (eds.) BrainLes 2018. LNCS, vol. 11384, pp. 311–320. Springer, Cham (2019). https://doi.org/10.1007/978-3-030-11726-9_28
22. Nalepa, J., Kawulok, M.: Adaptive genetic algorithm to select training data for support vector machines. In: Esparcia-Alcázar, A.I., Mora, A.M. (eds.) EvoApplications 2014. LNCS, vol. 8602, pp. 514–525. Springer, Heidelberg (2014). https://doi.org/10.1007/978-3-662-45523-4_42
23. Nalepa, J., Kawulok, M.: Adaptive memetic algorithm enhanced with data geometry analysis to select training data for SVMs. Neurocomputing **185**, 113–132 (2016)
24. Park, M.T.M., et al.: Derivation of high-resolution MRI atlases of the human cerebellum at 3T and segmentation using multiple automatically generated templates. NeuroImage **95**, 217–231 (2014)
25. Pinto, A., Pereira, S., Correia, H., Oliveira, J., Rasteiro, D.M.L.D., Silva, C.A.: Brain tumour segmentation based on extremely rand. forest with high-level features. In: Proceedings of IEEE EMBC, pp. 3037–3040 (2015). https://doi.org/10.1109/EMBC.2015.7319032
26. Pipitone, J., et al.: Multi-atlas segmentation of the whole hippocampus and subfields using multiple automatically generated templates. NeuroImage **101**, 494–512 (2014)
27. Rajendran, A., Dhanasekaran, R.: Fuzzy clustering and deformable model for tumor segmentation on MRI brain image: a combined approach. Procedia Eng. **30**, 327–333 (2012). https://doi.org/10.1016/j.proeng.2012.01.868
28. Ronneberger, O., Fischer, P., Brox, T.: U-net: convolutional networks for biomedical image segmentation (2015). CoRR abs/1505.04597
29. Saha, S., Bandyopadhyay, S.: MRI brain image segmentation by fuzzy symmetry based genetic clustering technique. In: Proceedings of IEEE CEC, pp. 4417–4424 (2007). https://doi.org/10.1109/CEC.2007.4425049
30. Sauwen, N., et al.: Comparison of unsupervised classification methods for brain tumor segmentation using multi-parametric MRI. Neuroimage Clin. **12**, 753–764 (2016)
31. Sauwen, N., Acou, M., Sima, D.M., Veraart, J., Maes, F., Himmelreich, U., Achten, E., Huffel, S.V.: Semi-automated brain tumor segmentation on multi-parametric mri using regularized non-negative matrix factorization. BMC Med. Imaging **17**(1), 29 (2017)
32. Simi, V., Joseph, J.: Segmentation of glioblastoma multiforme from MR images - a comprehensive review. Egypt. J. Radiol. Nucl. Med. **46**(4), 1105–1110 (2015)
33. Soltaninejad, M., et al.: Automated brain tumour detection and segmentation using superpixel-based extremely randomized trees in FLAIR MRI. Int. J. Comp. Assist. Radiol. Surg. **12**(2), 183–203 (2017)

34. Taherdangkoo, M., Bagheri, M.H., Yazdi, M., Andriole, K.P.: An effective method for segmentation of MR brain images using the ant colony optimization algorithm. J. Dig. Imaging **26**(6), 1116–1123 (2013)

35. Verma, N., Cowperthwaite, M.C., Markey, M.K.: Superpixels in brain MR image analysis. In: Proceedings of IEEE EMBC, pp. 1077–1080 (2013). https://doi.org/10.1109/EMBC.2013.6609691

36. Villanueva-Meyer, J.E., Mabray, M.C., Cha, S.: Current clinical brain tumor imaging. Neurosurgery **81**(3), 397–415 (2017). https://doi.org/10.1093/neuros/nyx103

37. Wu, W., Chen, A.Y.C., Zhao, L., Corso, J.J.: Brain tumor detection and segmentation in a CRF (conditional random fields) framework with pixel-pairwise affinity and superpixel-level features. Int. J. Comput. Assist. Radiol. Surg. **9**(2), 241–253 (2013). https://doi.org/10.1007/s11548-013-0922-7

38. Zhao, J., Meng, Z., Wei, L., Sun, C., Zou, Q., Su, R.: Supervised brain tumor segmentation based on gradient and context-sensitive features. Front. Neurosci. **13**, 144 (2019). https://doi.org/10.3389/fnins.2019.00144, https://www.frontiersin.org/article/10.3389/fnins.2019.00144

39. Zhao, X., Wu, Y., Song, G., Li, Z., Zhang, Y., Fan, Y.: A deep learning model integrating FCNNs and CRFs for brain tumor segmentation (2017). CoRR abs/1702.04528

40. Zhuge, Y., Krauze, A.V., Ning, H., Cheng, J.Y., Arora, B.C., Camphausen, K., Miller, R.W.: Brain tumor segmentation using holistically nested neural networks in MRI images. Med. Phys., 1–10 (2017). https://doi.org/10.1002/mp.12481

41. Zikic, D., et al.: Decision forests for tissue-specific segmentation of high-grade gliomas in multi-channel MR. In: Ayache, N., Delingette, H., Golland, P., Mori, K. (eds.) MICCAI 2012. LNCS, vol. 7512, pp. 369–376. Springer, Heidelberg (2012). https://doi.org/10.1007/978-3-642-33454-2_46

Multimodal Segmentation with MGF-Net and the Focal Tversky Loss Function

Nabila Abraham$^{(\boxtimes)}$ and Naimul Mefraz Khan

Ryerson University, Toronto, Canada
{nabila.abraham,n77khan}@ryerson.ca

Abstract. In neuro-imaging, MRI is commonly used to acquire multiple sequences simultaneously, including T1, T2 and FLAIR. Multimodal image segmentation involves learning an optimal, joint representation of these sequences for accurate delineation of the region of interest. The most commonly utilized fusion scheme for multimodal segmentation is early fusion, where each modality sequence is treated as an independent channel. In this work, we propose a fusion architecture termed the Moment Gated Fusion (MGF) network which combines feature moments from individual modality sequences for the segmentation task. We supervise our network with a variant of the focal Tversky loss function. Our architecture promotes explain-ability, light-weight CNN design and has achieved 0.687, 0.843 and 0.751 DSC scores on the BraTs 2019 test cohort which is competitive with the commonly used vanilla U-Net.

Keywords: Multimodal · Fusion · Focal Tversky loss

1 Introduction

Magnetic Resonance Imaging (MRI) is widely used in neuroimaging studies to obtain a variety of complementary scans to assist radiologists in contouring. T1-weighted MR scans and FLAIR sequences can both delineate basic brain anatomy but FLAIR can also detect white matter abnormalities associated with a variety of neuropathological conditions. Similarly, cerebral edema can be contoured from T2-weighted MRI but FLAIR is used to cross-check the extension and discriminate against healthy ventricular structures [9]. Figure 1 depicts the complementary information present within each MR modality and its utility in applications such as tumor delineation.

Glioma segmentation is a challenging problem as they present as varying size, heterogeneous structures with rich blood vessel circulation around them. Moreover, gliomas can be decomposed into histological sub-regions including necrotic tissues, fluid filled regions and enhancing tumorous structures. The variability in shape and intensity makes it harder for automatic segmentation approaches to generalize to the tumor signature. In this work, we attempt to utilize multimodal information in a deep learning setting with feature pruning blocks to enable the

© Springer Nature Switzerland AG 2020
A. Crimi and S. Bakas (Eds.): BrainLes 2019, LNCS 11993, pp. 191–198, 2020.
https://doi.org/10.1007/978-3-030-46643-5_18

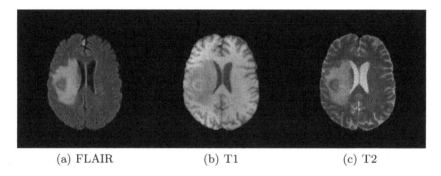

(a) FLAIR   (b) T1   (c) T2

Fig. 1. Mutli-parametric MR sequences of a patient with glioblastoma taken from the BraTs 2019 dataset [2–5,9]. Each MR modality highlights different types of intra-tumoral structures such as the tumor core visible in T1 and the cerebral edema visible in both FLAIR and T2 sequences.

network to pay attention to certain modality features and discard features that are not relevant for the segmentation task.

2 Methodology

Inspired by latent fusion models in the literature [6,7] and [8], we propose the Moment Gated Fusion Network (MGF-Net) to learn a relevant, joint multimodal feature representation for the segmentation task. Latent fusion architectures for multimodal imaging usually contain a modality specific encoder for each MR sequence and a common decoder to combine joint representations. We propose to

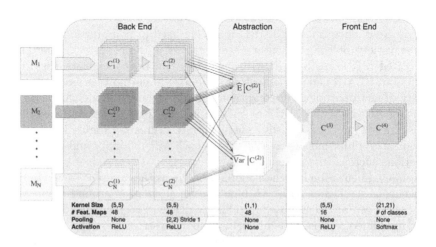

Fig. 2. HeMIS feature extraction [7]. The moments are computed per feature plane of each modality.

create a joint representation of the unimodal features by calculating the first and second image moments, as proposed by Havaei et al in the Hetero-modal image segmentation (HeMIS) algorithm [7], depicted in Fig. 2. Due to hetereogenity of the gliomas, we incorporate a recalibration block termed the Multimodal Gated Fusion (MGF) block to highlight important features across modalities, based on the joint mean and variance. The joint features are propagated to a common, simplistic decoder for the multi-class segmentation task. We utilize half the parameter space of the conventional U-Net and observe improved performance. Our proposed network is supervised with the focal Tversky loss function (FTL) [1] which has shown to be successful at small ROI based segmentation which proved useful for the enhancing tumor (ET) class in the BraTs 2019 dataset.

3   Network Architecture

The overall network architecture is depicted in Fig. 3. Each intermediate scale contains fused features computed with our MGF block. For brevity, we describe each of our network's significant architectural blocks in the following section.

3.1   Encoder Block

Each modality's features are extracted with a modality specific encoder. An encoder contains four 3×3 convolutional layers. Each convolution block is followed by a corresponding batch normalization and ReLU excitation. The encoder layers, similar to U-Net, serve to extract modality-specific, semantic information.

3.2   Decoder Block

The network uses one decoder which contains transpose convolution blocks to upsample coarse representations. The upsampled feature maps are concatenated with its corresponding MGF representation from the encoding half of the network. We adopt the strategy in BraTs 2018 winning submission [10] and train the network to predict the WT, TC and ET sub-regions. Therefore, the last convolution layer maps the decoded representations to these three classes using sigmoid activations. Similar to U-Net, the decoder localizes contextual information to predict segmentation maps.

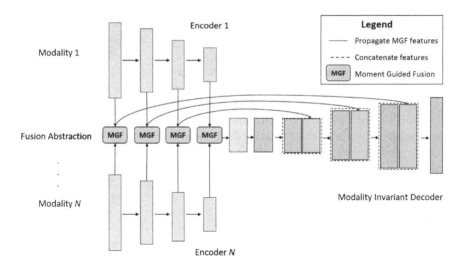

Fig. 3. Overview of our proposed Multimodal Gated Fusion Network (MGF-Net). The MGF block is used to extract and prune multimodal feature representations based on HeMIS feature abstraction. The blue blocks represent encoder layers and the grey blocks represent decoder layers. Skip connections are introduced into the decoder similar to the conventional U-Net and are concatenated with corresponding decoder blocks to improve feature localiztion. The final grey blocks utilize a Sigmoid activation across every class. (Color figure online)

3.3   MGF Block

Image moments computed from the HeMIS abstraction layer are gated using nonlinear excitation operations. Only the mean and variance are calculated as higher moments can be created by non-linear combinations of the mean and variance features. HeMIS feature extraction is depicted in Fig. 2. We recalibrate the mean and variance features to propagate important structural information (via the mean) and important outlier representation (via the variance). Depicted in Fig. 4, the variance features are independently gated using the squeeze and excitation (SE) pipeline. The non-linear activations include a global average pooling to reduce maps into a single descriptor which is then squeezed and excited using a multilayer perceptron and ReLU activations. The mean features are spatially excited using a similar workflow as the channel-wise SE blocks. As depicted in Fig. 4, this workflow includes a reduction using a 1×1 conv block, followed by a ReLU, 1×1 convolution and a sigmoid activation operation. Independently processing both spatial and channel-wise activations will encourage the network to learn relevance to structural features but also relevance to outlier features captured in the variance maps.

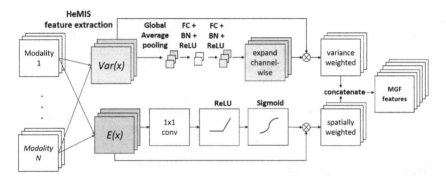

Fig. 4. Overview of our proposed MGF block. The upper portion of the MGF pipeline excites variance (Var(x)) features in order to capture possible outlier representations that are important for segmentation. The lower portion guides the mean ($E(x)$) feature extraction by following a similar spatial squeeze operation. By applying non-linear activations independently to feature moments, we discourage the network from being biased to only the mean features and therefore can account for outlier features in the fused representations.

4   Loss Function

We utilize the focal Tversky loss (FTL) proposed in [1] (Eq. 1). This loss function is a more generalized version of the Dice loss with tunable false negative and false positive penalizations. Moreover, a focal parameter γ is incorporated to help boost the model's focus to small regions of interest such as the enhancing tissue class. The FTL is defined as follows:

$$FTL = \frac{1}{C} \sum_c (1 - TI_c)^\gamma \tag{1}$$

where,

$$TI_c = \frac{2 * TP}{2 * TP + \alpha * FN + (1 - \alpha) * FP} \tag{2}$$

Based on the difficulty of segmentation for each class, we use a select value of alpha. Based on past experiments in [1], $\alpha = 0.7$ and $\gamma = 0.75$ are good starting points for the FTL parametrization. From cross-validation tests, we observe that the enhancing structures class, class 4, suffers more severe class imbalance and requires a stricture penalty on false negatives. Therefore, through numerous tests, we find the optimal α parameters for class 1, class 2 and class 4 to be $\alpha = 0.7$, $\alpha = 0.7$ and $\alpha = 0.75$, respectively. We experimented with different values of gamma and found the most stable training to occur at $\gamma = 0.75$.

5   Dataset

We utilize the BraTs 2019 dataset and validate the results on the online leaderboard [2–5,9]. This dataset consists of 335 patients with co-registered T1, T1c,

T2 and FLAIR MRI scans. All the scans are co-registered to a common template, resampled to 1 mm 3 and skull-stripped with a volume of $240 \times 240 \times 155$. The BraTs ground truth consist of three classes: non-enhancing/necrotic tissue, peritumoral edema and enhancing structures. A combination of each of these classes forms a sub-category for the evaluation board. ET denotes the enhancing class structures, TC constitutes the nonenhancing structures and edema classes and the WT refers too all classes in the annotation. We adopt the strategy in [10] and train the network to predict sub-regions ET, TC and WT.

6   Implementation Details

We normalize the dataset by min-max normalization of each modality. Since several slices are blank and contain no tumor information, we discard the first and last 12 slices and crop our volumes to $192 \times 192 \times 128$ and operate our 2D model at the slice level. The training cohort was split into a 80–20 ratio for training and validation, respectively. No additional data augmentation was used to supplement the training cohort. All our experiments were developed in PyTorch and trained on a single NVIDIA 1080Ti GPU. We use Stochastic Gradient Descent (SGD) and a learning rate of 1e-5 with momentum and weight decay. The model trained for a maximum of 200 epochs and we obtain the best model through early stopping based on constantly decreasing validation loss with a patience of 10 epochs. The output masks depict sub-regions WT, TC and ET where each class was predicted as a one-versus-all segmentation task. In order to prepare the final multi-class segmentation mask, we use an adhoc scheme to order the class pixels. The scheme essentially allows enhancing structures to take precedence over all other classes, then edema and finally, overall tumor core structure. In this way, we hope to not miss any enhancing structures class.

7   Results

Table 1 depicts a comparative performance between the U-Net and our MGF-Net on the validation set from the competition. Table 2 depicts our final score on the leaderboard (team name: RML). Figure 5 depicts our empirical comparison

Table 1. Mean DSC and Sensitivity scores for Validation set on leaderboard

Model	ET DSC	WT DSC	TC DSC	TPR ET	TPR WT	TPR TC	Parameters
U-Net	0.63809	0.85894	0.71711	0.73263	0.86843	0.70372	9.335 M
MGF-Net	0.63226	0.86046	0.71076	0.77968	0.88711	0.71326	4.449 M

Table 2. Mean DSC and Hausdorff 95% (HD95) scores for Test set on leaderboard

Model	ET DSC	WT DSC	TC DSC	HD95 ET	HD95 WT	HD95 TC
MGF-Net	0.68665	0.84381	0.7513	7.79124	17.56236	11.86324

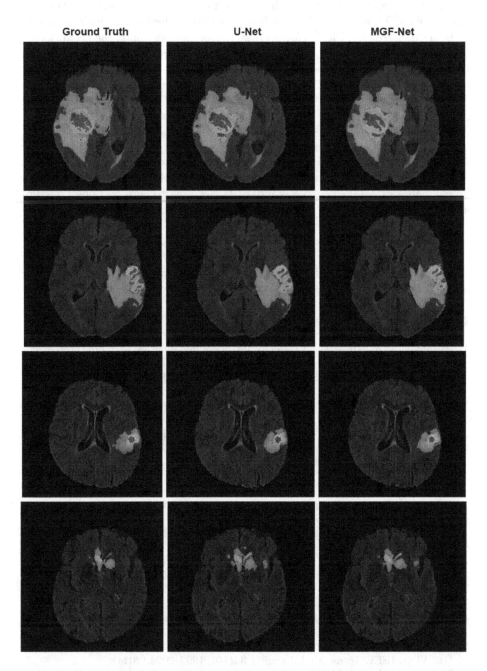

Fig. 5. Example results. From left to right, Ground Truth, U-Net segmentation, MGF-Net segmentation.

between U-Net and our proposed MGF-Net. A common characteristic of training with the focal Tversky loss involves false positive predictions as depicted across the board in all rows of Fig. 5. However, we see for the most part, the networks are competitive in performance but MGF-Net localizes small ROIs better due to the spatial weighting on the average features in the MGF block. This phenomenon can be depicted in rows 1, 2 and 4 of Fig. 5.

8   Conclusion

In this paper, we propose a novel network for brain tumor segmentation. The Moment Gated Fusion block intelligently prunes the mean and variance features of each specific modality using the a spatial and channel-wise excitation pipeline. We compare our MGF-Net to a standard U-Net architecture and observe competitive performance at 50% fewer parameters. When training with the focal Tversky loss with an alpha weighting on each class, we observe improved performance over U-Net. We hypothesize this is due to the added layers in the U-Net which hinders the overall gradient. We anticipate a 3D version of the MGF-Net could lead to improved performance but our 2D architecture sets the stage for simpler, more interpretable detection for glioma segmentation.

References

1. Abraham, N., Khan, N.M.: A novel focal tversky loss function with improved attention U-Net for lesion segmentation. In: 2019 IEEE 16th International Symposium on Biomedical Imaging (ISBI 2019), pp. 683–687. IEEE (2019)
2. Bakas, S., et al.: Segmentation labels and radiomic features for the pre-operative scans of the TCGA-GBM collection. The Cancer Imaging Archive (2017)
3. Bakas, S., et al.: Segmentation labels and radiomic features for the pre-operative scans of the TCGA-LGG collection. The Cancer Imaging Archive, **286** (2017)
4. Bakas, S., et al.: Advancing the cancer genome atlas glioma MRI collections with expert segmentation labels and radiomic features. Sci. Data **4**, 170117 (2017)
5. Bakas, S., et al.: Identifying the best machine learning algorithms for brain tumor segmentation, progression assessment, and overall survival prediction in the BRATS challenge. arXiv preprint arXiv:1811.02629 (2018)
6. Chartsias, A., Joyce, T., Giuffrida, M.V., Tsaftaris, S.A.: Multimodal MR synthesis via modality-invariant latent representation. IEEE Trans. Med. imaging **37**(3), 803–814 (2018)
7. Havaei, M., Guizard, N., Chapados, N., Bengio, Y.: HeMIS: hetero-modal image segmentation. In: Ourselin, S., Joskowicz, L., Sabuncu, M.R., Unal, G., Wells, W. (eds.) MICCAI 2016. LNCS, vol. 9901, pp. 469–477. Springer, Cham (2016). https://doi.org/10.1007/978-3-319-46723-8_54
8. Liang, X., Hu, P., Zhang, L., Sun, J., Yin, G.: MCFNet: Multi-layer concatenation fusion network for medical images fusion. IEEE Sens. J. **19**, 7107–7119 (2019)
9. Menze, B.H., et al.: The multimodal brain tumor image segmentation benchmark (BRATS). IEEE Trans. Med. Imaging **34**(10), 1993–2024 (2014)
10. Myronenko, A.: 3D MRI brain tumor segmentation using autoencoder regularization. In: Crimi, A., Bakas, S., Kuijf, H., Keyvan, F., Reyes, M., van Walsum, T. (eds.) BrainLes 2018. LNCS, vol. 11384, pp. 311–320. Springer, Cham (2019). https://doi.org/10.1007/978-3-030-11726-9_28

Brain Tumor Segmentation Using 3D Convolutional Neural Network

Kaisheng Liang[✉] and Wenlian Lu[✉]

Fudan University, Shanghai, China
{ksliang16,wenlian}@fudan.edu.cn

Abstract. Brain tumors segmentation is one of the most crucial procedures in the diagnosis of brain tumors because it is of great significance for the analysis and visualization of brain structures that can guide the surgery. With the development of natural scene segmentation model FCN, the most representative model U-net has been developed. An increasing number of people are trying to improve the encoder-decoder architecture to achieve better performance currently. In this paper, we focus on the improvement of the encoder-decoder network and the analysis of 3D medical images. We propose an additional path to enhance the encoder part and two separate up-sampling paths for the decoder part of the model. The proposed approach was trained and evaluated on BraTS 2019 dataset.

Keywords: Brain tumor segmentation · Convolutional neural network · U-Net

1 Introduction

Glioma, one of the most common brain tumors, has two types: low grade gliomas (LGG) and high grade glioblastomas (HGG). As an essential diagnostic tool for brain tumors, 3D magnetic resonance imaging (MRI) data using for tumor segmentation usually consist of four different kinds, including T1-weighted, a post-contrast T1-weighted, a T2-weighted and a Fluid-Attenuated Inversion Recovery (FLAIR), which emphasizing different tissue characteristics. The brain tumor segmentation challenge (BraTS) [1–5] aims to promote the development of advanced brain tumor segmentation methods by providing annotated, high-quality images of gliomas, including LGG and HGG. The training dataset of the BraTS 2019 challenge consists of 259 HGG and 76 LGG samples, which are manually marked by multiple raters. These MRI data were collected from 19 institutions using different protocols and magnetic field strength. Each tumor was divided into edema, necrosis, non-enhanced tumor, and enhanced tumor. The segmentation performance was measured by the Dice coefficient, sensitivity, specificity, and the 95th percentile of the Hausdorff distance. Besides, validation and testing datasets require participants to upload the segmentation results to the organizer's server for evaluation. Validation data sets (125 cases) allow

© Springer Nature Switzerland AG 2020
A. Crimi and S. Bakas (Eds.): BrainLes 2019, LNCS 11993, pp. 199–207, 2020.
https://doi.org/10.1007/978-3-030-46643-5_19

multiple submissions while the test dataset (166 cases) was only allowed to be committed once and was used to calculate the final challenge ranking.

There is no doubt that the segmentation and measurement of glioma are the keys to the treatment of glioma and related scientific research. If the location and shape of the tumor can be accurately and quickly identified and segmented, it can not only save doctors a lot of time but also potentially provide valuable information. However, the boundary between tumor and normal tissue is often blurred, which undoubtedly increases the difficulty of glioma segmentation. In recent years, the most effective segmentation algorithms are deep neural networks. Since the success of fully convolutional network (FCN) [6], a convolutional neural network algorithm has been proved to be effective in the field of image segmentation. After that, the convolutional neural network was used by Mask R-CNN [7] to completely replace the traditional segmentation method. Deeplab series [8,9] explores the use of dilated convolution in network models. Many scholars have developed a large number of deep learning networks for medical image segmentation, and the BraTS competition has greatly promoted the development of this field. In recent years, almost all the winning models in the BraTS challenge have been built using CNNs.

In this work, we propose a semantic segmentation method, an improved U-net model based on 3D CNN. We try to strengthen the connection between the encoder part and decoder part in the U-net model and also propose a novel loss function.

2   Related Work

While the data of natural scene segmentation is usually two-dimensional images, three-dimensional image data bring a challenge to the medical segmentation algorithm. Besides, the amount of medical data is too small to fit a normal semantic segmentation, so people have to figure out a suitable deep neural network for medical image segmentation. One of the most popular networks is U-net [10] which evolves from the famous FCN [6] is a light-weight convolutional neural network model. Compared with the original fully convolution neural network, its encoding part does not use the classical CNN structure such as ResNet to extract deep learning features. With a small special encoding part, U-net can be trained with a small number of training data. And also, U-net uses skip-connections between the encoder part and decoder part, which helps it have a very impressive performance although it has a small number of parameters. Other papers [11–13] show that the method of the convolutional neural network is effective in medical segmentation. For 3D data input, we can simply use 3D convolution to replace the normal 2D convolution in the U-net structure. 3D U-net [14], V-net [15], and other similar model [16–18]achieve very good results in 3D segmentation.

In the BraTS challenge 2017, Kamnitsas et al. [19] proposed a method ensembling several models for robust segmentation which takes advantage of an ensemble of several independently trained architectures including Kamnitsas et al. [13,20] and U-net [10]. Wang et al. [21] used dilated convolution to build a complicate encoder part and also decompose the 3*3*3 convolution kernel to save on both the computational time and the GPU memory. In the BraTS challenge 2018, Myronenko [16] proposed a novel encoder-decoder architecture using autoencoder regularization to improve the decoder part and use Group Normalization [22] to replace the common operation. Isensee et al. [17] proposed a generic U-net model that can achieve competitive performance with only a few modifications.

3   Methods

We mainly implemented U-net [10] and the winning models of the BraTS 2018 such as Myronenko [16] and Isensee et al [17]. We tried to improve some of the architecture of U-net, but the results were not surprising. We also try to optimize the loss function base on the modification of operations between channels. The details are detailed below.

3.1   Model

Our model is a kind of 3D U-net model. Group normalization would maintain the performance of the model while decreasing the number of parameters, so instead of using Batch Normalization or Instance Normalization, our model uses Group Normalization [22] and 3D CNN. To learn better deep features, we add a supervise path for the encoding process and enhance the decoding process using two paths for upsampling decoding. We also develop a skip connection, and the two different upsampling paths are also connected with the element-wise sum operation. Finally, the output layer followed by a personalized cascade loss block. Figure 1 show more details.

3.2   Data Preprocessing

We regularize all the input data to ensure the mean is zero and the variance is one. In this process, only non-zero values are considered. Since the region with a value of zero represents the extracranial region, so we have to avoid the influence of information from the extracranial region.

3.3   Training

For data generator, because the 3D data is relatively large, our graphics card can only transmit a complete single data at one time, which means batch size is 1. And also, we try another two methods of data generator during the training. One is to resize each data into 128*128*128 and input it to the network model. The other one is to crop each data into several 128*128*128 pieces and input them to the network model. The batch size for both these two methods is 2. When we change the size of the input data, the speed of the training procedure boost, so it's a valuable attempt. The results of the experiments are shown in the next section.

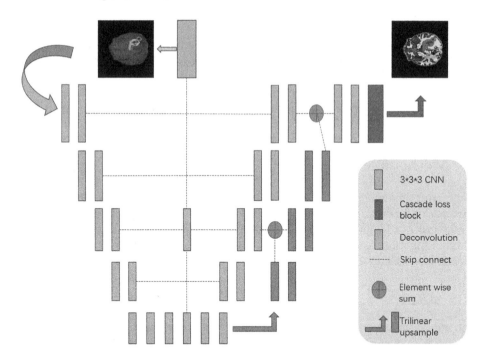

Fig. 1. Our U-shaped model using 3D convolution kernels (3*3*3) and Group Normalization. The left side of the model is the encoder section, we design a path (grey block) similar to the Autoencoder. In the decoder section, we design two decoder paths (blue and brown blocks, respectively), where the blue block path uses deconvolution upsampling, the brown block path uses trilinear upsampling. Finally, two decoder paths are merged followed by our loss block output (green block). (Color figure online)

During the training, considering that the data volume of the BraTS 2019 challenge is still insufficient, so we used a series of methods to avoid overfitting. Firstly, we used the early stop method. Specifically, if validation loss is not optimized after every 50 epochs, the network model would stop training. Secondly, our learning rate is gradually reduced linearly. Every 20 epochs, if validation loss has not been optimized, the learning rate will be reduced by half unless it came to 1e–6. We started the training by learning rate is 1e–3 and used Adam as the optimizer.

3.4   Prediction

When we use the full size of the training data for the training, the prediction is simple obviously because the output of the trained model is what we want directly. But when it comes to the issue of output having a different size from the original data, we have to figure out a few methods. For the case where each data is resized to 128*128*128, we use bilinear interpolation to restore the output to its original size. For the case where each data crop is 128*128*128, we also split the validation/test data crop into 128*128*128, and then finally assemble the output into its original size. When we crop the input data, each 128*128*128 patch may overshadow each other. So we determine the labels of the overshadowed pixes depend on the predicted labels of relevant patches or the mean of the relevant logits before the final output layer.

4   Experiments

4.1   Crop Data

Deep network segmentation tasks are very limited by the memory size of the GPU. We all know that the segmentation task is a dense output task, and most of Deep CNN models need to occupy a large amount of memory space. Though the U-net model already reduces the number of parameters of the original fully convolution network, the model proposed in here occupies larger GPU memory. Because we add some network layers that aim to promote the segmentation results. At the same time, the image data of BraTS competition is three-dimensional data (240*240*155) in which each patient has four modalities of image data. Also, it will consume far more memory space than a normal two-dimensional image. The larger data size also increases the pressure on the CPU while preparing training data which may decrease the speed of the training procedure.

We use different resize sizes and crop size sizes in the experiments. Unfortunately, we found that resizing is not a good way to change the size of input data, because the performance is pretty bad. For cropping data, the input data after cropping is a part of the original data called a patch. Each patch should have the same crop size, if the last few parts of the data cropped have not enough size, we will fill them with 0. During the evaluation, the validation data will be cropped into the fixed-size too. The prediction of each patch will be assembled to get the completed segmentation. Refer to Table 1.

Table 1. Results on BraTS 2019 training data and validation data. The crop size is 128*128*128.

method+evaluation dataset	ET	WT	TC
full+training data	0.8190	0.8812	0.8096
full+validation data	0.7912	0.8668	0.7802
crop+training data	0.7894	0.8803	0.8368
crop+validation data	0.6964	0.8723	0.7586

4.2   Sparse Label

Notice that each label may be missing in some data because the training data is unbalanced. Some labels like ET have very few pixels in image data, which is the

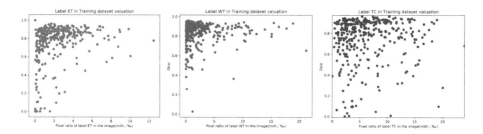

Fig. 2. These scatter plots show the relationship between the accuracy of segmentation and the sparseness of data. The horizontal axis is the ratio of the number of each label pixels to the amount of all pixels in the data. The vertical axis is dice accuracy.

Table 2. Results on BraTS 2019 validation data.

method	ET	WT	TC
Our model	0.791	0.867	0.780
Myronenko [16]	0.803	0.901	0.832
Isensee [17]	0.795	0.893	0.837

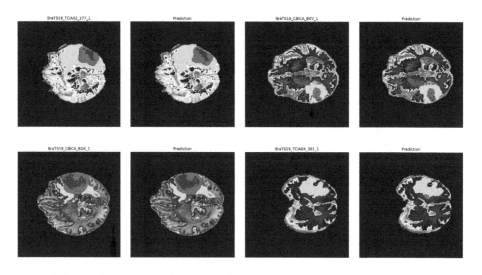

Fig. 3. The results predicted by our model is compared to the ground truth. The left of each prediction sub image is its ground truth.

sparse label, have a big impact on evaluation while others have a large number of pixels. The model is very well segmented on most of the data, but there are some "abnormal" data with very low dice accuracy. Most of these poor results are from the data with the sparse label. Figure 2 shows that the label with low dice accuracy is always relatively sparse. How to improve the performance of sparse label data is an interesting direction in the future.

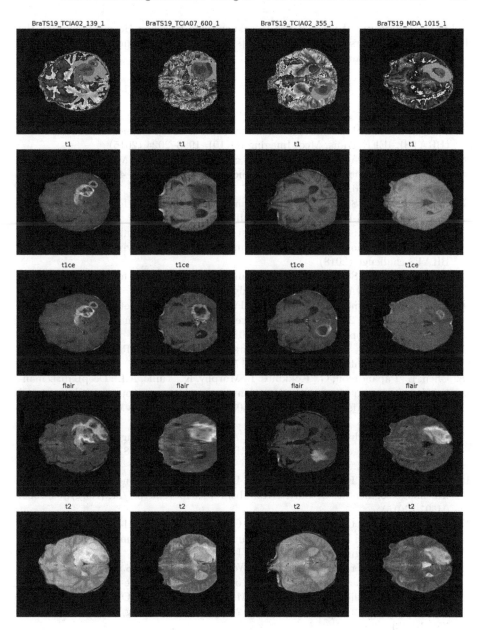

Fig. 4. Our results in validation data

4.3   Result

We evaluate our model in the validation dataset, and the mean of our dice accuracy of ET, WT, and TC is 79.1%, 86.7%, and 78.0% respectively. Figure 3 and Fig. 4 are demonstration of our results (Tabel 2).

Acknowledgement. We would like to acknowledge Huashan Hospital, through the discussion with doctors and medical students, we have a deeper understanding of brain tumor segmentation and Glioma.

References

1. Menze, B.H., et al.: The multimodal brain tumor image segmentation benchmark (BRATS). IEEE Trans. Med. Imaging **34**(10), 1993–2024 (2015)
2. Bakas, S., et al.: Advancing the cancer genome atlas glioma MRI collections with expert segmentation labels and radiomic features. Sci. Data **4**(1), 170117 (2017). https://doi.org/10.1038/sdata.2017.117
3. Bakas, S., et al.: Segmentation labels and radiomic features for the pre-operative scans of the TCGA-GBM collection, July 2017
4. Bakas, S., et al.: Identifying the best machine learning algorithms for brain tumor segmentation, progression assessment, and overall survival prediction in the BRATS challenge (2018)
5. Bakas, S., et al.: Segmentation labels and radiomic features for the pre-operative scans of the TCGA-LGG collection, July 2017
6. Long, J., Shelhamer, E., Darrell, T.: Fully convolutional networks for semantic segmentation. In: 2015 IEEE Conference on Computer Vision and Pattern Recognition (CVPR), pp. 3431–3440, June 2015
7. He, K., Gkioxari, G., Dollár, P., Girshick, R.B.: Mask R-CNN. CoRR, vol. abs/1703.06870. http://arxiv.org/abs/1703.06870 (2017)
8. Chen, L., Papandreou, G., Kokkinos, I., Murphy, K., Yuille, A.L.: DeepLab: semantic image segmentation with deep convolutional nets, atrous convolution, and fully connected crfs. CoRR, vol. abs/1606.00915. http://arxiv.org/abs/1606.00915 (2016)
9. Chen, L., Papandreou, G., Schroff, F., Adam, H.: Rethinking atrous convolution for semantic image segmentation. CoRR, vol. abs/1706.05587. http://arxiv.org/abs/1706.05587 (2017)
10. Ronneberger, O., Fischer, P., Brox, T.: U-net: convolutional networks for biomedical image segmentation. CoRR, vol. abs/1505.04597. http://arxiv.org/abs/1505.04597 (2015)
11. Li, X., Chen, H., Qi, X., Dou, Q., Fu, C.-W., Heng, P.-A.: H-DenseUNet: hybrid densely connected UNet for liver and tumor segmentation from CT volumes. IEEE Trans. Med. Imaging **37**(12), 2663–2674 (2018)
12. Isensee, F., Jaeger, P.F., Full, P.M., Wolf, I., Engelhardt, S., Maier-Hein, K.H.: Automatic cardiac disease assessment on cine-MRI via time-series segmentation and domain specific features. In: Pop, M., et al. (eds.) STACOM 2017. LNCS, vol. 10663, pp. 120–129. Springer, Cham (2018). https://doi.org/10.1007/978-3-319-75541-0_13
13. Kamnitsas, K., et al.: Efficient multi-scale 3D CNN with fully connected CRF for accurate brain lesion segmentation. Med. Image Anal. **36**, 61–78 (2017)
14. Çiçek, Ö., Abdulkadir, A., Lienkamp, S.S., Brox, T., Ronneberger, O.: 3D U-Net: learning dense volumetric segmentation from sparse annotation. CoRR, vol. abs/1606.06650. http://arxiv.org/abs/1606.06650 (2016)
15. Milletari, F., Navab, N., Ahmadi, S.: V-Net: fully convolutional neural networks for volumetric medical image segmentation. CoRR, vol. abs/1606.04797. http://arxiv.org/abs/1606.04797 (2016)

16. Myronenko, A.: 3D MRI brain tumor segmentation using autoencoder regularization. CoRR, vol. abs/1810.11654. http://arxiv.org/abs/1810.11654 (2018)
17. Isensee, F., Kickingereder, P., Wick, W., Bendszus, M., Maier-Hein, K.H.: No newnet. CoRR, vol. abs/1809.10483. http://arxiv.org/abs/1809.10483 (2018)
18. Kayalibay, B., Jensen, G., van der Smagt, P.: CNN-based segmentation of medical imaging data. CoRR, vol. abs/1701.03056. http://arxiv.org/abs/1701.03056 (2017)
19. Kamnitsas, K., et al.: Ensembles of multiple models and architectures for robust brain tumour segmentation. In: Crimi, A., Bakas, S., Kuijf, H., Menze, B., Reyes, M. (eds.) BrainLes 2017. LNCS, vol. 10670, pp. 450–462. Springer, Cham (2018). https://doi.org/10.1007/978-3-319-75238-9_38
20. Kamnitsas, K., et al.: Deepmedic for brain tumor segmentation. In: Crimi, A., Menze, B., Maier, O., Reyes, M., Winzeck, S., Handels, H. (eds.) BrainLes 2016. LNCS, vol. 10154, pp. 138–149. Springer, Cham (2016). https://doi.org/10.1007/978-3-319-55524-9_14
21. Wang, G., Li, W., Ourselin, S., Vercauteren, T.: Automatic brain tumor segmentation using cascaded anisotropic convolutional neural networks. In: Crimi, A., Bakas, S., Kuijf, H., Menze, B., Reyes, M. (eds.) BrainLes 2017. LNCS, vol. 10670, pp. 178–190. Springer, Cham (2018). https://doi.org/10.1007/978-3-319-75238-9_16
22. Wu, Y., He, K.: Group normalization. CoRR, vol. abs/1803.08494. http://arxiv.org/abs/1803.08494 (2018)

DDU-Nets: Distributed Dense Model for 3D MRI Brain Tumor Segmentation

Hanxiao Zhang[1,2(✉)], Jingxiong Li[3], Mali Shen[2], Yaqi Wang[4],
and Guang-Zhong Yang[1,2]

[1] The Institute of Medical Robotics, Shanghai Jiao Tong University, Shanghai, China
[2] The Hamlyn Centre for Robotic Surgery, Imperial College London, London, UK
hanxiao.zhang18@imperial.ac.uk
[3] Queen Mary University of London, London E1 4NS, UK
[4] Hangzhou Dianzi University, Hangzhou 310018, China

Abstract. Segmentation of brain tumors and their subregions remains a challenging task due to their weak features and deformable shapes. In this paper, three patterns (cross-skip, skip-1 and skip-2) of distributed dense connections (DDCs) are proposed to enhance feature reuse and propagation of CNNs by constructing tunnels between key layers of the network. For better detecting and segmenting brain tumors from multimodal 3D MR images, CNN-based models embedded with DDCs (DDU-Nets) are trained efficiently from pixel to pixel with a limited number of parameters. Postprocessing is then applied to refine the segmentation results by reducing the false-positive samples. The proposed method is evaluated on the BraTS 2019 dataset with results demonstrating the effectiveness of the DDU-Nets while requiring less computational cost.

Keywords: Brain tumor · Multi-modal MRI · 3D CNNs · Dense connection · Segmentation

1 Introduction

Gliomas are a kind of brain tumor developed from glial cells. It is one of the most threatening brain tumors as more than 40% of all tumors befall are malignant [12]. As a result, it is necessary to develop an accurate segmentation model for quantitative assessment of brain tumors, assisting early diagnosis and treatment planning. However, because of the diverse characteristics of tumor cells, reliable tumor segmentation remains a challenging task.

Focusing on the evaluation of state-of-the-art brain tumor segmentation methods, the annual Brain Tumor Segmentation Challenge (BraTS) provides datasets of brain magnetic resonance imaging (MRI) scans collected from multiple institutions [1–4,13]. The datasets include annotated MRI scans of low grade gliomas (LGG) and high grade glioblastomas (GMM/HGG), acquired under standard clinical conditions with different equipment and protocols. For each case, four 3D MRI modalities are provided consisting of a native T1-weighted

© Springer Nature Switzerland AG 2020
A. Crimi and S. Bakas (Eds.): BrainLes 2019, LNCS 11993, pp. 208–217, 2020.
https://doi.org/10.1007/978-3-030-46643-5_20

(T1), a post-contrast T1-weighted scan (T1Gd), a native T2-weighted scan (T2) and a T2 Fluid Attenuated Inversion Recovery (T2-FLAIR) scan. Each tumor is divided into 3 subregions for evaluation, which are enhancing tumor (ET), tumor core (TC) and whole tumor (WT), referred as complete tumor region extent [4]. All labels are evaluated manually by professional raters and approved by internationally recognized expert neuroradiologists.

Recently, convolutional neural networks (CNNs) with encoder-decoder structure have demonstrated their ability in segmenting biomedical images [5,6,11, 16], where the encoder down-samples and extracts features from the input data while the decoder rebuilds segmentation of the targets. As a result, methods based on CNNs are popular for brain tumor segmentation. In 2018, Myronenko [14], who won the 1st prize in BraTS18, proposed their ResNet based 3D encoder-decoder model and achieved the highest accuracy and robustness among others. Isensee et al. [9] also shown that well-trained U-Net without much modification could achieve competitive segmentation accuracy.

Current challenges of CNN-based methods include false predictions caused by weak features of tumors, gradient vanishing and overfitting problems when training on deep CNNs, slow training speed due to a large amount of training data, and low accuracy caused by false-positive predictions. To deal with these problems, we propose in this paper several patterns of distributed dense connections (DDCs), which reuse features in different strategies. CNN-based models with DDCs are designed to automatically segment brain tumor targets. In addition, postprocessing is applied to reduce the false-positives for more accurate delineation.

2   Methods

2.1   Distributed Dense Connections (DDCs)

Although deeper CNNs could reach a better performance than that of shallow ones, the problem of gradient vanishing can have a negative impact on network capacity and efficiency. It has been shown that this can be alleviated by shortcut connections between contextual layers [7,8,10,15].

The concept of residual learning network is used by ResNet [7] to address the degradation problem, which uses shortcut connections to skip one or more layers by summation, providing implicit deep supervision. DenseNet [8] proposes dense blocks with more shortcut connections, combining the feature maps of all the preceding layers as the input of the subsequent layer using a more efficient concatenation strategy. In practice, DenseNet provides better performance but consumes more GPU memory as the number of input channels grows dramatically towards deeper layers. The structure of DenseNet only allows dense connections being operated within each dense block and no shortcut connections are operated between dense blocks. It shows that GPU memory consumption can be reduced at the expense of training time by introducing an implementation strategy of shared memory allocations for storing feature maps [15]. Different from this memory-saving strategy, we aim to improve our network efficiency, while performing a better accuracy with much fewer dense connections.

We note that the gradient is unlikely to vanish very quickly in a few layers, so there is no need to relearn redundant features right after the preceding layers. The feature reuse can also be strengthened when the early feature maps are recalled by a reasonable skip distance.

Consequently, we propose a novel densely connected unit called distributed dense connections (DDCs), which only transmit features between critical intermediate layers without settled blocks. This approach reduces the total number of dense connections and extends the radiation scope of early features to the deeper layers, thus reducing the number of parameters and enhancing the global integration of information flow. The identity function of a neural network using distributed dense connections can be expressed as:

$$x_n = F_n \left([x_{n_i}, x_{n_{ii}}, \cdots, x_{n_{i \ldots ii}}] \right) \tag{1}$$

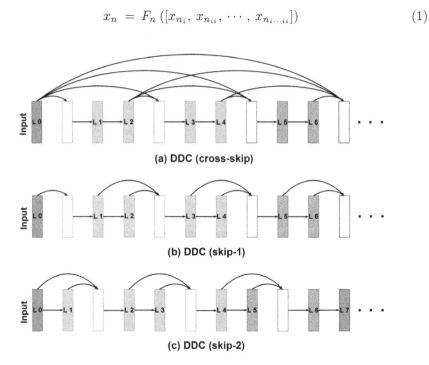

(a) DDC (cross-skip)

(b) DDC (skip-1)

(c) DDC (skip-2)

Fig. 1. Three architectures of distributed dense connections. The boxes with the same color denote hidden layers with the same size, known as the size-block. White boxes with colored borders mean concatenation operation for chosen layers. (a) stands for cross-skip, which reuses feature maps from the last layer of each size-block; (b) represents skip-1, which reuses the last 2 layers of preceding size-block; (c) represents skip-2, which reuses the last layer of preceding size-block and the first layer of the current block. (Color figure online)

where $F_n(\cdot)$ is non-linear transformation after each layer, n represents the nth layer. The output of the nth layer is represented as x_n. $[x_{n_i}, x_{n_{ii}}, \cdots, x_{n_{i\cdots ii}}]$ refers to the concatenation of the feature maps produced by the chosen layers n_i, n_{ii}, $\cdots$, $n_{i\cdots ii}$.

To match different sizes of feature maps, we consider three solutions: (a) Downsample the upper layer by performing 2×2 convolutions with the stride of 2; (b) 3×3 dilated convolutions with dilation 2 may increase the receptive field of feature maps when performing downsampling by the stride of 2; (c) Pooling is a simple way to halve the size of feature maps, which generates no additional parameter. In the network that is not deep enough, we recommend (average) pooling to extract different implicit features from former layers before passing to the back layers which may contain max pooling features.

Three patterns of distributed dense connections are designed in Fig. 1, varying in terms of choosing key layers and the methods for transmitting feature maps. In (a), DDC (cross-skip), each concatenation input consists of features from the final layer of all size-blocks. Due to the global transmission of chosen features, each size-block could reuse features from all the preceding blocks. For (b), each concatenation input includes features only chosen from the last two layers of preceding size-block where the information flow does not spread globally. We name this pattern as 'skip-1'. (c) reuses the features from 2 size-blocks. Each concatenation input consists of features chosen from the first layer of present size-block and the last layer of the preceding size block.

2.2 DDU-Net Architectures

Inspired by the encoder-decoder architecture, which is widely used for biomedical image segmentation, We modified U-Net [16] respectively by adding the above three patterns of distributed dense connections between each neighboring resolution stages in the encoder path. These proposed networks are named as DDU-Nets (distributed dense U-Nets), as shown in Fig. 2.

The networks in Fig. 2 inherit the encoder-decoder architecture with 4 resolution stages (levels) operating with different sizes and channels of layers. Every stage in the encoder path consists of two convolutional layers with $3 \times 3 \times 3$ kernels applied by 1 stride and 1 padding, each followed by a LeakyReLU ($alpha = 0.2$) and a 3D batch normalization. Max pooling with the stride of 2 is applied at each end of the encoder stage to downsize the feature maps. In the decoder path, each stage has an up-sampling operation and two convolutions each followed by a LeakyReLU and a batch normalization. In order to achieve the pixel-to-pixel localization, the feature maps at each end of the encoder stage are concatenated to the beginning of the corresponding decoder stage, which provides the high-resolution features to the decoder path. At the final layer, a $1 \times 1 \times 1$ convolution is used to produce the output with the required numbers of classes and the same image size as the input data.

At the encoder side of the network, we apply distributed dense connection to bridge over features between stages. For cross-skip pattern shown in Fig. 2 (a), the first layer of the upper stage in the encoder path is down-sampled by average

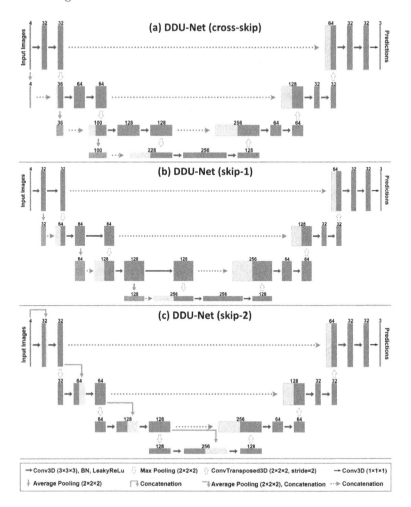

Fig. 2. DDU-Nets with different patterns of DDCs. (a), (b) and (c) represent DDU-Nets with cross-skip, skip-1 and skip-2 respectively. All the feature maps and translations are based on 3D volumes. Copied feature maps and concatenated feature maps can be distinguished by different colors of boxes. Operations are represented by arrows. (Color figure online)

pooling before concatenation, aiming to match the size of the first layer of the lower stage, which is also the output of the upper stage after max pooling. The input of each stage has direct access to all the previous representative feature maps on a global scale, thus enhancing the feature reuse and propagation with less redundant connections. Similarly, the DDU-Nets of skip-1 (b) and skip-2 (c) in Fig. 2 follow the patterns of their own distributed dense connections mentioned in Sect. 2.1. With the application of distributed dense connections, we empirically halve the feature channels of each stage comparing to the traditional U-Net [16].

Experiments (see Table 1) show that the proposed network architectures can effectively improve the performance of brain tumor segmentation tasks.

2.3   Loss Function

Our loss function includes two parts: the average of Dice loss and L2 regularization which are shown in Eqs. (2) and (3):

$$L_{Dice} = \sum_{class} (1 - DICE) \tag{2}$$

$$L = L_{Dice(mean)} + 0.01 L_{L2(total)} \tag{3}$$

To accurately represent the loss of inference, the Dice coefficient is used to represent the loss function, which is a frequently used measurement for pixel-wise segmentation tasks. One of the main challenges of brain tumor segmentation is the imbalance of each subregion. We try to reduce the impact by calculating the average value of three Dice loss functions for three output channels (predictions for each subregion), instead of calculating the Dice loss for the entire predictions directly. For the regularization part, L2 loss displays on the entire predictions and is assigned a hyper-parameter weight to prevent overfitting.

2.4   Training Configuration

BraTS 2019 dataset contains non-standardized 3D images with the size of $250 \times 250 \times 155$. Since the data is from different institutes, the value could vary due to different MRI machines or configurations. To ease these impacts and reduce the initial bias caused by the variations of cases, z-score standardization transform is applied to each of the four image modalities before concatenating them into an input with four channels. Then we reassigned four different labels (label 0, label 1, label 2, label 4) of ground truth into three combined subregions (see Fig. 3), representing enhancing tumor (label 4), tumor core (label 1 + label 4) and whole tumor (label 1 + label 2 + label 4), respectively, to optimize the segmentation accuracy for each region independently in the model. Therefore, the final layer of the network has three channels for the three subregions and we use sigmoid instead of softmax to output the segmentation predictions.

The network is implemented in Pytorch and trained using Adam optimizer with the learning rate of 3e–4. We run our operations parallelly on two GPUs (GeForce GTX 1080 Ti: 11G; TITAN Xp: 12G). In order to fit the capability of our network within GPU memory limits, we cropped all the data into a size of $192 \times 192 \times 128$, and then extracted three smaller overlapping volumes ($192 \times 192 \times 64$) by the stride of 32 in the third dimension. Partitioning images with overlapping area served as a type of data augmentation that ensures the seamless cohesion of separated small volumes, preventing information loss due to cropping. The batch size is 4 and trained the network for 100 epochs (335 cases for each epoch), taking 16 h in total.

2.5 Postprocessing

In some of the low grade gliomas (LGG) cases, there is no existence of enhancing tumor while the model may infer as existing, causing large error in Dice coefficient. Thus, if the number of voxels classified as the segmented enhancing tumor (ET) is less than 300 in a single case, those voxels are regarded as false-positive for ET (label 4) and replaced with the label of necrotic and non-enhancing tumor parts (label 1). Some independent small volumes disconnected with the largest tumor area are removed by connected component processing. If the voxel number of each small component is less than 30% of the total number of predicted class, those components were re-labeled as background.

3 Results

The proposed models are trained on the BraTS 2019 training dataset (335 cases) and initially evaluated on the validation dataset (125 cases). All the predicted results after reconstruction and post-processing are uploaded for the generalizability assessments by CBICA's Image Processing Portal (IPP). Example segmentation results are shown in Fig. 3.

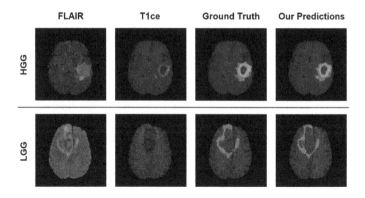

Fig. 3. Visualization results of two examples using DDU-Net (cross-skip) with corresponding FLAIR slices, T1ce slices and ground truth on BraTS 2019 Training dataset. Yellow: enhancing tumor (label 4); Red: necrotic and non-enhancing tumor core (label 1); Green: peritumoral edema (label 2). (Color figure online)

The performance of proposed models is evaluated and compared with baselines (U-Net and DU-net) by Dice score, Sensitivity, Specificity and Hausdorff distance (95%) for all subregions. Four metric average results of three subregions on the BraTS 2019 validation dataset are presented in Table 1.

To illustrate the effectiveness of postprocessing method, experiments on U-Net have been conducted. The results are shown in Table 1 (the 1st and the 2nd row). Comparing their performance on metrics, the effectiveness of the method is

Table 1. Models evaluated (mean value) on BraTS 2019 validation dataset. ET, WT and TC denote enhancing tumor, whole tumor and tumor core, respectively. P, cs, s1, s2 stand for postprocessing, cross-skip, skip-1, skip-2. DU-Net (dense U-Net) is an integrated model of the traditional dense connection and U-Net.

Models	Dice score			Sensitivity			Specificity			95 Hausdorff		
	ET	WT	TC	ET	WT	TC	ET	WT	TC	ET	WT	TC
U-Net	0.744	0.893	0.765	0.740	0.883	0.763	0.997	**0.995**	**0.997**	4.712	6.279	8.146
U-Net+P	0.776	0.894	0.768	0.795	0.883	0.770	**0.998**	**0.995**	**0.997**	3.470	4.968	7.882
DU-Net+P	0.767	0.887	0.773	0.758	0.883	0.763	**0.998**	**0.995**	**0.997**	3.843	5.908	7.989
DDU-Net(cs)+P	0.780	**0.898**	0.793	0.791	0.903	0.808	**0.998**	0.994	0.996	**3.376**	**4.874**	8.013
DDU-Net(s1)+P	0.765	**0.898**	0.793	0.787	**0.905**	**0.820**	**0.998**	0.994	0.996	4.058	5.225	8.127
DDU-Net(s2)+P	**0.784**	0.897	**0.794**	**0.804**	0.888	0.791	**0.998**	**0.995**	**0.997**	4.099	4.950	**7.399**

evident, especially for enhancing tumor. To explore the effects of DDCs, detailed experiments have been operated on each DDU-Net. Compared with U-Net baseline and dense U-net (DU-Net) using dense connection to bridge over features, the results show that the performance of the DDU-Nets surpasses that of baselines on Dice score and sensitivity in most of the subregions. Although DU-Net possesses the highest feature reuse rate, the redundant architecture makes it difficult to achieve better performance than that of the DDU-Nets, which reuse fewer features but achieve great improvement. For each model, Specificity has no obvious difference.

Within the DDU-Net models, different architectures dictate the relative performance that each metric of subregions displays. Cross-skip excels at edge characterization due to the global feature reuse of the localization information provided by original images. Skip-1 has the advantage in terms of Sensitivity at the expense of low Dice score in enhancing tumor. Skip-2 achieves a good result in Dice score, with a low feature reuse rate because of the specific design of DDU-Net (skip-2) that preserving neighborhood information in an adequate skip distance (2 stages). Overall, DDU-Nets with cross-skip pattern can be considered to acquire the best comprehensive performance with an easy model deployment.

Table 2. Performances (mean value) of DDU-Net (cross-skip) with different architectures on BraTS 2019 validation dataset.

Method	Dice			Sensitivity			Specificity			95 Hausdorff		
	ET	WT	TC	ET	WT	TC	ET	WT	TC	ET	WT	TC
Stage 5	**0.782**	0.896	0.784	**0.812**	0.905	0.799	0.998	**0.994**	0.996	4.392	5.616	8.023
Ch 64	0.772	0.873	0.786	0.780	**0.935**	**0.817**	0.998	0.988	0.995	4.699	7.984	8.260
Vol 128	0.674	0.862	0.640	0.661	0.872	0.617	**0.999**	0.993	**0.997**	6.370	6.379	10.846
Final	0.780	**0.898**	**0.793**	0.791	0.903	0.808	0.998	**0.994**	0.996	**3.376**	**4.874**	**8.013**

Apart from the approaches mentioned above, other potential architectures and postprocessing methods also deployed during our experiments. As Table 2 shown, we attempted to allocate DDU-Net (cross-skip) with more stages (5 stages) and feature channels (start with 64 channels in the first stage), but all led to worse results in general, which proves that the distributed dense connections can contribute better performance as well as decreasing in network depth and width. We tried to input with the larger volumes (192 × 192 × 128) without further cropping, but it didn't show better results as well. We also denied the opening operation solution used to denoise for the postprocessing which is replaced by connected component processing. Compared with those alternatives, the final proposed architecture achieved the best performance by balancing among the size of input data, the capability of network and the GPU memory consumption.

Table 3. Performance of DDU-Net (cross-skip) with postprocessing on BraTS 2019 testing dataset.

	Dice			95 Hausdorff		
	ET	WT	TC	ET	WT	TC
Mean	0.804	0.876	0.821	3.41	7.054	6.774
StdDev	0.193	0.117	0.237	7.44	11.600	13.238
Median	0.846	0.917	0.908	1.732	3.535	3.000
25quantile	0.774	0.854	0.836	1.414	2.236	1.946
75quantile	0.916	0.943	0.948	2.639	5.916	5.745

Table 3 presents mean value, standard deviation, median, 25 and 75 quantiles of two metrics on BraTS 2019 testing dataset (166 cases). Due to limited one submission chance, we only evaluated DDU-Net (cross-skip) on testing dataset. The results demonstrate that the performance of this model is highly competitive, with mean Dice scores of 0.804, 0.876, and 0.821 for enhancing tumor, whole tumor and tumor core, respectively.

4   Conclusion

In conclusion, this paper has shown a new network structure for brain tumor segmentation. Three distributed dense connections (DDCs) have been proposed for generic CNNs to inherit features efficiently. DDU-Nets are built to verify the effectiveness of DDCs. Postprocessing is deployed to eliminate false-positive pixels. The results show that the DDU-Nets can segment 3D MR images effectively by allocating DDCs to key layers, among which DDU-Net with cross-skip pattern achieved the competitive performance.

References

1. Bakas, S., et al.: Segmentation labels and radiomic features for the pre-operative scans of the TCGA-GBM collection. The Cancer Imaging Archive (2017)

2. Bakas, S., et al.: Segmentation labels and radiomic features for the pre-operative scans of the TCGA-LGG collection. The Cancer Imaging Archive 286 (2017)
3. Bakas, S., et al.: Advancing the cancer genome atlas glioma MRI collections with expert segmentation labels and radiomic features. In: Scientific Data (2017)
4. Bakas, S., et al.: Identifying the best machine learning algorithms for brain tumor segmentation, progression assessment, and overall survival prediction in the brats challenge. arXiv preprint arXiv:1811.02629 (2018)
5. Cao, H., Bernard, S., Heutte, L., Sabourin, R.: Improve the performance of transfer learning without fine-tuning using dissimilarity-based multi-view learning for breast cancer histology images. In: Campilho, A., Karray, F., ter Haar Romeny, B. (eds.) ICIAR 2018. LNCS, vol. 10882, pp. 779–787. Springer, Cham (2018). https://doi.org/10.1007/978-3-319-93000-8_88
6. Dong, H., Yang, G., Liu, F., Mo, Y., Guo, Y.: Automatic brain tumor detection and segmentation using U-Net based fully convolutional networks. In: Valdés Hernández, M., González-Castro, V. (eds.) MIUA 2017. CCIS, vol. 723, pp. 506–517. Springer, Cham (2017). https://doi.org/10.1007/978-3-319-60964-5_44
7. He, K., Zhang, X., Ren, S., Sun, J.: Deep residual learning for image recognition. In: Proceedings of the IEEE Conference on Computer Vision and Pattern Recognition, pp. 770–778 (2016)
8. Huang, G., Liu, Z., Van Der Maaten, L., Weinberger, K.Q.: Densely connected convolutional networks. In: Proceedings of the IEEE Conference on Computer Vision and Pattern Recognition, pp. 4700–4708 (2017)
9. Isensee, F., Kickingereder, P., Wick, W., Bendszus, M., Maier-Hein, K.H.: No new-net. In: Crimi, A., Bakas, S., Kuijf, H., Keyvan, F., Reyes, M., van Walsum, T. (eds.) BrainLes 2018. LNCS, vol. 11384, pp. 234–244. Springer, Cham (2019). https://doi.org/10.1007/978-3-030-11726-9_21
10. Larsson, G., Maire, M., Shakhnarovich, G.: FractalNet: ultra-deep neural networks without residuals. arXiv preprint arXiv:1605.07648 (2016)
11. Li, X., Chen, H., Qi, X., Dou, Q., Fu, C.W., Heng, P.A.: H-DenseUNet: hybrid densely connected UNet for liver and tumor segmentation from CT volumes. IEEE Trans. Med. Imaging 37(12), 2663–2674 (2018)
12. Mamelak, A.N., Jacoby, D.B.: Targeted delivery of antitumoral therapy to glioma and other malignancies with synthetic chlorotoxin (TM-601). Expert Opin. Drug Deliv. 4(2), 175–186 (2007)
13. Menze, B.H., et al.: The multimodal brain tumor image segmentation benchmark (brats). IEEE Trans. Med. Imaging 34(10), 1993–2024 (2014)
14. Myronenko, A.: 3D MRI brain tumor segmentation using autoencoder regularization. In: Crimi, A., Bakas, S., Kuijf, H., Keyvan, F., Reyes, M., van Walsum, T. (eds.) BrainLes 2018. LNCS, vol. 11384, pp. 311–320. Springer, Cham (2019). https://doi.org/10.1007/978-3-030-11726-9_28
15. Pleiss, G., Chen, D., Huang, G., Li, T., van der Maaten, L., Weinberger, K.Q.: Memory-efficient implementation of DenseNets. arXiv preprint arXiv:1707.06990 (2017)
16. Ronneberger, O., Fischer, P., Brox, T.: U-Net: convolutional networks for biomedical image segmentation. In: Navab, N., Hornegger, J., Wells, W.M., Frangi, A.F. (eds.) MICCAI 2015. LNCS, vol. 9351, pp. 234–241. Springer, Cham (2015). https://doi.org/10.1007/978-3-319-24574-4_28

Brain Tumor Segmentation Based on 3D Residual U-Net

Megh Bhalerao[1(✉)] and Siddhesh Thakur[2(✉)]

[1] National Institute of Technology, Karnataka, Surathkal, India
megh.bhalerao@gmail.com
[2] Shri Guru Gobind Singhji Institute of Engineering and Technology, Nanded, India
sid.cre8er@gmail.com

Abstract. We propose a deep learning based approach for automatic brain tumor segmentation utilizing a three-dimensional U-Net extended by residual connections. In this work, we did not incorporate architectural modifications to the existing 3D U-Net, but rather evaluated different training strategies for potential improvement of performance. Our model was trained on the dataset of the International Brain Tumor Segmentation (BraTS) challenge 2019 that comprise multi-parametric magnetic resonance imaging (mpMRI) scans from 335 patients diagnosed with a glial tumor. Furthermore, our model was evaluated on the BraTS 2019 independent validation data that consisted of another 125 brain tumor mpMRI scans. The results that our 3D Residual U-Net obtained on the BraTS 2019 test data are Mean Dice scores of 0.697, 0.828, 0.772 and Hausdorff$_{95}$ distances of 25.56, 14.64, 26.69 for enhancing tumor, whole tumor, and tumor core, respectively.

Keywords: Brain Tumor Segmentation · CNN · Glioblastoma · Segmentation · BraTS

1 Introduction

Gliomas are the most common type of adult brain tumors arising from glial cells. They are classified into High Grade Glioma (HGG - also referred to as Glioblastoma) and Lower Grade Glioma (LGG). Patients diagnosed with an LGG have better prognosis than an HGG patient. Multi-parametric Magnetic Resonance Imaging (mpMRI) is generally used by radiologists to detect the tumorous region in the brain to plan treatment and surgery, as well as postoperative monitoring of the patient.

The shape and structure of gliomas are highly variable making their detection and classification a tedious task, hence making the need for automatic segmentation algorithms imminent. Several machine learning approaches have been propounded in the past, with the most recent advent of deep learning (i.e., Convolutional Neural Networks) showing state-of-the-art performance in several segmentation tasks [1–5]. The crux of segmentation tasks lies in being able to extract global context as well as local information, which is effectively done by the encoder-decoder architecture of the 3D Residual U-net.

© Springer Nature Switzerland AG 2020
A. Crimi and S. Bakas (Eds.): BrainLes 2019, LNCS 11993, pp. 218–225, 2020.
https://doi.org/10.1007/978-3-030-46643-5_21

2   Materials and Methods

2.1   Data

To create our model we used the publicly available training dataset of the International Brain Tumor Segmentation (BraTS) challenge 2019 comprising mpMRI scans of 259 HGG and 76 LGG subjects (335 in total) [6–10]. For every subject, there are four available co-registered ans skull-stripped mpMRI modalities, namely native T1-weighted (T1), post-contrast T1-weighted (T1CE/T1Gd), T2-weighted (T2), and T2 Fluid Attenuated Inversion Recovery (FLAIR). Every subject was also accompanied by a corresponding ground-truth tumor segmentation label map. These label maps are manually-annotated by expert radiologists. The isometric view of the modalities along with the segmentation label is illustrated below.

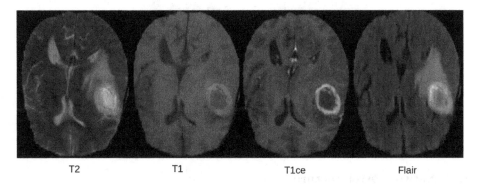

Fig. 1. An example of 4 mpMRI scans with the corresponding label map, comprising the tumor sub-structures. (Color figure online)

Specifically, the tumor label map is divided into 3 regions; non-enhancing tumor core and necrosis (label 1 - red color), enhancing tumor core (label 4 - yellow color), and peritumoral edema (label 2 - orange color), as shown in Fig. 1. The performance metrics used for the segmentation algorithms are the Dice score, the 95% of the Hausdorff Distance, as well as Sensitivity and Specificity.

2.2   Data Pre-processing

The original MRI scans have lots of background voxels with zero intensity value. For computational efficiency we focus on a region of interest (RoI), *i.e.*, the brain. To achieve this, we obtain a bounding box for all non-zero values across all four mpMRI modalities, and crop each of them according to the largest bounding box (amongst the four modalities), to accommodate the Brain-Region in all the four modalities.

After the bounding-box cropping of all mpMRI scans, we pad the images with zeros along every dimension, with the number of zeros chosen such that to make every dimension divisible by 16, and hence account for 4 downsampling layers in our U-Net architecture. We further normalize each image by $\frac{x-\mu}{\sigma}$, where μ and σ are the mean and standard deviation of that particular image's intensity values (it must be noted here that the normalization is done by calculating the μ & σ values only of the non-zero regions of the image). Furthermore, we apply N3 bias field correction to the BraTS data.

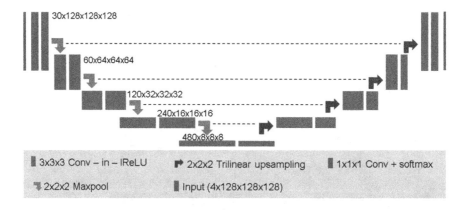

Fig. 2. The architecture followed by our model.

2.3 Network Architecture

We use the 3D U-Net from [2,11] with skip/residual connections as illustrated in Fig. 2.

Our proposed U-Net network is a fully convolutional architecture, meaning that the input image size to the network does not need to be constant (until it meets a certain criteria). The U-Net is an encoder-decoder architecture with 4 downsampling and upsampling modules. The encoder section of the network deciphers semantic information by down-sampling operations, but at the same time loses spatial information due to reduction of image size. This lost spatial information is recovered by the upsampling layers, which perform upsampling using transpose-convolution. A non-linear activation function - Leaky Rectified Linear Unit (leaky ReLU) [12] is applied before every convolution layer, with a leakiness of 10^{-2}. Also since Instance Normalization [13] has been shown to give better results for Image Related Tasks we use Instance Norm instead of Batch-Normalization. In order to avoid the computational bottleneck at the bottom of the "U", skip connections are used which simply concatenate the feature maps of the corresponding upsampling & downsampling layers, so enough contextual information is provided to the network.

Every convolutional module comprises of 2 convolutional layers with the following pipeline: Input → Instance Norm → Leaky ReLU → Convolution →

Dropout → Instance Norm → Leaky ReLU → Convolution → Output (+Input). As we see, we add the input to the output (residual connection), to give something as a shape prior.

At the beginning of the network we start off with 30 filters, successively doubling it with every down-sampling module, reaching a maximum of 480 filters at the bottom of the U. To reduce the computational requirement the number of feature maps is reduced to half, just before the first upsampling layer.

2.4   Training Procedure and Hyper-parameters

Feeding the entire 3D mpMRI scan to the network is very computationally expensive, requiring more than 12 GB of GPU memory, hence we experiment with different divisible-by-16 patch sizes and choose the largest one that fits in the memory. We use a batch-size of 1, again due to computational limitations. The largest image patch that can be employed given computational limitations we want to impose (i.e., Tesla P100 12 GB GPU) is ($128 \times 128 \times 128$). Randomly extracted patches are used to feed our network. 50 Images are randomly taken from the entire training dataset and are used for internal validation of the model during each epoch. Furthermore, we keep a track of which epoch gives us the best validation loss, and that epoch's model is used to generate the segmentations on the actual validation data. We start with a learning rate of 0.001, using a triangular schedule, with the minimum learning rate being 0.000001. Our learning rate varies in a triangular wave like fashion between these two maximum and minimum values, being updated after every iteration (*i.e.*, after every forward pass + back-propagation). This learning rate schedule is used to avoid the local minima spots where the weights might get stuck if we use a monotonically decreasing learning rate. The model is trained for 200 epochs with no-early stopping being done as of now. A stochastic gradient optimizer is used with a momentum of 0.9.

We use a 5-fold cross validation setup to train our models, by dividing our training dataset into 5 folds and training on 4 folds and validating on 1 fold. Hence our training generates 5 models.

In addition to the four modalities (channels - T1, T2, T1ce, Flair), we add an artificially generated fifth modality, with the aim of providing us with some more meaningful information about the tumor. This additional channel is generated by thresholding the flair modality at 0.2 intensity value which is determined heuristically. Thresholding flair provides better information about the whole tumor region and hence has the potential to improve whole tumor performance. Experiments were also carried out with substituting T1 with the artificial modality and also doing the same with T2. The best results were obtained from the experiment which added a fifth modality instead of substituting.

Our model segments the input into the 3 BraTS classes as mentioned above (in addition to 1 background class since we use a soft-max activation in the final layer). Since the most important metric used to quantitatively evaluate the segmentation performance is the Dice score, we continuously optimize towards maximizing the Dice. In the problem statement of BraTS there is an issue of class imbalance (i.e., the background/brain pixels are a lot more in number than the pixels of the segmentation labels). To partially tackle this issue of class imbalance, the Multi-Class dice loss function is used here. This calculates the Dice for individual classes and then averages them. The multi-class dice loss function used is as follows:

$$\ell_{mcd} = 1 - \frac{2}{K} \sum_K \frac{\sum_i u_i^k v_i^k}{\sum_i u_i^k + \sum_i v_i^k} \tag{1}$$

where K is the number of classes, k is the kth class, and i is the ith pixel, u is the predicted soft-max probability for the kth class, and v is the one hot encoded ground truth of the corresponding k^{th} class.

Since medical image datasets are generally smaller compared to other semantic segmentation tasks and the U-Net is a fairly deep network, there arises the issue of overfitting. This is tackled by data augmentation techniques - as of now we are using 90-degree rotation, mirroring, and 45-degree rotation. Furthermore, we add Gaussian noise to every training image, with a $\mu = 0$ and a $\sigma = 0.1$ so as to improve the generalization ability of the network on unseen data.

3   Results

We generate the predicted segmentation labels by thresholding our output probability map for the 125 validation cases and 166 test cases with 0.5 as the threshold. All the segmentations are generated using an ensemble of the 5 models generated by the 5 fold cross validation by using the process of voxel-wise majority voting. The results on the validation and test dataset are shown below (Tables 1, 2 and Figs. 3, 4):

Table 1. Results on BraTS 2019 validation data

	Dice			Haus		
	Enh.	Whole.	Core.	Enh.	Whole.	Core.
Mean	0.66677	0.85269	0.70912	7.27002	8.07931	9.57081
Std. Dev	0.29306	0.15486	0.27646	12.50261	13.56434	12.50147
Median	0.77811	0.90147	0.82969	2.44949	3.60555	4.89898

Table 2. Results on BraTS 2019 test data

	Dice			Haus		
	Enh.	Whole.	Core.	Enh.	Whole.	Core.
Mean	0.697289	0.828745	0.772904	25.56391	14.64394	26.69969
Std. Dev	0.239947	0.189408	0.271199	88.55071	57.10659	88.31207
Median	0.77039	0.894235	0.88047	2.23607	3.60555	3

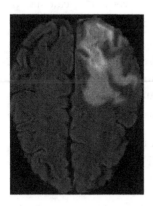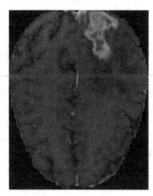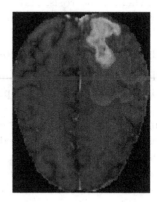

Fig. 3. This is the qualitative result of one of validation cases BraTS19_CBICA_AAM_1 and the corresponding segmentation. Left: Flair, Center: T1ce, Right: Predicted segmentation superimposed over T1ce. Orange is edema, yellow is enhancing tumor and red is necrosis. (Color figure online)

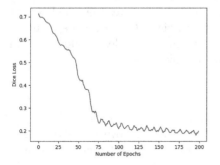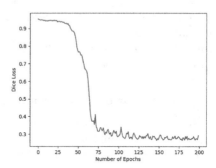

Fig. 4. Left: Training dice loss vs Number of epochs, Right: Validation dice loss vs Number of epochs

4   Discussion

Our model has the potential to provide better results with further improvements in our training process. Below we are listing some of the modifications which we are going to make to the existing process for improvement performance, in the future.

4.1   Post-processing

We further plan to use Isensee et al.'s [4] post-processing technique to improve the dice score of the enhancing tumor. It is known that LGG patients may have no enhancing tumor at all, hence even a single false-positive voxel prediction would make the ET dice score of that patient 0. To overcome this predicament we calculate an experiment-specific voxel threshold for enhancing tumor region. If the number of enhancing tumor voxels are less than this threshold we set all of them as necrosis, and if not, we leave it as it is. This threshold is calculated by mean-dice optimization over the BraTS 2019 training data.

4.2   Modification of Loss Functions

Though the Dice loss is an excellent metric to measure performance and for optimization purposes, it penalizes the False Positive and False Negative terms equally, but in reality, for medical image segmentation, a false negative term is much more dangerous than a false positive term since it means that the disease has not been detected at a place where it should have been. This issue can be addressed by using the Tversky Loss Function [14], which penalizes the false negative terms more than the false positive ones, and therefore improves performance. The mathematical expression for the Tversky Loss is given below:

$$\ell_{tv} = 1 - \frac{1}{K} \sum_K \frac{\sum_i u_i^k v_i^k}{\alpha \sum_i u_i^k + \beta \sum_i v_i^k} \tag{2}$$

All the terms are the same as Multi-class dice loss, except the additional parameters α and β, where $\alpha + \beta = 1$, and $\beta > 1$. The $\beta > 1$ ensures that the FN terms are penalized more than the FP ones. The Multi-class dice loss is a special case of the Tversky loss where $\alpha = \beta = 0.5$. Also, cross entropy loss could be used in conjunction with either the dice loss or the Tversky loss.

4.3   Architectural Modifications

Architectural modifications such as incorporation of inception modules instead of simple convolutional layers have the potential to improve the performance. Inception modules perform convolutions in-parallel with different filter sizes on the input feature map. Paralleled convolutions also help in reduction of bottlenecking of features as compared to convolutions performed sequentially, since multi-scale features can be detected by different filter sizes. The outputs of these parallel paths are then "depth"-concatenated.

5   Conclusions

From the experiments that were done, it can be concluded that, architectural modifications in the traditional 3-D U-Net do not significantly improve its performance on the Brain Tumor Segmentation dataset. Novel training strategies and procedures have a better potential to improve performance.

References

1. Kamnitsas, K., et al.: Efficient multi-scale 3D CNN with fully connected CRF for accurate brain lesion segmentation. MIA **36**, 61–78 (2017)
2. Isensee, F., Kickingereder, P., Wick, W., Bendszus, M., Maier-Hein, K.H.: Brain tumor segmentation and radiomics survival prediction: contribution to the BRATS 2017 challenge. In: Crimi, A., Bakas, S., Kuijf, H., Menze, B., Reyes, M. (eds.) BrainLes 2017. LNCS, vol. 10670, pp. 287–297. Springer, Cham (2018). https://doi.org/10.1007/978-3-319-75238-9_25
3. Li, X., Chen, H., Qi, X., Dou, Q., Fu, C.-W., Heng, P.A.: H-DenseUNet: hybrid densely connected UNet for liver and liver tumor segmentation from CT volumes. arXiv preprint arXiv:1709.07330 (2017)
4. Isensee, F., Jaeger, P.F., Full, P.M., Wolf, I., Engelhardt, S., Maier-Hein, K.H.: Automatic cardiac disease assessment on cine-MRI via time-series segmentation and domain specific features. In: Pop, M., et al. (eds.) STACOM 2017. LNCS, vol. 10663, pp. 120–129. Springer, Cham (2018). https://doi.org/10.1007/978-3-319-75541-0_13
5. Kamnitsas, K., et al.: Ensembles of multiple models and architectures for robust brain tumour segmentation. In: Crimi, A., Bakas, S., Kuijf, H., Menze, B., Reyes, M. (eds.) BrainLes 2017. LNCS, vol. 10670, pp. 450–462. Springer, Cham (2018). https://doi.org/10.1007/978-3-319-75238-9_38
6. Bakas, S., et al.: Advancing the cancer genome atlas glioma MRI collections with expert segmentation labels and radiomic features. Nat. Sci. Data **4**, 170117 (2017)
7. Bakas, S., et al.: Segmentation labels and radiomic features for the pre-operative scans of the TCGA-GBM collection. The Cancer Imaging Archive 286 (2017)
8. Bakas, S., et al.: Segmentation labels and radiomic features for the pre-operative scans of the TCGA-LGG collection. The Cancer Imaging Archive (2017)
9. Menze, B.H., Jakab, A., Bauer, S., et al.: The multimodal brain tumor image segmentation benchmark (BRATS). IEEE Trans. Med. Imaging **34**(10), 1993–2024 (2015). https://doi.org/10.1109/TMI.2014.2377694
10. Bakas, S., Reyes, M., Jakab, A., Bauer, S., Rempfler, M., Crimi, A., et al.: Identifying the best machine learning algorithms for brain tumor segmentation, progression assessment, and overall survival prediction in the BRATS challenge. arXiv preprint arXiv:1811.02629 (2018)
11. Çiçek, Ö., Abdulkadir, A., Lienkamp, S.S., Brox, T., Ronneberger, O.: 3D U-Net: learning dense volumetric segmentation from sparse annotation. In: Ourselin, S., Joskowicz, L., Sabuncu, M.R., Unal, G., Wells, W. (eds.) MICCAI 2016. LNCS, vol. 9901, pp. 424–432. Springer, Cham (2016). https://doi.org/10.1007/978-3-319-46723-8_49
12. Arora, R., Basu, A., Mianjy, P., Mukherjee, A.: Understanding deep neural networks with rectified linear units. arXiv preprint arXiv:1611.01491 (2016)
13. Ulyanov, D., Vedaldi, A., Lempitsky, V.: Instance normalization: the missing ingredient for fast stylization. arXiv preprint arXiv:1607.08022 (2016)
14. Salehi, S.S.M., Erdogmus, D., Gholipour, A.: Tversky loss function for image segmentation using 3D fully convolutional deep networks. In: Wang, Q., Shi, Y., Suk, H.-I., Suzuki, K. (eds.) MLMI 2017. LNCS, vol. 10541, pp. 379–387. Springer, Cham (2017). https://doi.org/10.1007/978-3-319-67389-9_44

Automatic Segmentation of Brain Tumor from 3D MR Images Using SegNet, U-Net, and PSP-Net

Yan-Ting Weng, Hsiang-Wei Chan, and Teng-Yi Huang[✉]

Department of Electrical Engineering, National Taiwan University of Science and Technology, Taipei, Taiwan
tyhuang@mail.ntust.edu.tw

Abstract. In the study, we used three two-dimensional convolutional neural networks, including SegNet, U-Net, and PSP-Net, to design an automatic segmentation of brain tumor from three-dimensional MR datasets. We extracted 2D slices from three slice orientations as the input tensor of the network in the training stage. In the prediction stage, we predict a volume several times with slicing along different angles. Based on the results, we learned that the result predicted more times has better outcomes than those predicted less times. Also, we implement two ensemble methods to combine the result of the three networks. According to the results, the above strategies all contributed to the improvement of the accuracy of segmentation.

Keywords: Gliomas · Deep learning · Image segmentation

1 Introduction

Brain glioma is the most common type of brain tumor associated with glial cells. Gliomas have different degrees of aggressiveness and several heterogeneous histological sub-regions. The subregions, i.e., peritumoral edema, enhancing tumor, necrotic core, and necrotic tumor, are portrayed across multimodal MRI scans. Manual segmentation is such a time-consuming and error-prone task that developing an automatic brain tumor segmentation is the critical issue nowadays. BraTs2019 hosts the competition for segmentation of brain gliomas with MR images [1–5]. In the study, we attempt to use and compare the recently advanced convolutional neural network architecture, SegNet [6], U-Net [7], and PSP Net [8], for our task. We separately train each architecture and then try to combine three results to improve the accuracy of the results. Because of the time limitation, we are still working on the BraTS 2019 data. In this article, the major findings are based on the BraTS 2018 data. Nonetheless, it presents the candidates of the methods that we are going to use during the final test stage.

© Springer Nature Switzerland AG 2020
A. Crimi and S. Bakas (Eds.): BrainLes 2019, LNCS 11993, pp. 226–233, 2020.
https://doi.org/10.1007/978-3-030-46643-5_22

2 Method

2.1 Data

In this article, we used the BraTS 2018 data as our training datasets and testing datasets. The BraTS 2018 data has 285 MRI datasets with the Nifty file format, including subjects with high-grade gliomas (HGG): 210, and low-grade gliomas (LGG): 75, as the training datasets. Each dataset consists of four types of MR image modalities, T1, T1 with contrast enhancement (T1ce), T2, and FLAIR, and corresponding segmentation map. The segmentation with four labels 1, 2, 4, 0 respectively are necrotic and nonenhancing tumor core (NCR/NET), peritumoral edema (ED), enhancing tumor (ET) and otherwise. The validation dataset includes 125 MRI datasets, which are similar to the training dataset without segmentation. Every dataset was all aligned to the same space, and the MR volume matrix was (X, Y, Z $= 240 \times 240 \times 155$).

2.2 Preprocessing

We padded zeros to each dataset to construct a 4D labeled volume ($256 \times 256 \times 256 \times 1$) and a 4D training volume ($256 \times 256 \times 256 \times 4$) which was combined with four contrasts. Also, we extracted two-dimensional slices ($256 \times 256 \times 4$)from three slice orientations, including axial, coronal, and sagittal. These slices without pixels labeled as gliomas were removed from the training datasets and the rest are normalized with the maximum intensity. With 285 subjects in the training datasets, we then had $256 \times 3 \times 285 = 218880$ images with a matrix size of 256×256. For enhancing the variability of the images, several methods of data augmentation are used in the training procedures, i.e., image flipping, rotation, transpose, and contrast adjustments.

2.3 Model Architecture (Segmentation)

We trained SegNet, U-Net, and PSP Net, respectively for the brain tumor segmentation task. All networks used the same input data and were implemented with the Tensor-flow framework (v1.8) and the Python (v3.6) environment.

SegNet. The design of our SegNet is illustrated in Fig. 1. On the encoding path, the max-pooling operation with stride two is displayed after two or three convolution layers which doubled the channel of the feature map. Each convolution layer uses a $3 \times 3 \times 3$ kernel and is followed by a rectified linear unit (ReLU). On the decoding path, the number of filters is reduced by half, and feature map size is increased by a factor of two every transpose convolution. At the last layer, a multi-class softmax classifier is used to map the multi-classes problem to four of classes (0, 1, 2, 4).

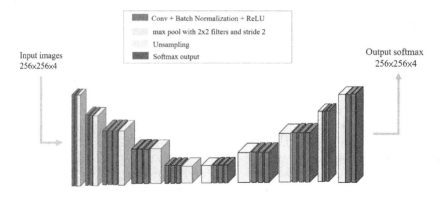

Fig. 1. The SegNet architecture

U-Net. The architecture of our U-net implementation is shown in Fig. 2. The U-net structure is similar to the SegNet one. There are three differences between them. First, there is dropout every two consecutive convolutions after the first two blocks. Second, the expanding path of U-net combines an up-convolution, a concatenation with an output of the decoder layer and the correspondingly feature map from the contracting path, two 3×3 convolutions, and dropout. Third, an 1×1 convolution layer is used before a softmax classifier.

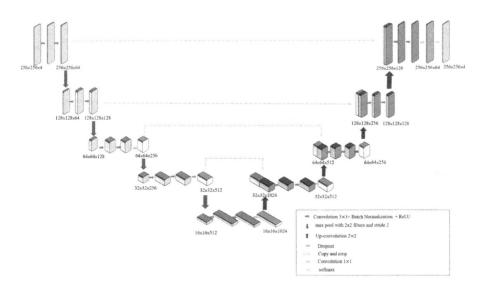

Fig. 2. The U-net architecture

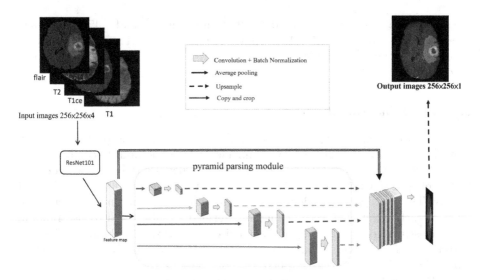

Fig. 3. The PSP-net architecture

PSP-Net. The architecture of our PSP-net implementation is shown in Fig. 3. We first use Res-Net101 [9] to extract a feature map and apply a four-level pyramid parsing module with bin sizes of 4×4, 8×8, 16×16, and 32×32, respectively. Then, we upsample the feature maps to get the same size feature as the original one via bilinear interpolation and concatenation them to the final feature representation. At the last step, the final feature representation was fed into a 1×1 convolution and was upsampled to form a final label prediction, which remained the same size as the original input images.

2.4   Volume Prediction Using Each Model

We used 66 validation datasets provided by BraTS2018 to evaluate the performance of the segmentation method. For each subject, we created multiple configurations of tensors as the input of the network. We compared three different configurations of the predicting procedures, termed Val-3, Val-12, and Val-21. In Val-3, the input volumes ($256 \times 256 \times 256$) were sliced along the X, Y, and Z dimension. We used each direction slices as input tensor to produce the network softmax output, respectively. And then, we merged the 2D output slices into three volumes, removed the zero-padded parts of every volume. The last step, we identified the class with the maximum value of each voxel as the prediction of the tumor subtype of the voxel. The Val-12 procedure was similar to the Val-3. However, the difference between was that the input volumes were rotated $45°$ along each axis (X, Y, Z) before being sliced. That is, there were three rotated volumes, and the original one as input volume and each volume would produce three softmax outputs because of three slicing orientationn. We then had $4 \times 3 = 12$ outputs volume to produce the final prediction result. As to the Val-21, we respectively rotated input volume with 45 and $-45°$ along each axis (X, Y, Z) and had $7 \times 3 = 12$ outputs softmax volume. The rest of the steps remain the same as in the Val-3 procedure. In Fig. 4, it is an overview of our prediction in Val-12.

2.5   Ensemble Model

At predicting stages, SegNet, U-Net, and PSP Net produced their softmax matrix of four tumors types. And we implemented two methods to combine these result. The first method, termed Com-1, was to calculate the summation of the three network soft-max volumes of four tumors types into a $240 \times 240 \times 155 \times 4$ matrix. The class with the highest value was selected as the final segmentation label of each voxel. The second method, termed Comb-2, was to use random forest classifier to identify the tumor label of each voxel. In the beginning, we flattened each network softmax volumes to 8928000×4 matrices and concatenated three of them to one 8928000×12 matrices as the input of our random forest [10]. Each pixel had 12 softmax value as input variables. In our random forest framework, there were twenty trees in the forest, and our feature selection criterion was Gini impurity.

3   Result

Table 1 lists the accuracy indexes of the validation results for the Val-3, Val12, and Val-21, with each network. Notice that every network seems to have the same trend. The average dice coefficients for the Val-3 are the worst, but they slightly increase for the Val-12 and the Val-21. That is the more slices extracted from different angles as input in the prediction stage, the better the segmentation result it will be. The result is inspiring that we can improve the accuracy of segmentation without re-training a network. Therefore, we can conclude that slicing the 3D volume in multiple orientations at the predicting stage is a possible way to improve the segmentation result for predicting 3D volume using two-dimensional convolutional neural networks. Table 2 lists the result of combining three networks with different ensemble strategy. Compared with the result in Table 1, without doubt, the result after combining with other networks has better outcomes than the one predicting with only one network. The dice coefficients for three type in Com-1 slightly increases from the Val-3 to Val-12, which has the same trend we mentioned above. Nonetheless, the dice coefficients of enhancing tumor in Com-2 decreases. We believe this drop is attributable to the imperfection of our random decision forests model for the Val-12.

Table 2 lists the result of combining three networks with different ensemble strategy. Compared with the result in Table 1, without doubt, the result after combining with other networks has better outcomes than the one predicting with only one network. The dice coefficients for three type in Com-1 slightly increases from the Val-3 to Val-12, which has the same trend we mentioned above. Nonetheless, the dice coefficients of enhancing tumor in Com-2 decreases.

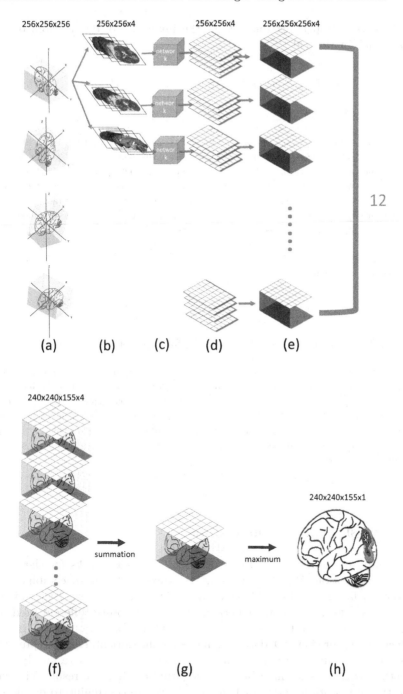

Fig. 4. Overview of our prediction in Val-12. We rotated the 3D input volume with 45° along three axis (a). Then, we extracted 2D slices from each axis (b) and fed the slices into a network (c) which produced the softmax matrix (d) as output. We obtained 12 softmax 3D volume after merging (e) and reshaping (f). Finally, we summed the softmax volume (g) and identified the class with the maximum value of each voxel (h).

We believe this drop is attributable to the immature of our random decision forests model for the Val-12.

Table 1. Average validation results for each model

	Model	Mean Dice			Mean Sensitivity			Mean Specificity			Mean Hausdorff_95		
		Enh.	Whole	Core	Enh.	Whole	Core	Enh.	Whole	Core	Enh.	Whole	Core
Val-3	Segnet	0.766	0.896	0.824	0.78	0.873	0.793	0.998	0.996	0.998	3.285	4.292	5.946
	U-net	0.761	0.891	0.811	0.79	0.862	0.787	0.998	0.997	0.998	3.23	4.652	6.613
	PSP-net	0.73	0.891	0.84	0.755	0.87	0.814	0.996	0.998	3.816	3.816	4.806	5.243
Val-12	Segnet	0.772	0.899	0.826	0.792	0.878	0.797	0.998	0.996	0.998	3.31	4.33	6.882
	U-net	0.765	0.894	0.811	0.8	0.868	0.788	0.998	0.996	0.998	3.160	4.572	7.631
	PSP-net	0.738	0.898	0.843	0.77	0.875	0.815	0.998	0.996	0.998	3.961	4.699	5.211
Val-21	Segnet	0.772	0.9	0.828	0.797	0.883	0.797	0.998	0.996	0.999	3.283	4.361	5.902
	U-net	0.766	0.896	0.811	0.804	0.873	0.788	0.998	0.996	0.998	3.193	4.66	7.407
	PSP-net	0.741	0.9	0.843	0.779	0.882	0.814	0.997	0.996	0.998	4.098	4.769	5.189

118 subjects for validation

Table 2. Average validation results for two ensemble methods

	Model	Mean Dice			Mean Sensitivity			Mean Specificity			Mean Hausdorff_95		
		Enh.	Whole	Core	Enh.	Whole	Core	Enh.	Whole	Core	Enh.	Whole	Core
Val-3	Com-1	0.774	0.902	0.823	0.798	0.879	0.791	0.998	0.996	0.999	3.216	4.001	6.933
	Com-2	0.777	0.903	0.825	0.806	0.88	0.796	0.998	0.996	0.998	3.26	4.201	5.604
Val-12	Com-1	0.779	0.906	0.840	0.811	0.906	0.81	0.998	0.995	0.999	3.208	4.358	5.270
	Com-2	0.765	0.908	0.839	0.809	0.91	0.812	0.998	0.995	0.999	3.2	4.458	5.481

118 subjects for validation

4   Discussions and Conclusions

In the study, we used three 2D convolutional neural networks for the task of glioma segmentation. We used different ensemble methods to combine them. However, so far, it hardly told which one is better. Now, we keep going on with the more effective ensemble method. Also, we proposed different prediction strategies. According to the segmentation results, we found out, for 2D network architectures to predict 3D datasets, multiple slice orientations in prediction stages truly contributed to the accuracy of the segmentation result. However, the method is time-consuming but only slightly improves the result. Therefore, during the competition period of BraTS2019, we are continuing to develop the method. We attempt to figure out the proper rotating angles and rotating times for the method, which makes it more efficient.

References

1. Bakas, S., et al.: Segmentation labels and radiomic features for the pre-operative scans of the TCGA-GBM collection. The Cancer Imaging Archive (2017). https://doi.org/10.7937/K9/TCIA.2017.KLXWJJ1Q
2. Bakas, S., et al.: Segmentation labels and radiomic features for the pre-operative scans of the TCGA-LGG collection. The Cancer Imaging Archive (2017). https://doi.org/10.7937/K9/TCIA.2017.GJQ7R0EF
3. Menze, B.H., et al.: The multimodal brain tumor image segmentation benchmark (BRATS). IEEE Trans. Med. Imaging **34**(10), 1993–2024 (2015)
4. Bakas, S., et al.: Advancing the cancer genome atlas glioma MRI collections with expert segmentation labels and radiomic features. Nat. Sci. Data **4**, 170117 (2017). https://doi.org/10.1038/sdata.2017.117
5. Bakas, S., Reyes, M., Jakab, A., Bauer, S., Rempfler, M., Crimi, A., et al.: Identifying the best machine learning algorithms for brain tumor segmentation, progression assessment, and overall survival prediction in the BRATS challenge. arXiv preprint arXiv:1811.02629 (2018)
6. Badrinarayanan, V., Kendall, A., Cipolla, R.: SegNet: a deep convolutional encoder-decoder architecture for image segmentation. CoRR abs/1511.00561 (2015). http://arxiv.org/abs/1511.00561
7. Ronneberger, O., Fischer, P., Brox, T.: U-Net: convolutional networks for biomedical image segmentation. In: Navab, N., Hornegger, J., Wells, W.M., Frangi, A.F. (eds.) MICCAI 2015. LNCS, vol. 9351, pp. 234–241. Springer, Cham (2015). https://doi.org/10.1007/978-3-319-24574-4_28. CoRR abs/1505.04597. http://arxiv.org/abs/1505.04597
8. He, K., Zhang, X., Ren, S., Sun, J.: Spatial pyramid pooling in deep convolutional networks for visual recognition. In: Fleet, D., Pajdla, T., Schiele, B., Tuytelaars, T. (eds.) ECCV 2014. LNCS, vol. 8691, pp. 346–361. Springer, Cham (2014). https://doi.org/10.1007/978-3-319-10578-9_23
9. He, K., Zhang, X., Ren, S., Sun, J.: Deep residual learning for image recognition. In: CVPR (2016)
10. Breiman, L.: Manual on setting up, using, and understanding random forests v4.0 (2003)

3D Deep Residual Encoder-Decoder CNNS with Squeeze-and-Excitation for Brain Tumor Segmentation

Kai Yan[1,2,3], Qiuchang Sun[1,2], Ling Li[2], and Zhicheng Li[1,2(✉)]

[1] University of Chinese Academy of Sciences (UCAS), Beijing, China
[2] Shenzhen Institutes of Advanced Technology, Chinese Academy of Science,
1068 Xueyuan Avenue, Shenzhen University Town, Shenzhen, China
{kai.yan,qc.sun,zc.li}@siat.ac.cn
[3] Peng Cheng Laboratory, No. 2, Xingke 1st Street, Nanshan, Shenzhen, China

Abstract. Segmenting brain tumors from multimodal MR scans is thought to be highly beneficial for brain abnormality diagnosis, prognosis monitoring, and treatment evaluation. Due to the highly heterogeneous appearance and shape, segmentation of brain tumors in multimodal MRI scans is a challenging task in medical image analysis. In recent years, many segmentation algorithms based on neural network architecture are proposed to address this task. Observing the previous state-of-the-art algorithms, not only did we explore multimodal brain tumor segmentation in 2D space, 2.5D space and 3D space respectively, we also made a lot of attempts in attention block to improve the segmentation result. In this paper, we describe a 3D deep residual encoder-decoder CNNS with Squeeze-and-Excitation block for brain tumor segmentation. In order to learn more effective image features, we have utilized an attention module after each Res-block to weight each channel, which emphasizes useful features while suppresses invalid ones. To deal with class imbalance, we have formulated a weighted Dice loss function. We find that 3D segmentation network with attention block which can enhance context features can significantly improve the performance. In addition, the results of data preprocessing have a great impact on segmentation performance. Our method obtained Dice scores of 0.70, 0.85 and 0.80 for segmenting enhancing tumor, whole tumor and tumor core, respectively on the testing data set.

Keywords: 3D-Resnet · U-net · Brain tumor · Attention block · CNNS

1 Introduction

Brain tumors are one of the major lethal types of tumor [1]. There are two categories of brain tumors, namely primary and secondary tumor types [4, 5]. Primary brain tumors originate from brain cells. The most common occurring primary brain tumors are gliomas [2, 3]. Gliomas are categorized into high-grade gliomas (HGG) and low-grade gliomas (LGG) [4, 5]. HGGs are an aggressive type of malignant brain tumor with a worse survival prognosis. Magnetic resonance imaging (MRI) is a key diagnostic tool used in the brain tumor diagnosis, characterization, and surgery planning. Usually,

© Springer Nature Switzerland AG 2020
A. Crimi and S. Bakas (Eds.): BrainLes 2019, LNCS 11993, pp. 234–243, 2020.
https://doi.org/10.1007/978-3-030-46643-5_23

different MRI modalities are acquired - such as T2-weighted (T2), Fluid Attenuated Inversion Recovery (FLAIR), T1-weighted (T1), and contrast-enhanced T1-weighted (T1ce) - to emphasize different tissue properties and areas of tumor spread.

Multimodal Brain Tumor Segmentation Challenge (BraTs19), which is making available a large dataset with accompanying delineations of the relevant tumor sub-regions as a benchmark, aims to encourage researchers to design state-of-the-art algorithms [6–10]. In this year, the training data consisting of 335 cases (259 HGG and 76 LGG). Each case including 4 modalities (T1, T1c, T2 and FLAIR) are pre-processed, co-registered to the same anatomical template, interpolated to the same resolution (1 mm^3) and skull-stripped, which are acquired with different clinical protocols and various scanners from multiple (n = 19) institution. Each tumor is annotated into 3 sub-regions: whole tumor (WT), tumor core (TC) and enhancing tumor (ET).

2   Related Work

The performance of semantic segmentation have been significantly improved by using deep convolutional neural networks [11–17]. One of the first notably successful neural network for semantic segmentation was FCN, which replaces the last fully connected layer in CNN with a convolutional layer. Recently, there have been two main approaches focusing on semantic segmentation: 1) the U-shape architecture such as RefineNet, GCN, DFN. 2) Dilation CNNs such as PSPNet and Deeplab series. In medical image segmentation task, 3D U-shape architecture is notably successful. In BraTs 2018, many methods with high performance were at least partially based on 3D encoder-decoder networks.

Top ranking methods on BraTS 2017 included DeepMedic, a 3D CNN introduced by Kamnitsas et al. [18]. They presented a fully automatic approach based on a dual pathway, 11-layers deep, multi-scale 3D CNN. In BraTS 2018, Andriy Myronenko proposed a 3D CNN including a dual pathway in decoder. Apart from a normal decoding structure, an additional branch auto-encoder architecture was added to guide the encoder part due to the limiting dataset size [19]. Apart from a normal decoding structure, an additional branch auto-encoder architecture was added to guide the encoder part in the case of the limiting dataset size [20].

3   Methods

Observing the previous state-of-the-art algorithms, we tried to use U-shape network and its variants to explore the multimodal brain tumor segmentation in 2D space, 2.5D space and 3D space respectively. In 2D space, we use 2D slices with lesions of all the training sets as the input data of the 2D-Unet [22]. In 2.5D space, we also extract all the slices with lesions of all the training sets and adopt the same network structure, but, in contrast to the 2D space, for each slice, we respectively extract two more layers above and below to make better use of the 3D-space information. So, after concatenate all of the models corresponding to the same layer for each case, the shape of the input data is $240 \times 240 \times 20$ and the convolution kernel in the network is 2-dimensional.

By comparing the three results, we find that 3D CNN is better than 2.5D while 2.5D is better than 2D. MRI volume data contains spatial coherence information, and adjacent regions with similar spatial feature states are likely to belong to the same tumor region or background. The spatial coherence of MRI images can help to separate tumor and background. On the other hand, recent studies witnessed that attention mechanism can improve the performance of deep semantic segmentation networks. We attempt to add attention mechanism modules in different places of the network structure. In this paper, we adopt a deep residual encoder-decoder CNN with Squeeze-and-Excitation block (SE-block) for brain tumor segmentation. In this architecture, we use a cropped patch of $160 \times 160 \times 128 \times 4$ as the input data and the batch size was set to as 10. The output layer has three-channels after sigmoid function and each channel represented a sub-region. The network architecture is built on a 3D Resnet with an encoder-decoder structure, enhanced by a SE-block after each Res-block to fully exploit image features while restrain useless feature (see also Fig. 1).

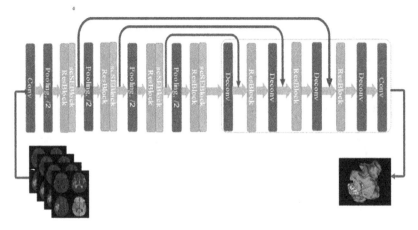

Fig. 1. Schematic visualization of the network architecture. The input of the network is a 3D MRI crop with a shape of $160 \times 160 \times 160 \times 4$. we adopt an encoder with 3D Resnet 34 structure. The squeeze-and-excitation residual blocks that are adopted after each convolutional block in encoding part. After each residual block, a SE-block is utilized before each skip connection.

3.1 Data Preprocessing and Augmentation

Data augmentation has served as an effective solution to he problem of over-fitting, particularly in the absence of large labeled training sets. In this paper, the image data is reshaped as a four-dimensional matrix with shape of $160 \times 160 \times 128 \times 4$. For each patient, four modality images are concatenate together to from a four-channel data. Same as last year's second-ranking method [20], we normalize each modality of each patient independently by subtracting the mean and dividing the standard deviation of the brain region (Figs. 2 and 3). The region outside the brain is set to 0 (Fig. 4).

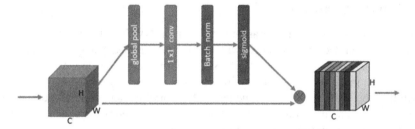

Fig. 2. Schematic visualization of the SE block architecture.

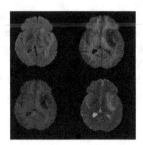

Fig. 3. BraTs19 data   **Fig. 4.** Standardization

3.2   Network Architecture

We adopt 3D-Resnet as backbone network due to its outstanding performance in computer vision tasks. Considering physical memory, we adopt a 3D resnet-34 as the backbone network of the encoding part, which includes four Res-blocks with SE-block [21] and four max pooling layers. After each down sampling the size of feature maps becomes one-half of the original. Each module is composed of several basic residual units, as shown in Fig. 5.

The Res-block in decoding part is the same as that in the encoding part. After each up-sampling step the size of feature maps becomes twice of the original. In encoding part, each Res-block followed by a SE-block as illustrated in Fig. 5.

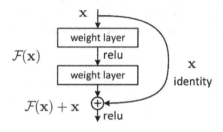

Fig. 5. Basic residual units

3.3   Training Procedure

We implement our network in Keras and trained it on NVIDIA Tesla V100 32G GPU. Our network architecture is trained with randomly sampled patches of size $160 \times 160 \times 128 \times 4$ voxels and batch size 10. We train the network using Adam optimizer with learning rate 0.0001. If the loss value does not decrease after 5 epochs of training, we will take the learning rate to be one tenth of the original, and the minimum learning rate is not less than 1e−6. Early stop strategy is also used. For each epoch, it will take about three minutes and ten seconds with 36 iterations.

3.4   Data Augmentation

Because we use 3D CNNS architecture, the amount of data is obviously insufficient. When training large neural networks from limited training data, we must take some strategies to avoid overfitting. In this work, data augmentation is one of strategies to solve this problem. We rotate the data on the (X, Y) coordinate plane and mirror rotation along the X-axis and Y-axis. We apply a random (per channel) intensity shift (−0.1 to 0.1 of image standard) and random scaling (0.9 to 1.1) on input image channels.

3.5   Loss Function

In order to tackle with the imbalanced data problem, we adopt the weighted Dice as the loss function. The two-class form of Weighted-Dice can be expressed as:

$$\text{Weighted\_Dice} = -\frac{2}{K^2}\left(\frac{P \wedge T}{|P| + |T|}\right) \tag{1}$$

Where P represents the predicted probability. T represents the true value. K represents the number of each category.

4   Result

Table 1 shows the results of our model on the BraTS 2019 dataset. The output value of each channel is between 0-1, so we need a cutoff value for each sub-region to output binary value. In this work, the cutoffs for WT, TC and ET are 0.55, 0.50, and 0.35. We find that the TC and ET are very sensitive to the choice of its cutoff value (Fig. 6).

According to our statistics, we find that some sub-regions are very small and even not present in some cases, but the model we have trained using the training data sets

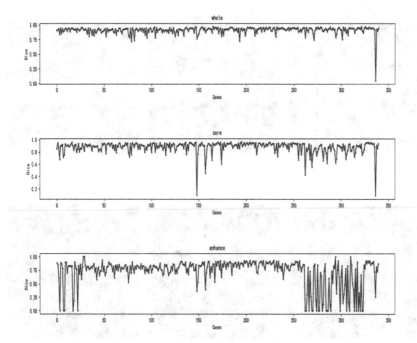

Fig. 6. Dice coefficient statistics for different regions in training data

with 3 labels may predict 3 sub-regions including whole, enhance, and core tumor. For example, if one case without enhance sub-region but the prediction mask image generated by the model includes enhance sub-region, which will make other sub-regions included in original data diminish and reduce dice value. Moreover, small sub-regions are more sensitive to the cutoffs. It is widely acknowledged that the small sub-region is very sensitive when using dice value as evaluation method, which needs to be emphasized. Recent studies witnessed that earlier layer features are typically more general while later layer features exhibit greater levels of specificity [21]. Therefore, we find that adding attention mechanism can improve the segmentation accuracy of small areas (Fig. 7).

Figure 8 shows a typical segmentation example with true and predicted labels which overlaid over T1c MRI axial, sagittal and coronal slices. The whole tumor (WT) is visible by a union area including green, yellow and red labels. The ET is shown in yellow. The TC entails the ET, as well as the necrotic (fluid-filled) and the non-enhancing (solid) parts of the tumor which is shown in red (Fig. 9).

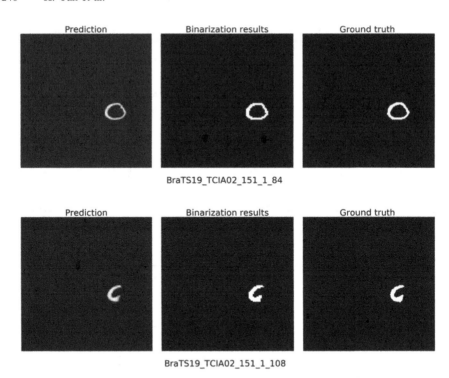

Fig. 7. A small subregion segmentation example (ET). The image on the left is prediction labels, whose values are between 0 and 1. The image on the right is true mask and the image in the middle of the figure is the final binary label.

Table 1. Results on BraTS2019 training set validation data and test set.

	Dice			Hausdorff (mm)		
	WT	TC	ET	ET	TC	ET
Training	0.91	0.88	0.72	4.73	3.60	3.75
Validation	0.86	0.73	0.66	40.31	10.4	18.53
Testing	0.85	0.80	0.70	31.98	11.47	26.28

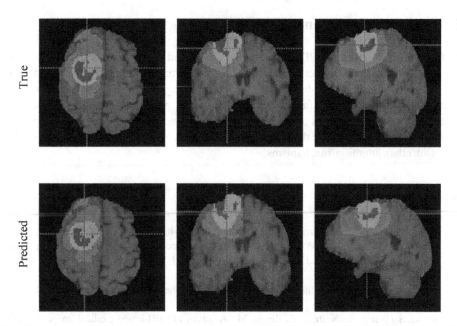

Fig. 8. A typical segmentation result example on BraTS2019 training set (Color figure online)

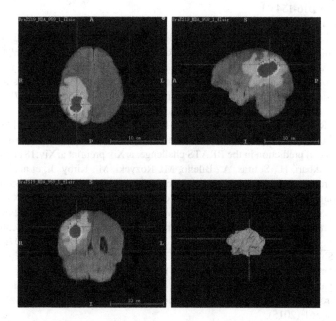

Fig. 9. A typical segmentation result example on BraTS2019 validation set. It shows a typical segmentation example only with predicted labels which overlaid over T1c MRI axial, sagittal and coronal slices. (Color figure online)

5 Discussion and Conclusion

In this work, we have tried 2D, 2.5D and 3D CNNs and many variants. Based on these networks, we find that 3D network has the best performance in all our experiments. On the other hand, adding SE-block can enhance the segmentation performance of the small area. However, it is far from the performance we expected. So next, we will further improve our network in these two aspects. Moreover, we will change the output layer of the network and the data structure of the label. We will continue to explore the effect of other attention mechanisms.

References

1. DeAngelis, L.M.: Brain tumors. New Engl. J. Med. **344**(2), 114–123 (2001)
2. Ohgaki, H., Kleihues, P.: Population-based studies on incidence, survival rates, and genetic alterations in astrocytic and oligodendroglial gliomas. J. Neuropathol. Exp. Neurol. **64**(6), 479–489 (2005)
3. Goodenberger, M.L., Jenkins, R.B.: Genetics of adult glioma. Cancer Genet. **205**(12), 613–621 (2012)
4. Bauer, S., Wiest, R., Nolte, L.P., Reyes, M.: A survey of MRI-based medical image analysis for brain tumor studies. Phys. Med. Biol. **58**(13), R97–R129 (2013)
5. Louis, D.N., Perry, A., Reifenberger, G., von Deimling, A., Figarella-Branger, D., Cavenee, W.K., et al.: The 2016 World Health Organization classification of tumors of the central nervous system: a summary. Acta Neuropathol. **131**(6), 803–820 (2016). https://doi.org/10.1007/s00401-016-1545-1
6. Menze, B.H., Jakab, A., Bauer, S., Kalpathy-Cramer, J., Farahani, K., Kirby, J., et al.: The multimodal brain tumor image segmentation benchmark (BRATS). IEEE Trans. Med. Imaging **34**(10), 1993–2024 (2015). https://doi.org/10.1109/TMI.2014.2377694
7. Bakas, S., Akbari, H., Sotiras, A., Bilello, M., Rozycki, M., Kirby, J.S., et al.: Advancing the cancer genome atlas glioma MRI collections with expert segmentation labels and radiomic features. Nat. Sci. Data **4** (2017). Article number: 170117. https://doi.org/10.1038/sdata.2017.117
8. Bakas, S., Reyes, M., Jakab, A., Bauer, S., Rempfler, M., Crimi, A., et al.: Identifying the best machine learning algorithms for brain tumor segmentation, progression assessment, and overall survival prediction in the BRATS challenge. arXiv preprint arXiv:1811.02629 (2018)
9. Bakas, S., Akbari, H., Sotiras, A., Bilello, M., Rozycki, M., Kirby, J., et al.: Segmentation labels and radiomic features for the pre-operative scans of the TCGA-GBM collection. The Cancer Imaging Archive (2017). https://doi.org/10.7937/K9/TCIA.2017.KLXWJJ1Q
10. Bakas, S., Akbari, H., Sotiras, A., Bilello, M., Rozycki, M., Kirby, J., et al.: Segmentation labels and radiomic features for the pre-operative scans of the TCGA-LGG collection. The Cancer Imaging Archive (2017). https://doi.org/10.7937/K9/TCIA.2017.GJQ7R0EF
11. Long, J., Shelhamer, E., Darrell, T.: The IEEE Conference on Computer Vision and Pattern Recognition (CVPR), pp. 3431–3440 (2015)
12. Badrinarayanan, V., Kendall, A., Cipolla, R.: SegNet: A deep convolutional encoder-decoder architecture for image segmentation, vol. 39, no. 12, pp. 2481–2495. CoRR abs/1511.00561 (2015)

13. Noh, H., Hong, S., Han, B.: Learning deconvolution network for semantic segmentation In: Proceedings of the IEEE International Conference on Computer Vision, pp. 1520–1528 (2015)
14. Lin, G., Milan, A., Shen, C., Reid, I.D.: RefineNet: multi-path refinement networks for high-resolution semantic segmentation. In: Proceedings of the CVPR, vol. 1, p. 5 (2017)
15. Zhao, H., Shi, J., Qi, X., Wang, X., Jia, J.: Pyramid scene parsing network. In: Proceedings of the IEEE Conference on Computer Vision and Pattern Recognition (CVPR), pp. 2881–2890 (2017)
16. Chen, L.-C., Papandreou, G., Kokkinos, I., Murphy, K., Yuille, A.L.: DeepLab: semantic image segmentation with deep convolutional nets, atrous convolution, and fully connected CRFs. IEEE Trans. Pattern Anal. Mach. Intell. **40**(4), 834–848 (2018)
17. Peng, C., Zhang, X., Yu, G., Luo, G., Sun, J.: Large kernel matters improve semantic segmentation by global convolutional network. In: Proceedings of the IEEE Conference on Computer Vision and Pattern Recognition (CVPR), pp. 1743–1751. IEEE (2017)
18. Kamnitsas, K., et al.: Efficient multi-scale 3D CNN with fully connected CRF for accurate brain lesion segmentation. MIA **36**, 61–78 (2017)
19. Myronenko, A.: 3D MRI brain tumor segmentation using autoencoder regularization. In: Crimi, A., Bakas, S., Kuijf, H., Keyvan, F., Reyes, M., van Walsum, T. (eds.) BrainLes 2018. LNCS, vol. 11384, pp. 311–320. Springer, Cham (2019). https://doi.org/10.1007/978-3-030-11726-9_28
20. Isensee, F., Kickingereder, P., Wick, W., Bendszus, M., Maier-Hein, K.H.: No new-net. In: Crimi, A., Bakas, S., Kuijf, H., Keyvan, F., Reyes, M., van Walsum, T. (eds.) BrainLes 2018. LNCS, vol. 11384, pp. 234–244. Springer, Cham (2019). https://doi.org/10.1007/978-3-030-11726-9_21
21. Hu, J., Shen, L., Sun, G.: Squeeze-and-excitation networks. In: The IEEE/CVF Conference on Computer Vision and Pattern Recognition (CVPR), pp. 7132–7141 (2018)
22. Ronneberger, O., Fischer, P., Brox, T.: U-Net: convolutional networks for biomedical image segmentation. In: Navab, N., Hornegger, J., Wells, W.M., Frangi, A.F. (eds.) MICCAI 2015. LNCS, vol. 9351, pp. 234–241. Springer, Cham (2015). https://doi.org/10.1007/978-3-319-24574-4_28

Overall Survival Prediction Using Conventional MRI Features

Yanhao Ren[1], Pin Sun[2], and Wenlian Lu[1(✉)]

[1] School of Mathematical Sciences, Fudan University,
No. 220, Handan Road, Shanghai, China
{18110840015,wenlian}@fudan.edu.cn
[2] Department of Neurosurgery, Huashan Hospital, Shanghai Medical College,
Fudan University, Shanghai, China

Abstract. Gliomas are common primary brain malignancies. The sub-regions of gliomas are depicted by MRI scans, reflecting varying biological properties. These properties have effect on the diagnosis of neurosurgeons on whether or what kind of resection should be done. The survival days after gross total resection is also of great concern. In this paper, we propose a semi-auto method for segmentation, and extract features from slices of MRI scans, including conventional MRI features and clinical features. 13 features of a subject are selected finally and a support vector regression is used to fit with the training data.

Keywords: MRI scans · Semi-auto segmentation · Survival prediction · Conventional MRI features · Support vector regression

1 Introduction

Gliomas are the most common primary brain malignancies, with different degrees of aggressiveness, variable prognosis and various heterogeneous histologic sub-regions, i.e., edematous tissue, necrotic core, active and non-enhancing core. The intrinsic heterogeneity of glioma is also portrayed in their radio-phenotype, as their sub-regions are depicted by intensity profiles disseminated across mpMRI scans, reflecting varying biological properties.

The MRI scans that are commonly utilized for study include 4 contrasts: native (T1), post-contrast T1-weighted (T1Gd), T2-weighted (T2) and Fluid Attenuated Inversion Recovery (FLAIR) volumes. In BraTS'19, 259 training subjects of high grade glioma (HGG) and 75 of low grade glioma (LGG) are included. The annotated sub-region of necrotic core and non-enhancing tumor (NCR), edematous tissue (ED), and active tumor (AT), as well as the resection status and patients' age are accessible [2, 8, 9]. The study includes three tasks: a) Segmentation of gliomas. b) Prediction of patient overall survival (OS). c) Quantification of Uncertainty in Segmentation.

© Springer Nature Switzerland AG 2020
A. Crimi and S. Bakas (Eds.): BrainLes 2019, LNCS 11993, pp. 244–254, 2020.
https://doi.org/10.1007/978-3-030-46643-5_24

Segmentation is a common problem in MRI. Convolutional neural networks (CNN), especially the structure of U-Net, is commonly used to deal with the problem [5–7]. 3D U-Net is commonly used in the previous BraTS' competition [5, 6]. But sometimes, the prediction of segmentation using the trained U-Net is not that accurate, especially in medical imaging. In this paper, we propose a semi-auto segmentation method, with the help of both neural networks and the neurosurgeon.

Some state-of-the-art studies on glioma using MRI scans show that machine learning algorithms, and also CNN, can be utilized in classification tasks, such as 1p/19q codeletion [1], GBM vs supra-bMET [10], H3K27M mutation [4] and IDH mutation [3]. Studies in these literatures often utilize not only features of MRI scans, but also several clinical features to build their model and achieve a good performance [1, 3–5, 10]. In these studies, some conventional MRI features are often extracted without the help of neural networks [4, 5]. These features include texture, shape, sharpness, pixel intensity, etc., considering both 3D and 2D slices. These features and machine learning methods can also be utilized to BraTS' overall survival task on patients with gross total resection. In this paper, we extract 13 features, including 10 MRI scan features and 3 clinical features for each subject, and use an SVR (support vector regression) model to fit the training data of 210 subjects and their label of survival days.

2 Materials and Methods

2.1 Datasets

During segmentation, all of the 335 subjects in the training data are included. While predicting the overall survival, only the 210 subjects of HGG (without the age label missing) are included. While dealing with the survival model, we also slice the MRI scans along the z-axis to get 2D slices for further feature extraction using the Python package Nibabel.

2.2 The Semi-auto Segmentation

The segmentation on the validation set is conducted using a semi-auto method. A segmentation model is trained using the 335 training data on a 3D U-Net. The training is end-to-end. The input size is $240 \times 240 \times 155 \times 4$ by fusing the four contrasts of the same subject and the output is the segmentation results on each voxel. On the validation set, the segmentation is firstly predicted using the trained 3D U-Net and then tuned or re-drawn by the neurosurgeon. We slice each MRI scan along the z-axis an get 155 slices for each MRI scan, and we only use one slice, called the 'most important' slice, and the final segmentation result on that slice, for survival prediction.

The approach to get the 'most important' slice and its segmentation result is conducted through the following two steps:

Step 1: Finding the 'most important' slice

We call the union of NCR and AT the tumor region (**TR**). Step 1 is conducted by the following sub steps:

1) The neurosurgeon checks the prediction of the segmentation by 3D U-Net on each slice (the check is done on FLAIR). The region of ED, NCR, AT are colored automatically by three different colors on FLAIR.
2) i) If the neurosurgeon judges that the segmentation in 1) is accurate enough, then we choose the slice that the area of TR is the largest, as the 'most important' slice.
3) ii) If the neurosurgeon judges the segmentation in 1) is not accurate enough, then he check all the slices (on FLAIR) artificially and choose a slice that is considered has the largest area of TR as the 'most important' slice.

After the 'most important' slice is chosen, we also take out the same slice of T1Gd and T2, and draw the segmentation result by the 3D U-Net automatically. We draw ED on FLAIR (in green), NCR on T1Gd (in yellow), and TR on T2 (in red), which are shown in Figs. 2, 3, 4, and 5. Then we move to Step 2.

Step 2: Getting the final segmentation result on the 'most important' slice for survival prediction

In this step, the neurosurgeon judges the segmentation result on the 'most important' slice. If it is accurate enough, then the neurosurgeon will not change the result or only just make a fine tune (Note that the judgment in Step 2 is much stricter than in Step 1 by the neurosurgeon. The judgment in Step 1 is general, but in Step 2 the result will be used for further calculation). If not, the neurosurgeon draws the region of ED, NCR and AT manually. The neurosurgeon involved in this study is P. Sun from Department of Neurosurgery, Huashan Hospital, Shanghai Medical College, Fudan University, Shanghai, China.

After Step 1 and Step 2, the segmentation result on the 'most important' slice is ready for the prediction of survival, including a FLAIR image with ED (green), a T1Gd image with NCR (yellow) and a T2 image with TR (red). A flow chart of getting the 'most important' slice and its segmentation result is shown in Fig. 1. The reason why we use the semi-auto segmentation and only one slice for survival prediction is that we take a synthetical consideration of the limit of time and accuracy. Among the cases that may happen in the flow chart, there are some examples, shown in Figs. 2, 3, 4, and 5, to explain how the neurosurgeon check and modify the results of the segmentation.

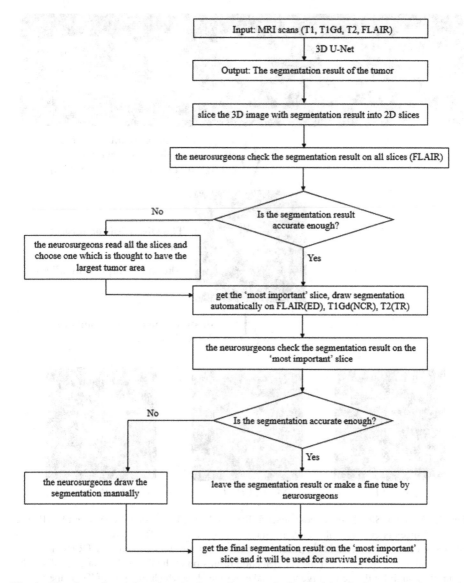

Fig. 1. The flow chart of getting the 'most important' slice and its final segmentation result according to Step 1 and Step 2.

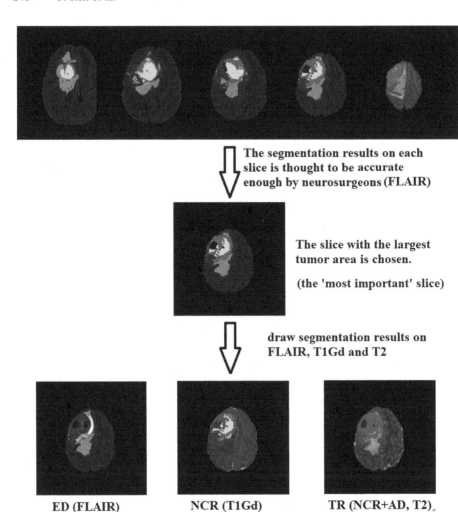

ED (FLAIR)   **NCR (T1Gd)**   **TR (NCR+AD, T2)**

Fig. 2. A case of Step 1. We choose several slices of the prediction of 3D U-Net on FLAIR. The image of slices is colored automatically, shown in the first row. The predicted ED (label 2) is colored green, the predicted NCR (label 1) is colored yellow and the predicted TR (NCR + AT, label 1 and 4) is colored red. If the segmentation results on each slice is thought generally to be accurate enough by the neurosurgeon, we take out the slice which the area of TR is the largest (shown in the second row). Then we take out the same slice in T1Gd and T2, and draw the segmentation (shown in the third row) separately with ED on FLAIR, NCR on T1Gd and TR on T2 automatically. The three images shown the third row are the ones that will be used for Step 2. (Color figure online)

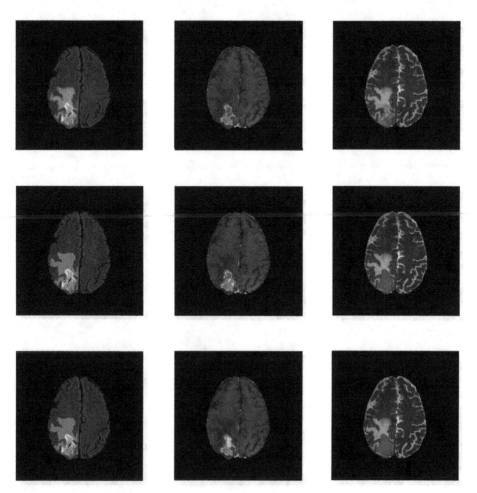

Fig. 3. A case of Step 2. The first row is the 'most important' slice without color. The second row is what we get in Step 1. In this case, the segmentation result is accurate enough both on all slices generally and on the 'most important' slice. The neurosurgeon only fine tune the results (in the case only the NCR, yellow), and the images in the third row is the final segmentation results of the 'most important' slice that will be used for survival prediction. (Color figure online)

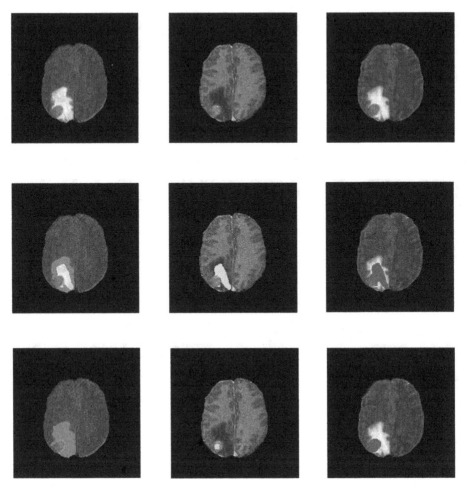

Fig. 4. A case of Step 2. In this case, the segmentation result of the 3D U-Net is thought to be accurate enough generally after reading all the slices by the neurosurgeon, but the segmentation result of the 'most important' slice is not accurate enough (shown in the second row). So the neurosurgeon draw the tumor region manually, shown in the third row, which will be used for survival prediction. (Color figure online)

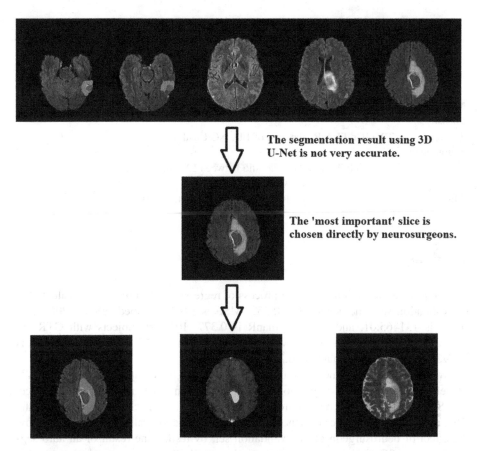

The segmentation result using 3D U-Net is not very accurate.

The 'most important' slice is chosen directly by neurosurgeons.

Fig. 5. A case of Step 1 and Step 2. In this case, the segmentation result of the 3D U-Net is thought to be not accurate enough generally (shown in the first row). The neurosurgeon then read all the slices again and choose one as the 'most important' slice (shown in the second row). Since the segmentation on this slice is not accurate, the neurosurgeon draw the segmentation manually (shown in the the third row), and the result is used for survival prediction. (Color figure online)

2.3   Overall Survival Prediction

To predict the survival days, we extract 13 features, including 10 conventional MRI features and 3 clinical features.

The 10 conventional MRI features are from the 'most important' slice, which includes the area of ED, TR and NCR, the pixel variance of ED, TR and NCR, the mean pixel intensity of TR and NCR, and in addition, the area ratio between ED and TR, and the area ratio between NCR and TR. While calculation, we use the results of Step 1 and Step 2 in Sect. 2.2, which means we calculate the pixel mean and variance of TR using T2, NCR using T1Gd, and so on. The 3 clinical features include GTR (1, 0), STR (0, 1), NA (0, 0), and the age of the patient. The details of the features we choose are listed in Table 1. All of the 210 patients with HGG are used for training, and an SVR model is used to fit the features and the survival days.

Table 1. Features selected in survival prediction.

Feature type	MRI scan concerned	Feature in detail	No. of feature
Pixel variance	T2/T1Gd/FLAIR	ED-FLAIR, NCR-T1Gd, TR-T2	3
Mean pixel intensity	T2/T1Gd	NCR-T1Gd, TR-T2	2
Tumor region area	T2/T1Gd/FLAIR	the area of ED, NCR and TR	3
Area ratio	T2/T1Gd/FLAIR	the area ratio between ED and TR, the area ratio between NCR and TR	2
Clinical	/	GTR (1, 0), STR (0, 1), NA (0, 0), age	3
Total features	13		

3 Results

210 subjects are used in the training process of regression. 28 subjects are evaluated in the validation set. The accuracy is 0.552, MSE is 84971.77, medianSE is 20877.69, stdSE is 131465.01, and the SpearmanR is 0.377. 107 test subjects with GTR are evaluated, the accuracy is 0.402, MSE is 444135.3, medianSE is 50565.43, stdSE is 1269290, and the SpearmanR is 0.34.

We also tried other models on the validation set, for comparison with the SVR model, including linear regression, linear regression with normalization (Ridge, LASSO), and RVR. The results are shown in Table 2. As is shown in the table, the SVR model has the best performance on the validation set overall. In addition, we show the effect of neurosurgeon on the validation set, by testing the result of an auto segmentation by 3D U-Net for segmentation plus an SVR model for survival prediction (without any help of neurosurgeon, the 'most important' slice is the slice with the largest TR area, and without Step 2 in Sect. 2.2). The results are also shown in Table 2. It is clear that the semi-auto segmentation method has some improvement to the survival prediction.

Table 2. Results on the validation set using different models

Model	Accuracy	MSE	medianSE	stdSE	SpearmanR
SVR	0.552	84971.77	20877.69	131465.01	0.377
Linear regression	0.517	114559.70	38298.11	188282.30	0.531
Ridge	0.552	121394.71	39213.48	198252.62	0.531
LASSO	0.517	114267.78	39466.44	186117.86	0.523
RVR	0.483	87749.99	24457.27	149333.11	0.496
SVR (without neurosurgeon)	0.448	125667.00	80902.05	143125.60	0.184

4   Discussion

In this paper, we propose a semi-auto segmentation method and an SVR model for predicting overall survival. The segmentation result is firstly predicted by a trained 3D U-Net model and is then partially modified by the neurosurgeon. For the survival task, with the help of segmentation results and the chosen 'most important' slice, we extract conventional MRI features and clinical features, and use an SVR model with 13 features as input.

There is a gap between the result of the validation set and test set, the reason may be considered as the following:

(1) In order to increase the number of training set, we use resection status of GTR, STR and NA subjects for training, that may lead to some problem when only the survival of GTR subjects are needed to be predicted.
(2) The SVR model may overfit. As the only way to see the accuracy of our model is the result on validation set, we tried many models, but the SVR model performs the best on the validation set.
(3) The validation set is small, so the result has some randomness.
(4) Some other objective reasons. For example, we often find it hard to distinguish short and mid survivors, and the mid and long survivors, which may be caused by some other individual differences that cannot be seen on MRI scans.

Because the training data is small, we finally remove some of the features and only choose the ones that are considered possible to affect the survival days under the suggestion of the neurosurgeon. Some other clinical features, such as the growth rate of glioma and the patient's genetic features, are also considered essential, but unfortunately, they are not accessible in the study. Our future work includes extracting more potential affective features, extracting features using neural networks, and also ensemble learning to deal with the survival prediction problem better.

References

1. Akkus, Z., et al.: Predicting deletion of chromosomal arms 1p/19q in low-grade gliomas from MR images using machine intelligence. J. Digit. Imaging **30**(4), 469–476 (2017). https://doi.org/10.1007/s10278-017-9984-3
2. Menze, B.H., Jakab, A., Bauer, S., Kalpathy-Cramer, J., Farahani, K., Kirby, J., et al.: The multimodal brain tumor image segmentation benchmark (BRATS). IEEE Trans. Med. Imaging **34**(10), 1993–2024 (2015). https://doi.org/10.1109/TMI.2014.2377694
3. Carrillo, J., Lai, A., Nghiemphu, P., et al.: Relationship between tumor enhancement, edema, IDH1 mutational status, MGMT promoter methylation, and survival in glioblastoma. Am. J. Neuroradiol. **33**, 1349–1355 (2012)
4. Pan, C.-C., et al.: A machine learning-based prediction model of H3K27M mutations in brainstem gliomas using conventional MRI and clinical features. Radiother. Oncol. **130**, 172–179 (2018). https://doi.org/10.1016/j.radonc.2018.07.011

5. Feng, X., Tustison, N., Meyer, C.: Brain tumor segmentation using an ensemble of 3D U-nets and overall survival prediction using radiomic features. In: Crimi, A., Bakas, S., Kuijf, H., Keyvan, F., Reyes, M., van Walsum, T. (eds.) BrainLes 2018. LNCS, vol. 11384, pp. 279–288. Springer, Cham (2019). https://doi.org/10.1007/978-3-030-11726-9_25

6. Myronenko, A.: 3D MRI brain tumor segmentation using autoencoder regularization. In: Crimi, A., Bakas, S., Kuijf, H., Keyvan, F., Reyes, M., van Walsum, T. (eds.) BrainLes 2018. LNCS, vol. 11384, pp. 311–320. Springer, Cham (2019). https://doi.org/10.1007/978-3-030-11726-9_28

7. Ronneberger, O., Fischer, P., Brox, T.: U-Net: convolutional networks for biomedical image segmentation. arXiv preprint arXiv:1505.04597 (2015)

8. Bakas, S., Akbari, H., Sotiras, A., Bilello, M., Rozycki, M., Kirby, J.S., et al.: Advancing the cancer genome atlas glioma MRI collections with expert segmentation labels and radiomic features. Nat. Sci. Data **4** (2017). Article number: 170117. https://doi.org/10.1038/sdata.2017.117

9. Bakas, S., Reyes, M., Jakab, A., Bauer, S., Rempfler, M., Crimi, A., et al.: Identifying the best machine learning algorithms for brain tumor segmentation, progression assessment, and overall survival prediction in the BRATS challenge. arXiv preprint arXiv:1811.02629 (2018)

10. Zhou, C., et al.: Segmentation of peritumoral oedema offers a valuable radiological feature of cerebral metastasis. Br. J. Radiol. (2016). https://doi.org/10.1259/bjr.20151054

A Multi-path Decoder Network for Brain Tumor Segmentation

Yunzhe Xue[1], Meiyan Xie[1], Fadi G. Farhat[1], Olga Boukrina[2], A. M. Barrett[3], Jeffrey R. Binder[4], Usman W. Roshan[1(✉)], and William W. Graves[5]

[1] Department of Computer Science, New Jersey Institute of Technology, Newark, NJ, USA
usman@njit.edu
[2] Stroke Rehabilitation Research, Kessler Foundation, West Orange, NJ, USA
[3] Emory University and Atlanta VA Medical Center, Atlanta, GA, USA
[4] Department of Neurology, Medical College of Wisconsin, Milwaukee, WI, USA
[5] Department of Psychology, Rutgers University – Newark, Newark, NJ, USA

Abstract. The identification of brain tumor type, shape, and size from MRI images plays an important role in glioma diagnosis and treatment. Manually identifying the tumor is time expensive and prone to error. And while information from different image modalities may help in principle, using these modalities for manual tumor segmentation may be even more time consuming. Convolutional U-Net architectures with encoders and decoders are state of the art in automated methods for image segmentation. Often only a single encoder and decoder is used, where different modalities and regions of the tumor share the same model parameters. This may lead to incorrect segmentations. We propose a convolutional U-Net that has separate, independent encoders for each image modality. The outputs from each encoder are concatenated and given to separate fusion and decoder blocks for each region of the tumor. The features from each decoder block are then calibrated in a final feature fusion block, after which the model gives it final predictions. Our network is an end-to-end model that simplifies training and reproducibility. On the BraTS 2019 validation dataset our model achieves average Dice values of 0.75, 0.90, and 0.83 for the enhancing tumor, whole tumor, and tumor core subregions respectively.

Keywords: Convolutional neural networks · Multi-modal · Brain MRI

1 Introduction

Gliomas are the most commonly occurring tumor in the human central nervous system [1]. They fall into low-grade (LGG) and high-grade (HGG) subtypes and have three subregions: the enhanced tumor, tumor core, and whole tumor. These regions show up with different intensities and areas across different image modalities [2]. The tumor core (TC) subregion shows the bulk of the tumor and is

A. Crimi and S. Bakas (Eds.): BrainLes 2019, LNCS 11993, pp. 255–265, 2020.
https://doi.org/10.1007/978-3-030-46643-5_25

typically removed. The TC contains the enhanced tumor and necrotic fluid-filled (NCR) and the non-enhancing solid parts (NET) of the tumor. These also show up with different intensities across image modalities. The whole tumor subregion describes the entire tumor since it contains the TC and the peritumoral edema (ED), which is typically depicted by hyper-intense signal in FLAIR.

Given the complexity of the tumor and different image modalities, manual identification of the tumor subregions takes time and is prone to error. Automated methods would facilitate physician diagnosis and lead to better overall patient treatment. A step towards this is the Multimodal Brain Tumor Segmentation (BraTS) challenge [2–5] that invites automated solutions to predict the three tumor subregions from images across four different modalities. It provides 335 patient samples, include 260 HGG cases and 75 LGG cases, each with four MRI modalities: T1, T1 contrast-enhanced (T1ce), T2, and FLAIR. Each image in this dataset has been pre-processed by the same method and rescaled to $1 \times 1 \times 1$ mm isotropic resolution and skull-stripped. The dataset also provides ground truth segmentations of the three subregions. Two additional datasets whose ground truth are unavailable to us are used for validation and test.

Inspired by the success of convolutional neural networks in image recognition tasks, we present a convolutional U-neural network (U-Net) solution to this problem. Our model has multiple encoders for each modality and multiple decoders for each tumor subregion. Below we describe our model in detail, followed by variants of our model and final accuracies on the challenge's validation dataset.

2   Related Work

The Convolutional U-Net [6] is the basic architecture for end-to-end semantic segmentation. In previous years of the BraTS competition, researchers have improved upon the basic UNet and addressed training overfitting problems on small datasets. For example, we have 2.5D multi-stage segmentation of different anatomical views [7] and 3D segmentation models [8–10]. The winning entry in the BraTS 2018 contest used an auto-encoder to regularize their network's encoder to prevent over-fitting caused by small sample size [11]. Most teams in the previous contest performed model ensembling or second phase correction to integrate the outputs of multiple models for better final results.

The multi-path approach that has separate encoders for different image types has been explored previously [12]. We have taken this approach further in our previous work [7,8] with weighted feature fusion blocks to combine features from different modality encoders. That approach works well for binary segmentation of the brain into healthy and non-healthy tissue. For multiple regions (also known as multi-class segmentation), however, a single fusion block may not work because it uses one set of shared weights for multiple subregions of the brain. The squeeze-and-excite block [13] that we have also used in previous work assigns weights to channel features from different encoders. However, those encoders still fed into the same fusion block. Considering that the different subregions of glioma have different intensities in different modal images, we expect that different subregions will require different weights in the feature fusion stage. Thus we propose a new model with separate fusion and decoder blocks for each subregion of the tumor to be segmented.

3 Methods

3.1 3D Convolutional Multi-encoder Multi-decoder Neural Network

In Fig. 1 we show the overview of our model. We see separate encoders for each of the four image modalities. Each decoder consists of convolutional and transposed convolutional (also called deconvolutional) blocks. We use three decoders corresponding to the three categories: enhanced tumor (ET), tumor core regions (NCR/NET), and the peritumoral edema (ED) within the whole tumor subregion.

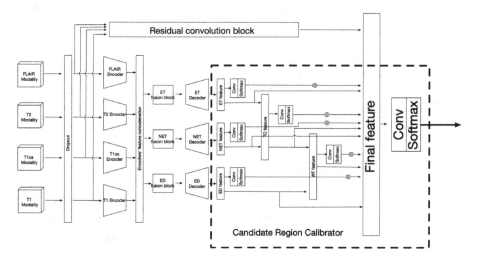

Fig. 1. Our multi-path network with independent encoders for each modality and independent decoders for each tumor subregion

In Fig. 1 we see a candidate region calibrator after the outputs of each subregion's decoder. Here we account for the hierarchical relationship between different tumor subregions. We send the output of each decoder to a simple convolutional block (without activation) that outputs a probability. We then multiply the probability by the decoder's output and concatenate it to the outputs of the other three decoders. We concatenate features from the ET and NCR/NET decoders to account for the TC subregion features. We also concatenate TC and ED subregion features to get features for the whole tumor (WT) subregion. We run the outputs from the TC and WT layers through simple convolutional blocks that output probabilities and multiply their outputs by the probabilities.

The purpose of our calibrator is to give specific attention to individual sub-regions as well as their larger combined parts, which we achieve by multiplying their outputs by probabilities as described above. The different subregions of gliomas do not necessarily appear in all four modal images. Therefore, we randomly set one of the four modality inputs to zero with probability 0.25 during the training process. We call this modality dropout that we describe in detail below.

Encoder. Our overall encoder shown in Fig. 2 downsamples the input image four times. In each downsampling we double the number of output filters and send the output to the feature fusion component. The encoder consists of residual convolution blocks shown in Fig. 3(c) and downsampling modules in Fig. 3(b). We achieve independent encoder training for different modality inputs using group convolution. We use instance normalization as the last component in the residual block so that the output follows a normal distribution. This avoids gradient problems in feature addition part in the decoder (that we identified and solved previously [8]).

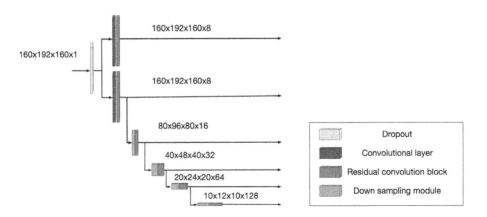

Fig. 2. We downsample the input image four times and collect the output after each residual convolutional block to give to the fusion block.

Feature Fusion. The output from each downsampling level of the encoder is given to the feature fusion block as shown in Fig. 4. The feature fusion block shown in Fig. 3(a) integrates features from different modal encoders with a $1 \times 1 \times 1$ convolution. Prior to integration we use the squeeze and effect module to give different weights to the input feature channels.

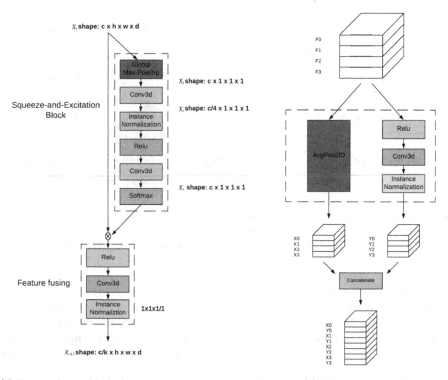

(a) Feature fusion block that contains squeeze and excite   (b) Downsampling block

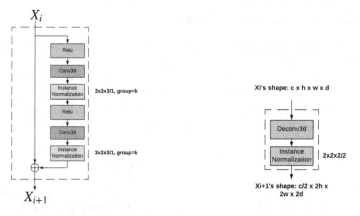

(c) Residual convolutional block   (d) Decoder upsampling block

Fig. 3. We show a detailed description of different components of our network as used in the overall network, encoder, and decoder.

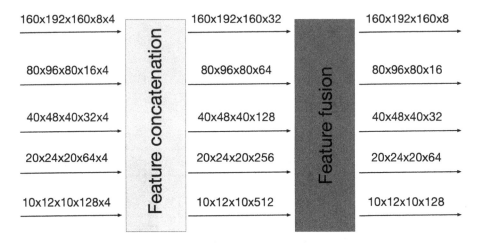

Fig. 4. Features from each downsampling layer of the encoder are collected for fusion. The first set on the left are features from the four encoders for each modality. These are first concatenated and then fused with a 3D convolutional kernel as shown in Fig. 3(a).

Decoder. The decoder in Fig. 5 consists of a residual convolutional block shown in Fig. 3(b) followed by a transposed convolutional block (also called upsampling or deconvolutions) shown in Fig. 3(d). We add features from the upsampling module to the output of the fusion block before sending them to further upsampling blocks. The output of the decoder is given to the candidate region calibrator.

3.2   Model Training and Parameters

Loss Function. We measure the Dice loss of predictions of each the three subregions ET, NCR/NET, and ED after their output from their respective decoder is given to the convolutional block followed by softmax. We also measure the loss of predictions of the TC and WT subregions in the candidate region calibrator after their outputs are passed through the convolutional block followed by softmax. For a given segmentation the Dice loss is defined to

$$D(p) = \frac{2 \sum_i p_i r_i}{\sum_i p_i^2 + \sum_i r_i^2},$$

where p_i are the predicted softmax outputs and r_i is 1 if the voxel has a lesion and 0 otherwise. We also have a final multi-class Dice loss for the three subregions. Our overall loss is the sum of the five Dice losses for the ET, NCT/NET, ED, TC, and WT subregions plus the final multi-class Dice.

Implementation and Optimization. We implement our network using the Pytorch library [14]. We use stochastic gradient descent (SGD) with Nesterov

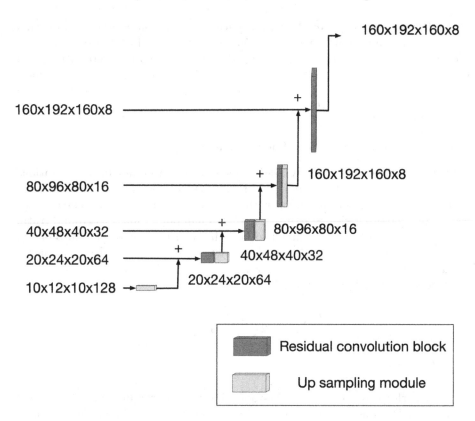

Fig. 5. Features from the fusion block are added to the output of the upsampling layer.

momentum. We set momentum to 0.9, our initial learning rate to 0.01 and number of epochs at 240. We decrease the learning rate if the current epoch's average loss is no less than the previous epoch's loss. We also use the Pytorch extension library called NVIDIA-apex for mixed precision (16-bit and 32-bit floats) and distributed training across multiple GPUs [15].

Dropout. Dropout is a popular technique to prevent overfitting in neural networks [16]. We perform an image-modality dropout: we randomly pick an image modality and set its input to all zeros. In other words we randomly ignore one out of the four image modalities. One reason for doing this is that some of the tumor subregions are visible only in some modalities. For example enhanced tumor (ET) subregion is visible in T1ce images while the ED subregion in T2 or FLAIR only. We achieve this dropout with the 3D dropout function in Pytorch [14]. Alternatively we could ignore more than one modality but we found this to lower the cross-validation accuracy than when zeroing just one.

Data Preprocessing and Augmentation. We randomly crop each original size image volume from $240 \times 240 \times 155$ to $160 \times 192 \times 144$. We then perform a mean 0 variance 1 normalization (including background zero intensity pixels) for each modality and each 3D image volume. We randomly flip each 3D image in the three view directions with probability 0.5. In the inference part, we do a center crop of size $160 \times 192 \times 160$ and zero pad to original size without flip.

3.3   Measure of Accuracy: Dice Coefficient

The Dice coefficient is typically used to measure the accuracy of segmentations in MRI images [17]. The output of our network is a binary mask of the same dimensions as the input image, but with a 1 for each voxel prediced to be a tumor region, and a 0 otherwise. Starting with the human binary mask as ground truth, each predicted voxel is determined to be either a true positive (TP, also one in true mask), false positive (FP, predicted as one but zero in the true mask), or false negative (FN, predicted as zero but one in the true mask). The Dice coefficient is formally defined as

$$DICE = \frac{2TP}{2TP + FP + FN}.$$

4   Results

We first performed a five-fold cross-validation on the training dataset provided by the BraTS consortium. This dataset contains four modality images along with the segmentations of the three tumor subregions. In Table 1 we see the average Dice accuracies and other statistics of our training samples obtained under cross-validation.

Table 1. Average Dice values of our model for each of the three tumor regions after 5-fold cross validation on the training dataset (total of 335 patients).

	Dice			Sensitivity			Specificity			Hausdorff95		
	ET	WT	TC	ET	WT	TC	ET	WT	TC	ET	WT	TC
Mean	0.75	0.90	0.84	0.81	0.90	0.83	1.00	0.99	1.00	5.34	5.67	6.06
Std. dev	0.26	0.07	0.17	0.21	0.09	0.18	0.00	0.01	0.00	10.34	8.31	9.11
Median	0.85	0.92	0.90	0.88	0.93	0.90	1.00	1.00	1.00	1.73	3.38	3.00

We then evaluated our model on the validation dataset provided by BraTS. This dataset contains only the four modality images without ground truth segmentations. To obtain the validation accuracies we uploaded our predicted segmentations to the BraTS server. In Table 2 we see our statistics for the validation dataset returned by the BraTS server. We see that the Dice mean and median values are similar in both training cross-validation and the validation dataset, which shows that our model is generalizing.

To obtain a sense of our method's performance relative to others we evaluate the rank of our validation data Dice accuracies on the BraTS 2019 challenge leaderboard on their web page. At the time of writing of our paper there were a 113 submissions on all 125 validation samples. We found that on the tumor core (TC) Dice measure our method stood at rank 16 from the top. On the whole tumor (WT) and enhanced tumor (ET) Dice measures our method stood at ranks 22 and 34 respectively. Thus, while not amongst the top three, our method obtained segmentations better than most other submissions.

Table 2. Average Dice values of our model of each of the three tumor regions on the validation dataset provided by BraTS (total of 125 patients)

	Dice			Sensitivity			Specificity			Hausdorff95		
	ET	WT	TC	ET	WT	TC	ET	WT	TC	ET	WT	TC
Mean	0.75	0.90	0.83	0.76	0.93	0.81	1.00	0.99	1.00	5.07	6.13	6.77
Std. dev	0.28	0.08	0.16	0.27	0.07	0.20	0.00	0.01	0.00	12.64	12.18	11.44
Median	0.85	0.92	0.89	0.85	0.95	0.89	1.00	0.99	1.00	2.24	3.16	3.39

In Table 3 we show the average Dice accuracies and the Hausdorff distance [18] on the test data. These were provided to us by the conference organizers since the test data is unavailable to all participants. We see that the average and median Dice accuracies are similar to what we obtained in cross-validation and validation testing above, thus further supporting our model's generalizability.

Table 3. Average Dice values of our model of each of the three tumor regions on the test dataset provided by BraTS

	Dice			Hausdorff95		
	ET	WT	TC	ET	WT	TC
Mean	0.8	0.88	0.83	2.2	4.89	4.09
Std. dev	0.21	0.12	0.24	2.1	5.8	6.8
Median	0.85	0.91	0.92	1.41	3.08	2.45

5   Discussion and Conclusion

We present a multi-encoder and multi-decoder convolutional neural network to handle different image modalities and predict different subregions of the input image (multi-class segmentation). We build upon previous work [8] where we used multiple encoders and squeeze-and-excite blocks [13] to give weights to different modalities. However, in that previous work we used a single feature fusion block that shares weights for different subregions, which is not as accurate as the separate fusion and decoder blocks we developed in this study.

The squeeze-and-excite blocks that we use here are designed for classification. Our global average pooling considers the entire feature map. For tumor subregion segmentation this may not be the best approach since the tumor region is given by just a small region of the full feature map. Thus in future work we plan to develop squeeze-and-excite to give better modality weights based on just the tumor region instead of the entire feature map.

With 3D components our model is more challenging to train than a 2D one. The additional parameters introduced by 3D typically require more data as well as memory and runtime to train. With the additional parameters the model may overlift and so careful training is required. In comparison a 2D model would be easier and faster to train and require less memory but may not be as accurate as a 3D one.

References

1. Goodenberger, M.L., Jenkins, R.B.: Genetics of adult glioma. Cancer Genet. **205**(12), 613–621 (2012)
2. Bakas, S., et al.: Identifying the best machine learning algorithms for brain tumor segmentation, progression assessment, and overall survival prediction in the brats challenge. arXiv preprint arXiv:1811.02629 (2018)
3. Menze, B.H., et al.: The multimodal brain tumor image segmentation benchmark (BRATS). IEEE Trans. Med. Imaging **34**(10), 1993–2024 (2014)
4. Bakas, S., et al.: Advancing the cancer genome atlas glioma MRI collections with expert segmentation labels and radiomic features. Sci. Data **4** (2017). Article number: 170117
5. Bakas, S., et al.: Segmentation labels and radiomic features for the pre-operative scans of the TCGA-GBM collection. The Cancer Imaging Archive 2017 (2017)
6. Ronneberger, O., Fischer, P., Brox, T.: U-Net: convolutional networks for biomedical image segmentation. In: Navab, N., Hornegger, J., Wells, W.M., Frangi, A.F. (eds.) MICCAI 2015. LNCS, vol. 9351, pp. 234–241. Springer, Cham (2015). https://doi.org/10.1007/978-3-319-24574-4_28
7. Xue, Y., et al.: A multi-path 2.5 dimensional convolutional neural network system for segmenting stroke lesions in brain MRI images. arXiv preprint arXiv:1905.10835 (2019)
8. Xue, Y., et al.: A fully 3D multi-path convolutional neural network with feature fusion and feature weighting for automatic lesion identification in brain MRI images (2019, Submitted)
9. Kamnitsas, K., et al.: Efficient multi-scale 3D CNN with fully connected CRF for accurate brain lesion segmentation. Med. Image Anal. **36**, 61–78 (2017)
10. Chen, H., Dou, Q., Lequan, Y., Qin, J., Heng, P.-A.: VoxResNet: deep voxelwise residual networks for brain segmentation from 3D MR images. NeuroImage **170**, 446–455 (2018)

11. Myronenko, A.: 3D MRI brain tumor segmentation using autoencoder regularization. In: Crimi, A., Bakas, S., Kuijf, H., Keyvan, F., Reyes, M., van Walsum, T. (eds.) BrainLes 2018. LNCS, vol. 11384, pp. 311–320. Springer, Cham (2019). https://doi.org/10.1007/978-3-030-11726-9_28

12. Tseng, K.-L., Lin, Y.-L., Hsu, W., Huang, C.-Y.: Joint sequence learning and cross-modality convolution for 3D biomedical segmentation. In: 2017 IEEE Conference on Computer Vision and Pattern Recognition (CVPR), pp. 3739–3746. IEEE (2017)

13. Hu, J., Shen, L., Sun, G.: Squeeze-and-excitation networks. In: Proceedings of the IEEE Conference on Computer Vision and Pattern Recognition, pp. 7132–7141 (2018)

14. Paszke, A., et al.: Automatic differentiation in pytorch. In: NIPS-W (2017)

15. Micikevicius, P., et al.: Mixed precision training. arXiv preprint arXiv:1710.03740 (2017)

16. Srivastava, N., Hinton, G., Krizhevsky, A., Sutskever, I., Salakhutdinov, R.: Dropout: a simple way to prevent neural networks from overfitting. J. Mach. Learn. Res. **15**(1), 1929–1958 (2014)

17. Zijdenbos, A.P., Dawant, B.M., Margolin, R.A., Palmer, A.C.: Morphometric analysis of white matter lesions in MR images: method and validation. IEEE Trans. Med. Imaging **13**(4), 716–724 (1994)

18. Rockafellar, R.T., Wets, R.J.-B.: Variational Analysis, vol. 317. Springer, Cham (2009)

The Tumor Mix-Up in 3D Unet
for Glioma Segmentation

Pengyu Yin, Yingdong Hu, Jing Liu, Jiaming Duan, Wei Yang,
and Kun Cheng[✉]

Beijing University of Posts and Telecommunications, Beijing, China
kcheng@bupt.edu.cn

Abstract. Automated segmentation of glioma and its subregions has
significant importance throughout the clinical work flow including diagnosis, monitoring and treatment planning of brain cancer. The automatic delineation of tumours have draw much attention in the past few
years, particularly the neural network based supervised learning methods. While the clinical data acquisition is much expensive and time consuming, which is the key limitation of machine learning in medical data.
We describe a solution for the brain tumor segmentation in the context
of the BRATS19 challenge. The major learning scheme is based on the
3D-Unet encoder and decoder with intense data augmentation followed
by bias correction. At the moment we submit this short paper, our solution achieved Dice scores of 76.84, 85.74 and 74.51 for the enhancing
tumor, whole tumor and tumor core, respectively on the validation data.

Keywords: Tumor segmentation · Data augmentation · 3D Unet ·
Mixup

1 Introduction

Glioblastoma is the most common and lethal intracranial tumor, the delineation
of tumor volume has great clinical importance in the disease diagnosis and treatment planning and prognosis [4]. While manual definition of tumor volumes is an
expert only task, which is also time consuming and somehow objective. Thanks to
the development of high-performance computing and advances in medical imaging industry, the automatic segmentation algorithms particularly neural network
based machine learning methods, have achieved promising results in the past
decades. Magnetic Resonance Imaging (MRI) provides good soft tissue contrast
of brain tumors without exposing the patient to radiations, consequently widely
used in the brain tumor diagnosis in the clinical [11]. Since 2012 the brain tumor
segmentation challenge (BraTS) offers great benchmark MRI datasets for evaluating the state-of-the-art brain tumor segmentation algorithms [1,2,13]. The
segmentation task is to identify glioma subregions including tumor core (TC),
enhancing tumor (ET) and whole tumor (WT), while these subregions can only
be separated when several modalities are combined with complementary information. T1-weighted (T1), Gadolinium-DTPA contrast enhanced T1-weighted

A. Crimi and S. Bakas (Eds.): BrainLes 2019, LNCS 11993, pp. 266–273, 2020.
https://doi.org/10.1007/978-3-030-46643-5_26

(T1c), T2-weighted (T2) and Fluid Attenuation Inversion Recovery (FLAIR) sequences are given with ground truth labels. T1 is the most commonly used sequence for structural analysis, T1c reveals the tumor boarder by with the contrast agent which breaks through the blood brain barrier and accumulates at the tumor region. The edema region, which surrounds the tumor appears brighter on T2. FLAIR distinguishes the edema and the Cerebrospinal Fluid (CSF) by suppressing the free water signal. It is convincing that the combination of the four mentioned sequences will provide necessary clinical evidence in order to accurately segment glioma subregions TC, ET and WT.

2   Related Work

In the past BRATS 2018 challenge, top 3 solutions were all encoder-decoder based neural networks, also knowns as Unet. The Unet is a symmetric shape convolutional neural network structure with skip connections between the encoder path and the decoder path, which provides local information to the semantic information while upsampling [15]. The Unet and its variants outperformed a series of popular conventional medical image analysis methodology such as random forest, atlas based, texture analysis and deformable models [5, 7, 10, 17]. The first place of BRATS18 adds variational auto-encoder (VAE) branch to regularize the shared decoder [14]. We agree that overfitting is very commonly occurred when training with medical data, especially the cancerous target volume segmentation due to much larger shape and appearance than health organs. In clinical practice, it is not possible to acquire millions labelled training data as Imagenet [3]. Consequently, the data augmentation and regularization are the key factors influencing the performance of model trained with medical data.

No New-net (nnUnet) [9] took the second place which only fine tuned the 3D Unet. Further attempt in [8] augmented the Unet with residual and preactivation residual blocks on kidney data, achieving promising results. It is most notably that these Unet variants suggest again that a well trained encoder and decoder architecture is quite efficient in the context of brain tumor segmentation.

Hence we choose the 3D Unet as our baseline for the remaining description of our solution. Our major work lands on the data augmentation and regularization, we found that the training of the VAE regularised Unet in [14] is very tricky, same case happened when we tried to add a generative adversarial branch to the vanilla Unet. 'Entities should not be multiplied without necessity' as Occam's razor said, since the lack of training data is the headache, the data augmentation might worth a revisit. Based on our experience the data augmentation process of medical data should be very careful, since the inappropriate augmentation might destroy the anatomical information of the training data. The detailed solution is presented in the next section.

3 Method

3.1 Data Description

This year challenge training set consists of total of 135 glioblastoma (GBM) and 108 low grade glioma (LGG) pre-operative scans, from multiple (n = 19) institutions. The training set was well preprocessed, isotropic interpolated, skull-removed and co-registered to an anatomical template. The official pre-precessing has saved us a lot of effort, the inter-patient variance have been reduced significantly, so we can focus more on the meaningful pathological features. Only gray level normalization (−1 to 1) was applied before we feed data into our model.

3.2 Data Augmentation

From our perspective, the training data provided is much smaller than 'sufficient' date set, in order to train a well performed model for abnormal growth (the tumor). Most of our experiments are trying to make the most from the given training data. We first reproduced the VAE Unet in [14] and found it tricky to fine tune due to our limited computing power. We also tried to put the Unet within a framework of generative adversarial framework, regularizing the Unet output as fake labels, while the performance was not stable as we expected. We applied a few widely used data augmentation, including flip, rotation, scaling and intensity shift, with minor increase in Dice. With a few attempts, we found the mixup is an interesting while effective data augmentation method [18], as shown in Eqs. (1) and (2),

$$\widetilde{x} = \lambda x_i + (1 - \lambda)x_j \tag{1}$$

$$\widetilde{y} = \lambda y_i + (1 - \lambda)y_j \tag{2}$$

where x_i and x_j are raw input vectors, y_i and y_j are one hot labels. By doing so, the mixup regularizes the neural network to favor simple linear behavior in-between training examples. We implemented the mixup augmentation in two ways, we firstly mixed up the whole brain volume. However, directly applying the mix up to brain tumor segmentation does not achieve significant improvements, particularly on WT and TC. In another way, we are looking at the tumor region only, the mixup only operates on the tumor region on interest (ROI). The x_i and x_j are now the label masked input vector as shown in Fig. 1. Since the images are co-registered before distributed, the overlay of the image are reasonable and visually natural.

We take two patient scan as the reference image x_i and target image x_j, tumor pixels in reference image are zeroed out, and then calculate the tumor grey level as in Eq. (1) to overlay on the target image. This process is only convincing when the two scans are registered beforehand. The mixed up image are gaussian filtered with kernel size 3 and bias corrected to create more natural image. By feeding the neural network the tumor mixed up input, we are expecting it to learn the tumor characteristics of the two patients on one training sample. The randomness induced by the straight forward mixup somehow improves the generalization particularly in the ET segmentation, as shown in Fig. 2.

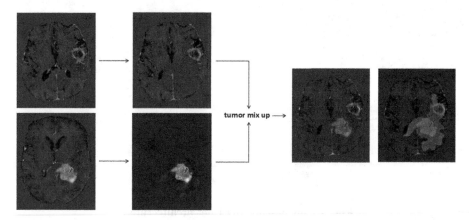

Fig. 1. The validation Dice of the ET segment. The tumor-mixup outperformed the baseline model on our validation set during training.

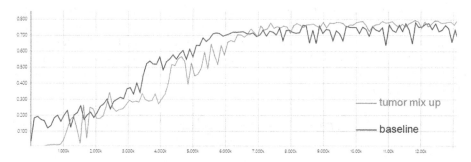

Fig. 2. The validation Dice of the ET segment. The tumor-mixup outperformed the baseline model on our validation set during training.

3.3   Network Architecture

An attempted modification was made to the nnU-net architecture by incorporating the hypercolumns [6], which uses the vector of activations of all network units above a specific pixel as the feature representation. After a few attempts, we found the nnU-net with hypercolumns learns faster than vanilla nnU-net while tends to overfit more easily, which we would like investigate further. At this moment, we stick the nnU-net wiht minor modification shown in Fig. 3.

3.4   Training Procedure

For the training of baseline, we randomly clip the data into 128*128*128 voxels, to be able to set the batch size to 2 on a GTX Titan XP (12 GB). The optimizer used is ADAM with an initial learning rate of 0.0001 with learning rate decay. For mixup augmentation training, the mixup is enabled until the loss researches plateau, afterwards mixup is disabled for fine tuning. In the entire BraTS2018

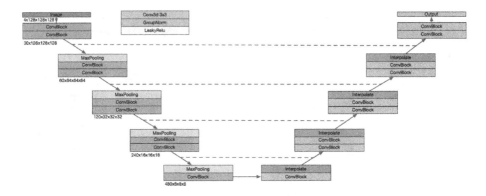

Fig. 3. Our baseline 3D Unet architecture.

dataset, approximately 98.88 % of voxels are background, 1.12% of voxels are WT, 0.48% of voxels are TC, and only 0.20% of voxels are ET as claimed in [12]. In order to solve this highly imbalanced problem between tumor subregions and background, we use the of Generalised dice loss from [16]. Let g_{ci} be the ground truth for the i-th voxel belonging to the class c and p_{ci} be the corresponding prediction, N be the total number of voxels, and C be the total number of classes. Then, we:

$$GDL = 1 - 2\frac{\sum_{c=1}^{l} w_c \sum_N g_{ci}p_{ci}}{\sum_{c=1}^{L} w_c \sum_{i=1}^{N} g_{ci} + p_{ci}} \tag{3}$$

where $w_l = 1/\sum_{i=1}^{N} g_{ci}$ is the corresponding weight assigned to class c. Lager weight will hence be assigned to the class with less pixels.

4 Preliminary Results

The preliminary result we achieved is not very competitive on the leaderboard, while we want to share our interesting finding, as summarized in Table. 1.

Table 1. Summarized preliminary results.

	Dice_ET	Dice_WT	Dice_TC	Sensitivity_ET	Sensitivity_WT	Sensitivity_TC
Baseline	0.72343	0.85243	0.75673	0.75394	0.91428	0.80001
Whole mixup	0.74053	0.84583	0.74453	0.75312	0.90132	0.8132
Tumor mixup	**0.76843**	**0.85743**	0.74513	**0.82394**	**0.91628**	**0.81351**
Tumor mixup ⋆	**0.78319**	**0.86735**	**0.79132**	NA	NA	NA

⋆ Our latest model enabled the tumor mixup for a pre-training, and fine tuning the model with tumor mixup disabled after loss stop decreasing. The

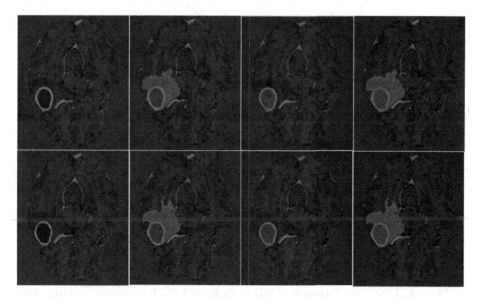

Fig. 4. Demo of tumor mixup prediction, top row prediction, bottom row are ground truth labels. Geen, yellow and red masks are WT, ET and TC respectively. (Color figure online)

performance was evaluated on our validation set, CBICA IPP platform submission jobs not return results since later 17th Aug when we have the tumor mixup predictions available (Fig. 4).

The whole brain mixup gives limited improvement on performance, we found that could be resulting from anatomical structure cancelling the tumour features. While the tumor mixup significantly improved both the Dice and sensitively for ET, which is the most challenging (lowest Dice) segmentation. It is reported the difficulty in ET is because of misclassified small blood vessel [9]. We believe that the imbalance in between tumor subregions should be resolved by the weight Dice loss, while the voxel number difference between background and whole tumor (100:1) is taken care of by the tumor mixup. Many '0' Dice for ET class is observed in our baseline prediction, which makes the ET prediction very unstable. The tumor mixup basically narrowing down the feature extraction within the tumor ROI, with reduced impact from the healthy brain tissue, to improve the ET prediction scores dramatically.

5   Future Work

As we discussed earlier, the data augmentation for medical data should be careful, the mixup method produce data with brute force which we agree might destroy anatomical information in brain structure. On the other hand, it might be viewed as a implicitly guided attention on the tumor tissue, by incorporating the randomness or noise brought in by the mixup operation. There is a major concern, the WT segment (the mode outside segment)obviously do not benefit from the mixup operation, because of the boundary between the WT and surrounding healthy tissue are mostly destroyed. We are considering using a model ensemble scheme to reduce this side effect caused by tumour mixup.

Acknowledgement. This work was supported by the National Natural Science Foundation of China (Grant No. 61605014) and the Fundamental Research Funds for the Central Universities (Grant No. 2018RC17 and 2018RC18).

References

1. Bakas, S., et al.: Advancing the cancer genome atlas glioma MRI collections with expert segmentation labels and radiomic features. Sci. Data **4**(July), 1–13 (2017). https://doi.org/10.1038/sdata.2017.117
2. Bakas, S., et al.: Identifying the best machine learning algorithms for brain tumor segmentation, progression assessment, and overall survival prediction in the BRATS challenge. CoRR abs/1811.02629 (2018). http://arxiv.org/abs/1811.02629
3. Deng, J., Dong, W., Socher, R., Li, L.J., Li, K., Fei-Fei, L.: ImageNet: a large-scale hierarchical image database. In: 2009 IEEE Conference on Computer Vision and Pattern Recognition, pp. 248–255. IEEE (2009)
4. Furnari, F.B., et al.: Malignant astrocytic glioma: genetics, biology, and paths to treatment. Genes Dev. **21**(21), 2683–2710 (2007). https://doi.org/10.1101/gad.1596707
5. Hamamci, A., Kucuk, N., Karaman, K., Engin, K., Unal, G.: Tumor-Cut: segmentation of brain tumors on contrast enhanced mr images for radiosurgery applications. IEEE Trans. Med. Imaging **31**(3), 790–804 (2011)
6. Hariharan, B., Arbeláez, P., Girshick, R., Malik, J.: Hypercolumns for object segmentation and fine-grained localization. In: Proceedings of the IEEE Conference on Computer Vision and Pattern Recognition, pp. 447–456 (2015)
7. Havaei, M., Larochelle, H., Poulin, P., Jodoin, P.-M.: Within-brain classification for brain tumor segmentation. Int. J. Comput. Assist. Radiol. Surg. **11**(5), 777–788 (2015). https://doi.org/10.1007/s11548-015-1311-1
8. Isensee, F., Maier-Hein, K.H.: An attempt at beating the 3D U-Net. arXiv preprint arXiv:1908.02182 (2019)
9. Isensee, F., Petersen, J., Kohl, S.A., Jäger, P.F., Maier-Hein, K.H.: nnU-Net: breaking the spell on successful medical image segmentation. arXiv preprint arXiv:1904.08128 (2019)
10. Khotanlou, H., Colliot, O., Atif, J., Bloch, I.: 3D brain tumor segmentation in MRI using fuzzy classification, symmetry analysis and spatially constrained deformable models. Fuzzy Sets Syst. **160**(10), 1457–1473 (2009)
11. Liang, Z.P., Lauterbur, P.C.: Principles of Magnetic Resonance Imaging: A Signal Processing Perspective. SPIE Optical Engineering Press, Bellingham (2000)
12. Mazumdar, I.: Automated brain tumour segmentation using deep fully convolutional residual networks. arXiv preprint arXiv:1908.04250 (2019)
13. Menze, B.H., et al.: The multimodal brain tumor image segmentation benchmark (BRATS). IEEE Trans. Med. Imaging **34**(10), 1993–2024 (2015). https://doi.org/10.1109/TMI.2014.2377694

14. Myronenko, A.: 3D MRI brain tumor segmentation using autoencoder regularization. In: Crimi, A., Bakas, S., Kuijf, H., Keyvan, F., Reyes, M., van Walsum, T. (eds.) BrainLes 2018. LNCS, vol. 11384, pp. 311–320. Springer, Cham (2019). https://doi.org/10.1007/978-3-030-11726-9_28

15. Ronneberger, O., Fischer, P., Brox, T.: U-Net: convolutional networks for biomedical image segmentation. In: Navab, N., Hornegger, J., Wells, W.M., Frangi, A.F. (eds.) MICCAI 2015. LNCS, vol. 9351, pp. 234–241. Springer, Cham (2015). https://doi.org/10.1007/978-3-319-24574-4_28

16. Sudre, C.H., Li, W., Vercauteren, T., Ourselin, S., Jorge Cardoso, M.: Generalised dice overlap as a deep learning loss function for highly unbalanced segmentations. In: Cardoso, M.J., et al. (eds.) DLMIA/ML-CDS -2017. LNCS, vol. 10553, pp. 240–248. Springer, Cham (2017). https://doi.org/10.1007/978-3-319-67558-9_28

17. Tustison, N.J., et al.: Optimal symmetric multimodal templates and concatenated random forests for supervised brain tumor segmentation (simplified) with ANTsR. Neuroinformatics **13**(2), 209–225 (2015). https://doi.org/10.1007/s12021-014-9245-2

18. Zhang, H., Cisse, M., Dauphin, Y.N., Lopez-Paz, D.: mixup: Beyond empirical risk minimization. arXiv preprint arXiv:1710.09412 (2017)

Multi-branch Learning Framework with Different Receptive Fields Ensemble for Brain Tumor Segmentation

Cheng Guohua[1]([✉]), Luo Mengyan[2], He Linyang[2], and Mo Lingqiang[2]

[1] Fudan University, Shanghai, China
17110850005@fudan.edu.cn
[2] Jian Pei Technology, Hangzhou, China

Abstract. Segmentation of brain tumors from 3D magnetic resonance images (MRIs) is one of key elements for diagnosis and treatment. Most segmentation methods depend on manual segmentation which is time consuming and subjective. In this paper, we propose a robust method for automatic segmentation of brain tumors image, the complementarity between models and training programs with different structures was fully exploited. Due to significant size difference among brain tumors, the model with single receptive field is not robust. To solve this problem, we propose our own method: i) a cascade model with a 3D U-Net like architecture which provides small receptive field focus on local details. ii) a 3D U-Net model combines VAE module which provides large receptive field focus on global information. iii) redesigned Multi-Branch Network with Cascade Attention Network, which provides different receptive field for different types of brain tumors, this allows to scale differences between various brain tumors and make full use of the prior knowledge of the task. The ensemble of all these models further improves the overall performance on the BraTS2019 [10] image segmentation. We evaluate the proposed methods on the validation DataSet of the BraTS2019 segmentation challenge and achieved dice coefficients of 0.91, 0.83 and 0.79 for the whole tumor, tumor core and enhanced tumor core respectively. Our experiments indicate that the proposed methods have a promising potential in the field of brain tumor segmentation.

Keywords: Brain tumor segmentation · Multi-branch · Receptive field

1 Introduction

The recent success of Deep Neural Networks has greatly improved the accuracy of semantic segmentation, object detection, classification, and other computer vision problems. In the medical image segmentation field, deep learning algorithms has gradually become the mainstream because its advantage over traditional machine learning algorithms. Methods like U-Net [12] and 3D U-Net [6] have shown their advanced ability in feature extraction and representation.

© Springer Nature Switzerland AG 2020
A. Crimi and S. Bakas (Eds.): BrainLes 2019, LNCS 11993, pp. 274–284, 2020.
https://doi.org/10.1007/978-3-030-46643-5_27

Magnetic resonance imaging (MRI) is commonly used in radiology to portray the phenotype and intrinsic heterogeneity of brain tumor. Accurate segmentation and quantitative analysis of brain tumor are critical for diagnosis and treatment planning. Developing a segmentation method, which helps automatic identify tumor regions from MRIs, will potentially improve diagnostics and follow-up treatment. However, it is a challenge task to do brain tumor image segmentation due to the heterogeneity in the size, location, and shape. In addition, the largely overlap between profiles of tumor regions and healthy parts also makes task more difficult. Many semantic segmentation models [4] based on using convolutional neural networks have been developed for brain tumor segmentation through past few years. The first successful neural network for brain tumor segmentation is DeepMedic which was proposed in 2016 [9]. It implements a 3D CNN architecture with multi-scale, residual connections and fully connected conditional random field. Later, Guotai et al. [7] proposed to segment tumor subregions in cascade using anisotropic convolutions, which become the winner of the BraTS2017 segmentation task [1,2]. In BraTS2018 challenge, Myronenko [11] presented an encoder-decoder architecture with a added auto-encoder branch added to reconstruct the input image itself in order to regularize the shared decoder and impose additional constraints on its layers. This approach won 1st place in the BraTS 2018 challenge. Later, Isensee [8] demonstrated that a generic U-Net architecture with a few minor modifications is enough to achieve competitive performance.

2 Method

2.1 Preprocessing

The training DataSet comprises of 259 glioblastoma (GBM/HGG) images and 76 low-grade gliomas volume images. The images were registered to a common space, resampled to isotropic 1 mm × 1 mm × 1 mm resolution with image dimensions 240 × 240 × 155 and were skull stripped by the organizers. Each MRI case comprises of 4 MR sequences [3,10], named T1-weighted (T1), contrast enhanced T1-weighted (T1ce), T2-weighted (T2) and Fluid Attenuation Inversion Recovery (FLAIR) images respectively, each sequence provides complementary proles for different sub-regions of brain tumor. We simply concatenate 4 MR sequences and normalize all sequences to have zero mean and unit standard deviation (based on non-zero voxels only) as our network input. In order to solve the class imbalance problem in the DataSet, we employ a few data augmentation techniques, including adding new synthetic images by performing operations and transformations on the input images and the corresponding ground truth images. Specifically, we apply random crops and random flip in our main augmentation methods. In addition, we also apply some random rotations and random scale.

2.2 Model

To make better use of the advantages of different model structures while avoiding the defects of certain models on certain tasks, we decide to separately train different models with different network structures, and fusion these trained models for the test. The models we use are as follows: 1. We employ a cascade method with a 3D U-Net like architecture, training with a sliding window method. 2. We borrow the model design ideas from NvNet [11], at the same time we add a VAE module structure and a multi-scale adaptive module to assist segmentation in the U-Net 3. An own designed model, which uses dilation convolutions to increase the receptive field and reduce the number of downsample operations.

Cascade Model. As described above, the goal of the challenge competition is to segment the whole tumor (WT), tumor core (TC) and enhancing tumor (ET) from the MRI sequences. The logical rules between these three sub-regions are that ET is a subset of TC, and TC is a subset of WT. A natural way to segment these sub-regions is training three networks for three sub-region respectively, which is also called the Model Cascade strategy [7]. The cascade method can be very powerful in segmenting the brain tumor. Due to the large volume size of the image, we consider using the sliding window method to segment each region in each sub-task in this paper.

The main network architecture is mainly based on the 3D U-Net. It consists of an encoding network and a decoding network. We particularly decrease the feature map size for some layers. Also, we use random dropout in the last layer of the encoding network to prevent the model from overfit. The overall number of downsample layers is 3. In each down-sample layer, the feature map size is down-sampled by a factor of 2. Train ET-Network, WT-Network, TC-Network for design for, the WT-Network designed for whole tumor (WT) and TC-Network designed for tumor core (TC) have bigger receptive field. In addition, we reduced one down-sample layer to increase the feature map size at the end of the encoding network. About ET-Network, we insert an Atrous Spatial Pyramid Pooling (ASPP) module [5] to extract the multi-scale features through different dilation convolution in ET-Network. Instead of training on the whole image size, we crop the whole image according to the bounding box with margins that derived from the ground truth segmentation. For the WT network, we use $80 \times 80 \times 48$ cubes sliding from the cropped image, the stride of the sliding window is $40 \times 40 \times 24$. The TC network uses $64 \times 64 \times 32$ cube size with $32 \times 32 \times 16$ stride size. The cube size of the ET network is $64 \times 64 \times 32$ and the stride size is $32 \times 32 \times 16$. In the test stage, the WT network segments the whole tumor from the original 3D volume. Then a bounding box is obtained based on the WT network's segmentation. The cropped region of the input image which based on the bounding box was feed to the TC network. The TC network segments the tumor core in the predicted WT bounding box. Similarly, the bounding box of predicted tumor core was feed to the ET network to segment the enhancing tumor core. Therefore, the networks run sequentially in the test stage.

U-Net with VAE. We imitated BraTS2019 champion scheme to design the model structure. On the basis of 3D U-Net, a VAE branch was added to provide regularization effect after training, this also help to optimize the encode part of the model and improve the fitting ability of the model. Due to the video memory limitation, the model structure is simplified to some extent (reduce two convolution block between the encode and decode) and the input patch size is reduced at the same time. In addition, in order to improve the adaptability of the model to multi-scale targets, we added the Atrous Spatial Pyramid Pooling (ASPP) [5] module into the encode and decode of the model, and set the dilate rates to be 1, 2, 4 and 8 in the ASPP [5] module structure. Because of limitations of video storage, batch normalization results are poor when batch size is small (batch size is 1 in our case). We compared the results of Group Normalization with Instance Normalization through experiments, and finally choose Instance Normalization as the replacement of batch Normalization.

Multi-branch Network. Although the down-sample operation can increase the receptive field of the model, it does great damage to the detailed information in the input image, especially for some tumors with a small range, such as ET and TC. When using U-Net-based model for segmentation, there are some defects in the segmentation of marginal details and small areas. As the area of WT tumor is large, large receptive fields are required to be able to take advantage of global information, and different types of tumors require different receptive fields. Thus, different models should be designed for training.

We designed an end to end model structure to better utilize the interrelationship between different types of tumors. There are two branches in order to meet different sensory needs. For WT, we used three down-sampling operations to obtain a larger receptive field. For TC and ET, we only used one down-sample operation, and combined with empty convolution to indicate the receptive field. The two branches share the encode module but the loss was calculated respectively. In this model, the residual module structure is adopted in each convolution module to facilitate better information fusion. We use Instance Normalization instead of Batch Normalization according to our experimental comparison result. Our model can take into account the detail characteristics and satisfy requirements of receptive field. Therefore, a single model with Multi-Branch can have the ability to match multiple target sizes. Further, in order to better capture the characteristics of different levels and prompt the convergence rate, we add a skip connection structure in the model, which is used to fuse highlevel semantic information with low-level semantic information (Fig. 1).

Multi-branch Network with Cascade Attention Module. Because of the inclusion relationship among three types of tumors, many programs use multiple model cascades for prediction. However, this scheme needs to train multiple models separately and the process is relatively tedious. In order to ensure the high efficiency of the training and prediction process, we designed the cascade attention module, which could take advantage of the hierarchical property of brain tumor structure.

It would be easy to add cascade attention module to our Multi-Branch Network, the output results of the second branch were calculated through cascade attention module, which was used to perform weighted operation on the input

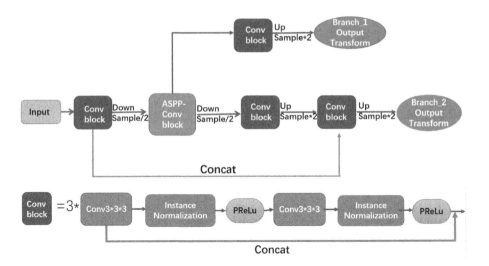

Fig. 1. Architecture of the Multi-Branch Network with different receptive field. (The diagram above shows the overall model structure, the diagram below is used to explain the Conv block structure.

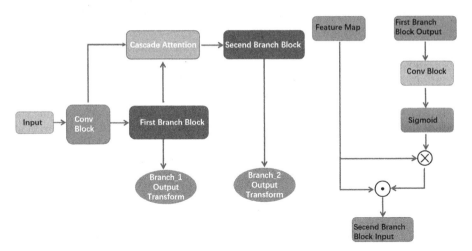

Fig. 2. Left figure is the architecture of the Multi-Branch Network with the Cascade Attention Module. Right figure is the structure of Cascade Attention Module.

feature map of the first branch. We used the first branch to predict WT and the second branch was designed to predict both TC and ET. This module use the prediction results of WT to weight the feature map and reduce the prediction difficulty of TC and ET. The Fig. 2 briefly describes the model structure we designed.

2.3 Loss Function

In this challenge competition, we combine the dice coefficient and the weighted cross-entropy loss into a hybrid loss as our main loss function in all the experiments.

Dice. The dice similarity coefficient measures the agreement between the predict and the ground truth. Also, it is one of the evaluation metrics on the validation phase. The equation is as follows, where S means the segmentation and R means the reference.

$$DSC = \frac{2\,|S \cap R|}{|S| + |R|} \tag{1}$$

2.4 Training

We designed our training scheme by running 5-fold cross-validation on the training set. The 335 cases of the BraTS2019 training set are randomly split into five folds with 67 patients each. Each fold contains the almost equal number of high-grade patient and low-grade patient. In order to further improve the generalization of the model and enhance the robustness of the model, we designed several data augmentation operations to increase the diversity of data. We use the data after random cropping as the training data, and carry out random flip, random rotation and elastic deformation operations. We use the ADAM optimizer with an initial learning rate 0.0001, and then reduce the learning rate by factor 5 for every 50 epochs. For full training, each model trains 450 epochs on its own cross.

3 Experiment and Results

3.1 Evaluation

The total number of validation cases of the BraTS 2019 segmentation challenge competition is 125. All the reference segmentation of tumor sub-regions in the validation set are derived from multiple experts' manual segmentation. We submit our prediction to the official online evaluation system to test our models and the final evaluation results are calculated separately in the three sub-regions, WT, TC, and ET.

3.2 Metrics

Three metrics are used to evaluate the prediction result in the segmentation of brain tumors:

Dice Coefficient. The dice coefficient measures the similarity between predicted segmentation and ground truth segmentation. It penalizes false positive and false negative of the prediction and is very commonly used in image segmentation.

Hausdorff Distance (HD). Hausdorff distance measures how far two subsets of a metric space are from each other. HD is one of the most informative and useful criteria because it is an indicator of the largest segmentation error. The 95% Hausdorff distance indicates the distance that point in one set to its closest point in the other set is greater or equal to exactly 95% of the other points.

$$d_H(A, B) = max \left\{ \sup_{a \in A} \inf_{b \in B} d(a, b), \sup_{b \in B} \inf_{a \in A} d(a, b) \right\} \tag{2}$$

Sensitivity and Specificity. Sensitivity measures the proportion of actual positives that are correctly identified as such. And Specificity measures the proportion of actual negatives that are correctly identified as such. They are used since brats'17 to determine potential over or under-segmentation of the tumor sub-regions by participating methods.

$$Sensitivity = \frac{TP}{TP + FN} \tag{3}$$

$$Specificity = \frac{TN}{TN + FP} \tag{4}$$

3.3 Results and Discussion

Each model segments brain tumor individually and outputs its prediction in the test phase. We ensemble all the predictions to yield the final segmentation. About the fusion method, we average the probabilistic results of multiple models as the prediction result of the final integration model. The final results are shown in Table 1, 2.

Table 1. 2019 Brats 2019 validation submit results

	Dice ET	Dice WT	Dice TC
Cascade model	0.75046	0.90043	0.79411
U-Net + VAE	0.75937	0.90134	0.81085
Multi-branch network	0.76419	0.90493	0.82
Multi-branch network with cascade attention	0.76879	0.90609	0.82431
Ensemble model	0.79161	0.90848	0.83545

From the result of the above table, we can see that the Multi-Branch model we designed has certain advantages over the traditional Cascade model and U-Net model. The addition of Cascade Attention to Multi-Branch Network further improves its performance. Using a single end-to-end model, which takes full advantage of the relevance and prior knowledge between different tumor

Table 2. 2019 Brats 2019 validation submit results

	Hausdorff95 ET	Hausdorff95 WT	Hausdorff95 TC
Cascade model	5.19945	5.49878	7.97158
U-Net + VAE	4.39729	5.6109	7.67346
Multi-branch network	3.40238	5.40841	7.37504
Multi-branch network with cascade attention	3.32457	5.35498	6.96617
Ensemble model	4.9227	5.27584	6.52394

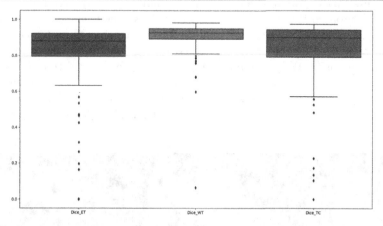

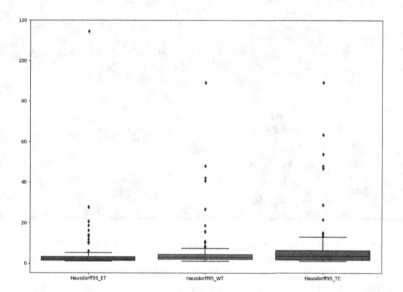

Fig. 3. Box plot of evaluation index. The above is about dice, the below is about hausdorff distance.

categories, is more convenient and efficient during the training and prediction stages. Because the four models we trained are complementary to some extent, the ensemble model displays better performance in prediction accuracy and robustness.

In order to better analyze the shortcomings of the model, so as to provide ideas for future improvement. We performed statistical analysis on the predicted results of our ensemble model. The Fig. 3 shows a box diagram of the predicted results of the integrated model. As we can see from the graph, the median of each indicator is better than the mean. Some hard example has a great influence on the overall indicators of the model. We visualized the prediction result labels and then compared the difference. The Fig. 4 shows the visualization.

In Fig. 4, we are able to distinguish intratumoral regions by color-code: enhancing tumor (yellow), peritumoral edema (green) and necrotic and non-enhancing tumor (red). By analyzing Fig. 4 and 5, we can find that TC and ET are prone to missing detection, especially when the range of ET is small. These characteristics result in the poor performance of the model in some data, which affects the overall performance of the model. This is also a direction we continue to improve.

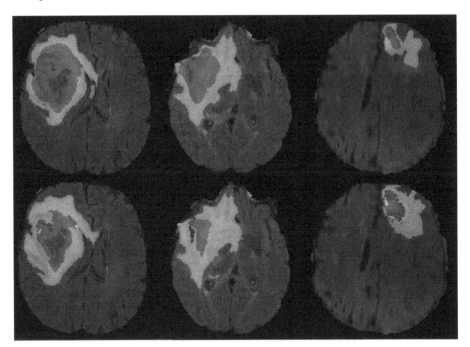

Fig. 4. Some segmentation examples which are difficult to segment with true and predicted labels. The visualization of segmentation results about data which are difficult to be segmented. The whole tumor (WT) class includes all visible labels (a union of green, yellow and red labels), the tumor core (TC) class is a union of red and yellow, and the enhancing tumor core (ET) class is shown in yellow (a hyperactive tumor part). (Color figure online)

4   Conclusion

In this paper, we improved the traditional cascade model and U-Net model by training the modified cascade model and the U-Net model with VAE. At the same time, according to the specific characteristics of the task, we designed our two-branch network to adapt to a variety of sensory field needs. In order to make better use of prior knowledge, Cascade Attention Module was added to further improve the accuracy of the model. It is known that neural networks can make unpredictable errors which may result in the un-robustness. An ensemble of three models improves the robustness of the neural network in prediction. Also, the overall accuracy of the individual prediction is improved when using the ensemble model. Due to the time limit, other methods are not tried in our experiment. Good post-processing may good for reducing the Hausdorff distance and a better hyperparameter tuning method may improve the dice score, which are both worth to be explored in the future experiment.

References

1. Bakas, S., et al.: Segmentation labels and radiomic features for the pre-operative scans of the TCGA-GBM collection (2017). https://doi.org/10.7937/K9/TCIA. 2017.KLXWJJ1Q
2. Bakas, S., et al.: Segmentation labels and radiomic features for the pre-operative scans of the TCGA-LGG collection. Cancer Imaging Archive (2017). https://doi.org/10.7937/K9/TCIA.2017.GJQ7R0EF
3. Bakas, S., et al.: Advancing the cancer genome atlas glioma MRI collections with expert segmentation labels and radiomic features. Nat. Sci. Data **4**, 170117 (2017). https://doi.org/10.1038/sdata.2017.117
4. Bakas, S., et al.: Identifying the best machine learning algorithms for brain tumor segmentation, progression assessment, and overall survival prediction in the brats challenge. arXiv preprint arXiv:1811.02629 (2018)
5. Chen, L.C., Papandreou, G., Kokkinos, I., Murphy, K., Yuille, A.L.: DeepLab: semantic image segmentation with deep convolutional nets, atrous convolution, and fully connected CRFs. IEEE Trans. Pattern Anal. Mach. Intell. **40**(4), 834–848 (2017)
6. Çiçek, Ö., Abdulkadir, A., Lienkamp, S.S., Brox, T., Ronneberger, O.: 3D U-Net: learning dense volumetric segmentation from sparse annotation. In: Ourselin, S., Joskowicz, L., Sabuncu, M.R., Unal, G., Wells, W. (eds.) MICCAI 2016. LNCS, vol. 9901, pp. 424–432. Springer, Cham (2016). https://doi.org/10.1007/978-3-319-46723-8_49
7. Wang, G., Li, W., Ourselin, S., Vercauteren, T.: Automatic brain tumor segmentation using cascaded anisotropic convolutional neural networks. In: Crimi, A., Bakas, S., Kuijf, H., Menze, B., Reyes, M. (eds.) BrainLes 2017. LNCS, vol. 10670, pp. 178–190. Springer, Cham (2018). https://doi.org/10.1007/978-3-319-75238-9_16
8. Isensee, F., Kickingereder, P., Wick, W., Bendszus, M., Maier-Hein, K.H.: No new-net. In: Crimi, A., Bakas, S., Kuijf, H., Keyvan, F., Reyes, M., van Walsum, T. (eds.) BrainLes 2018. LNCS, vol. 11384, pp. 234–244. Springer, Cham (2019). https://doi.org/10.1007/978-3-030-11726-9_21

9. Kamnitsas, K., et al.: Deepmedic for brain tumor segmentation. In: Crimi, A., Menze, B., Maier, O., Reyes, M., Winzeck, S., Handels, H. (eds.) BrainLes 2016. LNCS, vol. 10154, pp. 138–149. Springer, Cham (2016). https://doi.org/10.1007/978-3-319-55524-9_14

10. Menze, B.H., et al.: The multimodal brain tumor image segmentation benchmark (brats). IEEE Trans. Med. Imaging, 1993–2024 (2015). https://doi.org/10.1109/TMI.2014.2377694

11. Myronenko, A.: 3D MRI brain tumor segmentation using autoencoder regularization. In: Crimi, A., Bakas, S., Kuijf, H., Keyvan, F., Reyes, M., van Walsum, T. (eds.) BrainLes 2018. LNCS, vol. 11384, pp. 311–320. Springer, Cham (2019). https://doi.org/10.1007/978-3-030-11726-9_28

12. Ronneberger, O., Fischer, P., Brox, T.: U-Net: convolutional networks for biomedical image segmentation. In: Navab, N., Hornegger, J., Wells, W.M., Frangi, A.F. (eds.) MICCAI 2015. LNCS, vol. 9351, pp. 234–241. Springer, Cham (2015). https://doi.org/10.1007/978-3-319-24574-4_28

Domain Knowledge Based Brain Tumor Segmentation and Overall Survival Prediction

Xiaoqing Guo[1], Chen Yang[1], Pak Lun Lam[2], Peter Y. M. Woo[3], and Yixuan Yuan[1(✉)]

[1] Department of Electrical Engineering, City University of Hong Kong, Kowloon, Hong Kong SAR, China
yxyuan.ee@cityu.edu.hk
[2] Department of Diagnostic and Interventional Radiology, Kwong Wah Hospital, Yaumatei, Hong Kong SAR, China
[3] Department of Neurosurgery, Kwong Wah Hospital, Yaumatei, Hong Kong SAR, China

Abstract. Automatically segmenting sub-regions of gliomas (necrosis, edema and enhancing tumor) and accurately predicting overall survival (OS) time from multimodal MRI sequences have important clinical significance in diagnosis, prognosis and treatment of gliomas. However, due to the high degree variations of heterogeneous appearance and individual physical state, the segmentation of sub-regions and OS prediction are very challenging. To deal with these challenges, we utilize a 3D dilated multi-fiber network (DMFNet) with weighted dice loss for brain tumor segmentation, which incorporates prior volume statistic knowledge and obtains a balance between small and large objects in MRI scans. For OS prediction, we propose a DenseNet based 3D neural network with position encoding convolutional layer (PECL) to extract meaningful features from T1 contrast MRI, T2 MRI and previously segmented sub-regions. Both labeled data and unlabeled data are utilized to prevent over-fitting for semi-supervised learning. Those learned deep features along with handcrafted features (such as ages, volume of tumor) and position encoding segmentation features are fed to a Gradient Boosting Decision Tree (GBDT) to predict a specific OS day.

1 Introduction

The annual incidence of primary brain tumors is increasing and poses a significant burden on public health [7]. Glial cells comprise approximately half of the total volume of the brain with a glial cell-to-neuron ratio of 1:1 [5]. They are principally responsible for maintaining homeostasis, providing support and protecting neurons. Gliomas are the most common primary brain tumors in humans and originate from glial cells, accounting for 35% to 60% of all intracranial tumors [7]. The age-standardized incidence rate of gliomas is 4.7 per 100000

© Springer Nature Switzerland AG 2020
A. Crimi and S. Bakas (Eds.): BrainLes 2019, LNCS 11993, pp. 285–295, 2020.
https://doi.org/10.1007/978-3-030-46643-5_28

person-years and in clinical practice the diagnosis of such tumors requires neurosurgery to obtain a tissue biopsy, which entails considerable risks for patients. Moreover, according to the World Health Organisation (WHO), gliomas can be histologically classified to into four grades, with each resulting in distinctly different durations of overall survival (OS). Although high-grade gliomas (HGG), i.e. WHO grade III or IV, are considered more aggressive, there is a growing body of evidence that such a histopathological classification is inadequate to prognosticate OS due to the nuanced variations in molecular profile from one tumor to the next [6]. By incorporating the glioma pathological diagnosis data, segmenting tumor sub-regions exhibited by magnetic resonance imaging (MRI) has been known to provide additional quantitative information for OS prediction. However, the process of manual image segmentation is highly time-consuming, often requires experienced neuro-radiologists and can be subject to inter-observer variations. To address these issues, developing an automated accurate segmentation tool that can reliably detect OS-relevant imaging biomarkers is urgently needed.

Gliomas, especially HGGs, often possess intratumoral heterogeneity that could represent different MRI signal intensity profiles across multi-modality imaging sequences [12]. Over the last decade, a number of scholars have proposed algorithms to automatically segment these glioma sub-regions in order to determine an accurate preoperative prognosis with varying degrees of success [9,10,13,16,18]. Promising progress has been made using traditional machine learning methods [16,18], which calculated low-level handcrafted, radiological features to describe images and trained a classifier or a regression for tumor segmentation and OS prediction. These handcrafted features were usually defined by experienced neuro-radiologists founded on prior knowledge of the exact histological diagnosis of the glioma and could have been a potential source of bias. This simple, straight-forward feature analysis approach potentially also disregarded a great deal of useful information embedded within the MR images, prohibiting the full effective utilization of sub-region segmentation for OS prediction. Recently, deep learning methods have demonstrated superior image processing capabilities that have been proven to effectively overcome these limitations [9–11,13]. Instead of defining handcrafted features, Convolutional Neural Network (CNN) methods jointly trains feature extractor and classifier to adaptively derive high-yield information and enhance model performance [11]. Inspired by the superior outcomes of this methodology, researchers are increasingly applying CNN for brain tumor segmentation and patient OS prediction [9,10,13]. Nie et al. [13] adopted a VGG-based network to automatically extract high-yield features from gadolinium contrast-enhanced T1 (T1ce) and diffusion tensor imaging sequences, and then utilized the extracted features together with tumor volume data to train a support vector machine for final OS prediction. Kao et al. [10] incorporated location information from a brain parcellation atlas to obtain accurate glioma segmentation results. Kao et al. [10] also analyzed connectome tractography information to identify potentially tumor-induced damaged brain regions and demonstrated that incorporation of this feature dataset resulted in superior OS prediction than including customary age, volumetric, spatial and morphological features alone.

Despite the relatively good performance of automatic tumor segmentation, the results of OS prediction are far from satisfactory [4,12]. For example, the patient OS prediction model of the first-ranking team [9] of the Brain Tumor Segmentation (BraTS) 2018 challenge, an international competition with open-source MRI Digital Images in Communications in Medicine (DICOM) data organized by the School of Medicine of the University of Pennsylvania, only resulted in a accuracy of 0.62 [4]. Two factors may have resulted in the limited predictive capacity of previous deep learning methodologies. First, [10] was the first effort to incorporate tumor location-based information in the CNN for brain tumor segmentation. However, such location data, which is crucial for OS prediction, has not been considered to predict OS. Secondly, pre-existing algorithms for OS prediction were usually based on supervised CNN models. However, the limited available number of datasets, led to considerable over-fitting problems [9,10,13]. In contrast, unlabeled MRI DICOM data are readily accessible in the clinical setting. Therefore, making adequate use of such data during the training process could be a promising strategy to improve OS prediction.

In this paper, we present a 3D dilated multi-fiber network (DMFNet) trained with weighted dice loss to segment glioma sub-regions from MRI scans. Then these predicted segmentation results are combined with T1 contrast and T2 MRI together as inputs for the proposed PECL-DenseNet to extract high-level and meaningful features that is trained with unlabeled as well as labeled data. In addition, we combine the extracted deep features from PECL- DenseNet with handcrafted features (age, tumor volume, volume ratio, surface area, surface area to volume ratio, location of the tumor's epicenter, its corresponding brain parcellation, relevant location of the tumor epicenter to the brain epicenter and resection status) and position encoding segmentation features to train the Gradient Boosted Decision Tree (GBDT) regression for patient OS prediction.

2   Methodology

2.1   Dataset

The Brain Tumor Segmentation (BraTS) 2019 dataset [1–4,12] provides 335 training subjects, 125 validation subjects and 167 testing ones, each with four MRI modality sequences (T1, T1ce, T2 and FLAIR). All the training data have corresponding pixel-level annotations, including necrosis and non-enhancing tumor, edema, and enhancing tumor sub-regions. Partial training data have corresponding subject-level annotations, indicating the OS duration and the resection status, respectively. In particular, only HGG patients with gross total resection (GTR) were evaluated, since resection status is the only consistent modifiable treatment predictor for OS, and only 101 (30%) training subjects are eligible. There remaining 109 brain tumor training subjects do not undergo GTR have OS data and 49 subjects miss OS labels. 29 validation subjects and 107 testing ones have complete subject-level annotations.

2.2   Brain Tumor Segmentation

A major difficulty with the existing BraTS segmentation challenge is the high computational cost required, since each subject has four modality MRI scans. To tackle this dilemma, our segmentation model is primarily based on DMFNet [8], which can significantly reduce the computational cost of 3D networks by an order of magnitude. It slices a complex neural network into an ensemble of lightweight networks or fibers, and further incorporates multiplexer modules to facilitate information flow between fibers. To enlarge the respective field and to capture the multi-scale 3D spatial correlations, DMFNet adds dilated convolution to multiplexer modules (Fig. 1).

The accuracy of contrast-enhancing tumor segmentation is usually the worst, compared with peri-tumoral edema and intratumoral necrosis regions, since tumor tissue enhancement often constitutes the smallest volume of the entire tumor. Therefore, we introduce prior volume knowledge to traditional dice loss to resolve this imbalanced class problem, namely by weighted dice loss. In particular, we apply the reciprocal of each tumor volume as our dice weight, given as 0.38, 0.15, 0.47 for necrosis, edema and enhancing tumor, respectively.

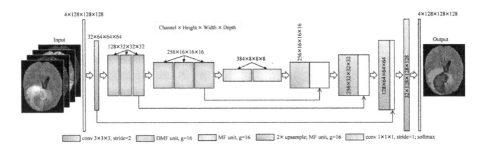

Fig. 1. Illustration of DMFNet framework for brain tumor segmentation.

2.3   Overall Survival Prediction

Features extracted from deep CNN, handcrafted features and position encoding segmentation features are incorporated for the OS prediction. The geometry and location of tumor are crucial for the OS prediction [14]. Therefore, we propose a PECL-DenseNet with considering the location information to extract meaningful features and make adequate use of unlabeled data to prevent over-fitting. With the calculated segmentation result, we define 36 handcrafted features that involves geometry and location information for accurate OS prediction. Moreover, we apply the pooling operator to predicted segmentation to accurately obtain the tumor location information. GBDT regression is performed to fit with normalized features of 210 training data that is with OS labeling. The source code for extracting the handcrafted features and implementing GBDT regression is available at https://github.com/Guo-Xiaoqing/BraTS_OS.

PECL-DenseNet. From T1ce, T2 sequence MRI images and the predicted sub-region segmentation from DMFNet, deep features are extracted by alternate-cascaded 3D dense blocks and transition layers. The dense connectivities in dense blocks can combine information from different convolutional layers, therefore encourage feature reuse and ensure maximum information flow between layers. Specifically, our proposed framework includes four dense blocks as shown in Fig. 2. Each block is comprised of 7 densely connected layers, and every layer consists of a batch normalization, a ReLU, and the proposed PECL module. Then the deep features are concatenated with the resection status to derive a five-class OS prediction classification. Specifically, we utilize a digit to represent resection status, given as GTR (2), STR (1) and NA (0). The dimension of extracted image features are reduced to 50 by principal components analysis for further processing.

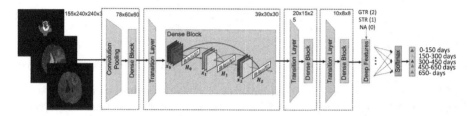

Fig. 2. Illustration of the proposed PECL-DenseNet for OS prediction. T1 contrast, T2 MRI images and the predicted sub-regions segmentation from DMFNet are concatenated as the input of the PECL-DenseNet. Deep features extracted from the PECL-DenseNet are then combined with resection status to make a five-classes prediction.

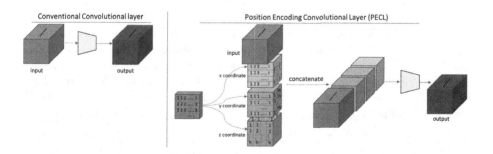

Fig. 3. Comparison of 3D convolutional layer and the proposed PECL.

We propose a position encoding convolutional layer (PECL) to incorporate location knowledge for OS prediction. Due to the translation invariance of convolution and global average pooling operator, the extracted deep feature vector usually ignores the spatial information. However, the tumor location is essential for the diagnosis and prognosis of gliomas, especially for HGG. In this regard, we extend the conventional 3D convolutional layer to PECL by incorporating the position information as in Fig. 3. In contrast from the standard convolutional layer, PECL introduces three extra channels (x, y, z) to derive a 3D Cartesian coordinate. The introduced channels are individually normalized by dividing their maximum value. Then input feature maps are concatenated with these additional channels for further processing.

To make adequate use of the limited available labeled data and additional unlabeled data in our hand, we develop an effective loss function for semi-supervised learning. Assuming the training set is $\mathcal{D}$ consisting of N samples. Denoting $\mathcal{L} = \{(x_i, y_i)\}_{i=1}^{L}$ is labeled dataset and $\mathcal{U} = \{x_i\}_{i=L+1}^{N}$ is unlabeled dataset. We aim to learn an OS prediction network parameterized by Θ through optimizing the following loss function:

$$L = \frac{1}{N} \sum_{i=1}^{N} (\alpha \cdot \frac{(n-1) \|z_i - \mathbf{c}_{y_i}\|}{\sum_{j \neq y_i}^{n} \|z_i - \mathbf{c}_j\|} \underset{x \in \mathcal{L}}{} - \beta \cdot \underset{x \in \mathcal{L}}{y_i \log p_i} - \gamma \cdot \underset{x \in \mathcal{U}}{p_i \log p_i}), \tag{1}$$

where $p_i = \frac{e^{W^\top z_i + b}}{\sum_{j=1}^{n} e^{W_j^\top z_i + b_j}}$. W_j is the weight for j^{th} class in fully connected layer and b_j is bias. z_i denotes extracted features of ith samples and $\mathbf{c}_j$ is the jth class feature centroid. N and n represent batch size and number of classes. α, β, γ are set as 0.5, 1 and 0.1, respectively. The first term is inspired by [17] and aimed to enforce the extracted features to approximate their corresponding feature centroid and to distance away from other centroids. The accumulative feature centroids are updated by formulation: $\mathbf{c}_j^{t+1} = \mathbf{c}_j^t - 0.5 \cdot \frac{\sum_{i=1}^{N} \delta(j=y_i) \cdot (\mathbf{c}_j - x_j)}{1 + \sum_{i=1}^{N} \delta(j=y_i)}$, where t denotes sequential iterations. $\delta(\cdot) = 1$ if condition is satisfied, and $\delta(\cdot) = 0$ if not. The second term is softmax cross entropy loss for labeled data, and the third one is an information entropy loss for unlabeled data.

Handcrafted Feature. We define 36 handcrafted features that involves non-image features and image features. Non-image features includes age and resection status. In particularly, a two dimensional feature vector is used to represent resection status, given as GTR (1, 0), STR (0, 1) and NA (0, 0). With the calculated segmentation from DMFNet, we calculate 34 image features including volume (V_{whole}, $V_{necrosis}$, V_{edema}, $V_{enhancing}$), volume ratio ($\frac{V_{whole}}{V_{brain}}$, $\frac{V_{necrosis}}{V_{brain}}$, $\frac{V_{edema}}{V_{brain}}$, $\frac{V_{enhancing}}{V_{brain}}$, $\frac{V_{necrosis}}{V_{enhancing}}$, $\frac{V_{edema}}{V_{enhancing}}$, $\frac{V_{necrosis}}{V_{edema}}$), surface area ($S_{whole}$, $S_{necrosis}$, S_{edema}, $S_{enhancing}$), surface area to volume ratio ($\frac{S_{whole}}{V_{whole}}$, $\frac{S_{necrosis}}{V_{necrosis}}$, $\frac{S_{edema}}{V_{edema}}$, $\frac{S_{enhancing}}{V_{enhancing}}$), position of the whole tumor epicenter (3 coordinates and its corresponding brain parcellation), position of the enhancing tumor epicenter (3 coordinates and its corresponding brain parcellation), relevant location of the whole tumor epicenter to brain epicenter (3 coordinates) and relevant location of the enhancing tumor epicenter to brain epicenter (3 coordinates). Note that V and S indicate volume and surface area. *whole, necrosis, edema, enhancing* and *brain* denote the entire tumor, necrosis and non-enhancing tumor, edema, enhancing tumor and the entire brain, respectively. To obtain the brain parcellation for tumor epicenter location, we register all the data to LPBA40 atlas [15], and 56 different brain parcellations are delineated.

Position Encoding Segmentation. To reserve the tumor location information, we apply a pooling operator to the predicted segmentation, where the kernel of the pooling operator is $5 \times 12 \times 12$ and the resolution of predicted segmentation is $155 \times 240 \times 240$. Thus, a 12400-dimensional feature vector is obtained.

3   Experiments and Results

3.1   Experiment Setup

In the brain tumor segmentation experiment, we trained the DMFNet with all the 335 training subjects and evaluated on 125 validation subject data and 167 testing data. We used a batch size of six and trained the DMFNet on two parallel Nvidia GeForce 1080Ti GPUs for 300 epochs. The initial learning rate was set as 0.001. During the training phase, we randomly cropped the data into $128 \times 128 \times 128$ for training data augmentation. In the testing phase, we utilized zero padding to make the resolution of input MRI data $240 \times 240 \times 160$.

In the OS prediction experiment, we made subjects with OS labels (101 + 109 = 210 subjects) as labeled data, and regarded the remaining 49 subjects without OS labels and 96 validation subjects as unlabeled data for the 3D PECL-DenseNet training in a semi-supervised strategy. Results of OS prediction were evaluated on 29 validation subjects and 107 testing ones. During 3D PECL-DenseNet training, the initialized learning rate was set to 0.1, and was dropped by 0.1 at 150 and 250 epochs, respectively. All training steps for labeled data and unlabeled data both use batch size of four. Both handcrafted features and location encoding segmentation features are extracted from 210 training subjects, 29 validation subjects and 107 testing ones, which are then combined with deep features extracted from PECL-DenseNet to feed into the GBDT for training and testing, respectively.

3.2   Results

Brain Tumor Segmentation. For brain tumor segmentation, we first conducted five-fold cross-validation evaluation on the training set, and our DMFNet achieved average dice scores of 80.12%, 90.62% and 84.54% for enhancing tumor (ET), the whole tumor (WT) and the tumor core (TC), respectively. The segmentation results are shown in Fig. 4, and our results match well with ground truth. Besides, 125 validation cases were evaluated after submitting to the CBICA's Image Processing Portal, achieving average dice scores of 76.88%, 89.38% and 81.56% for ET, WT and TC, respectively. The 3rd *row* in Table 1 shows the performance metrics that the segmentation network achieved on the testing data.

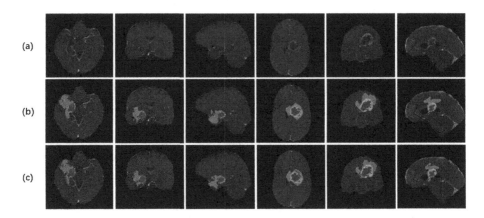

Fig. 4. Prediction of DMFNet for BraTS 2019 cross-validation on training data. (a) MRI (T1ce), (b) predicted segmentation (c) ground truth.

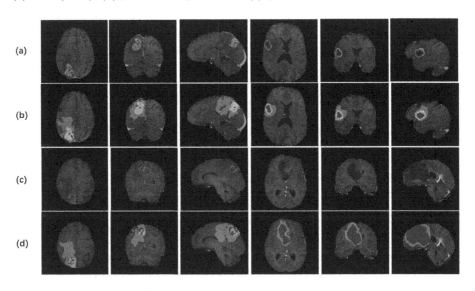

Fig. 5. Prediction of DMFNet for BraTS 2019 validation and testing data. (a) Validation data; (b) predicted segmentation on (a); (c) testing data; (d) predicted segmentation on (c).

It obtains average dice scores of 78.99%, 86.71% and 82.09% for ET, WT and TC, respectively. We also visualized the segmentation results of DMFNet, as shown in Fig. 5.

Overall Survival Prediction. We extracted different features to solve OS prediction problems as follows:

(a) Features extracted from PECL-DenseNet
(b) handcrafted features
(c) Position encoding segmentation features

Firstly, we trained GBDT regression with 36 handcrafted features. Subsequently, 18 important features were selected by their regression weight and fed into GBDT regression model for training. A comparison of the results obtained from training with 36 and 18 handcrafted features are shown in Table 2. It is clear that feature selection improves the performance of OS prediction.

Table 1. Dice and Hausdorff for BraTS 2019 validation and testing dataset.

Dataset	Dice_ET	Dice_WT	Dice_TC	Hausdorff_ET	Hausdorff_WT	Hausdorff_TC
Validation	76.88%	89.38%	81.56%	4.50841	5.03648	6.58191
Testing	78.99%	86.71%	82.09%	20.24	12.45	26.99

Table 2. 36 handcrafted VS 18 handcrafted.

Method	Accuracy	MSE	medianSE	stdSE	SpearmanR
36 handcrafted	0.31	152597.065	76777.495	192410.851	−0.091
18 handcrafted	0.448	142485.235	64070.727	192720.964	0.061

Table 3. Validation results of OS prediction with different methods.

Method	a	b	c	Accuracy	MSE	medianSE	stdSE	SpearmanR
Method 1	√			0.379	431949.975	270585.384	488132.042	−0.347
Method 2		√		0.448	142485.235	64070.727	192720.964	0.061
Method 3			√	0.379	105019.348	47771.914	139093.436	0.07
Method 4	√	√		0.483	118374.49	68989.292	132897.288	0.238
Method 5	√		√	0.448	120356.082	52497.44	186701.876	0.012
Method 6		√	√	0.586	104985.694	86581.049	117638.724	0.218
Method 7	√	√	√	0.517	200169.575	51368.509	309567.261	0.142

Moreover, we arranged and combined features (a), (b), (c) to train the GBDT regression model, and the corresponding results on validation data were shown in Table 3. It is obvious that handcrafted features is of great importance compared with the deep features (a) and position encoding features (c) (2nd to 4th rows). Selecting and combining these three features groups, it is observed that incorporating both handcrafted features and location encoding segmentation data achieved the highest accuracy of 0.586 (Method 6). We then saved the parameters obtained from the regression model that yielded the best result and applied it on testing data (Table 4). We achieved an accuracy of 0.523 for OS time prediction.

Table 4. Testing results of OS prediction with the method 6.

Accuracy	MSE	medianSE	stdSE	SpearmanR
0.523	407196.811	55938.713	1189657.961	0.281

4   Conclusion

In this paper, we utilize DMFNet with weighted dice loss for brain tumor segmentation, which significantly reduces the computation cost and obtains a balance between small and large objects from MRI scans. Segmentations predicted from DMFNet are further utilized to provide explicit tumor information for patient OS prediction. As for OS prediction, GBDT regression is implemented by combining of deep features derived from the proposed PECL-DenseNet, handcrafted features and position encoding segmentation features. Specifically, we propose a PECL-DenseNet to extract meaningful features, which makes adequate use of unlabeled data and to prevent over-fitting issues. Besides, several clinical features are defined and combined with the deep features from PECL-DEnseNet to train GBDT regression for OS days prediction. Although our methods reveals promising performances for both the brain tumor segmentation and OS prediction tasks, we believe that the performance will be further improved by integrating more MRI modality data and brain tumor molecular information.

References

1. Bakas, S., et al.: Segmentation labels and radiomic features for the pre-operative scans of the TCGA-GBM collection. Cancer Imaging Archive (2017)
2. Bakas, S., et al.: Segmentation labels and radiomic features for the pre-operative scans of the TCGA-LGG collection. Cancer Imaging Archive **286** (2017)
3. Bakas, S., et al.: Advancing the cancer genome atlas glioma MRI collections with expert segmentation labels and radiomic features. Sci. Data **4**, 170117 (2017)
4. Bakas, S., et al.: Identifying the best machine learning algorithms for brain tumor segmentation, progression assessment, and overall survival prediction in the brats challenge. arXiv preprint arXiv:1811.02629 (2018)
5. von Bartheld, C.S., Bahney, J., Herculano-Houzel, S.: The search for true numbers of neurons and glial cells in the human brain: a review of 150 years of cell counting. J. Comp. Neurol. **524**(18), 3865–3895 (2016)
6. van den Bent, M.J.: Interobserver variation of the histopathological diagnosis in clinical trials on glioma: a clinician's perspective. Acta Neuropathol. **120**(3), 297–304 (2010)
7. Bray, F., Ferlay, J., Soerjomataram, I., Siegel, R.L., Torre, L.A., Jemal, A.: Global cancer statistics 2018: Globocan estimates of incidence and mortality worldwide for 36 cancers in 185 countries. CA Cancer J. Clin. **68**(6), 394–424 (2018)
8. Chen, C., Liu, X., Ding, M., Zheng, J., Li, J.: 3D dilated multi-fiber network for real-time brain tumor segmentation in MRI. arXiv preprint arXiv:1904.03355 (2019)

9. Feng, X., Tustison, N., Meyer, C.: Brain tumor segmentation using an ensemble of 3D U-Nets and overall survival prediction using radiomic features. In: Crimi, A., Bakas, S., Kuijf, H., Keyvan, F., Reyes, M., van Walsum, T. (eds.) BrainLes 2018. LNCS, vol. 11384, pp. 279–288. Springer, Cham (2019). https://doi.org/10.1007/978-3-030-11726-9_25

10. Kao, P.-Y., Ngo, T., Zhang, A., Chen, J.W., Manjunath, B.S.: Brain tumor segmentation and tractographic feature extraction from structural MR images for overall survival prediction. In: Crimi, A., Bakas, S., Kuijf, H., Keyvan, F., Reyes, M., van Walsum, T. (eds.) BrainLes 2018. LNCS, vol. 11384, pp. 128–141. Springer, Cham (2019). https://doi.org/10.1007/978-3-030-11726-9_12

11. LeCun, Y., Bengio, Y., Hinton, G.: Deep learning. Nature **521**(7553), 436 (2015)

12. Menze, B.H., et al.: The multimodal brain tumor image segmentation benchmark (brats). IEEE Trans. Med. Imag. **34**(10), 1993–2024 (2014)

13. Nie, D., et al.: Multi-channel 3D deep feature learning for survival time prediction of brain tumor patients using multi-modal neuroimages. Sci. Rep. **9**(1), 1103 (2019)

14. Pérez-Beteta, J., et al.: Glioblastoma: does the pre-treatment geometry matter? A postcontrast T1 MRI-based study. Eur. Radiol. **27**(3), 1096–1104 (2017)

15. Shattuck, D.W., et al.: Construction of a 3D probabilistic atlas of human cortical structures. Neuroimage **39**(3), 1064–1080 (2008)

16. Wang, K., et al.: Radiological features combined with IDH1 status for predicting the survival outcome of glioblastoma patients. Neuro-oncology **18**(4), 589–597 (2015)

17. Wen, Y., Zhang, K., Li, Z., Qiao, Y.: A discriminative feature learning approach for deep face recognition. In: Leibe, B., Matas, J., Sebe, N., Welling, M. (eds.) ECCV 2016. LNCS, vol. 9911, pp. 499–515. Springer, Cham (2016). https://doi.org/10.1007/978-3-319-46478-7_31

18. Zhao, Z., Yang, G., Lin, Y., Pang, H., Wang, M.: Automated glioma detection and segmentation using graphical models. PLoS one **13**(8) (2018). e0200745

Encoder-Decoder Network for Brain Tumor Segmentation on Multi-sequence MRI

Andrei Iantsen$^{(\boxtimes)}$, Vincent Jaouen, Dimitris Visvikis, and Mathieu Hatt

LaTIM, INSERM, UMR 1101, University Brest, Brest, France
andrei.iantsen@inserm.fr

Abstract. In this paper we describe our approach based on convolutional neural networks for medical image segmentation in a context of the BraTS 2019 challenge. We use the conventional encoder-decoder architecture enhanced with residual blocks, as well as spatial and channel squeeze & excitation modules. The present paper describes the general pipeline including the data pre-processing, the choices regarding the model architecture, the training procedure and the chosen data augmentation techniques. Our final results in the BraTS 2019 segmentation challenge are Dice scores equal to 0.76, 0.87 and 0.80 for enhanced tumor, whole tumor and tumor core sub-regions, respectively.

Keywords: Medical imaging · Tumor segmentation · Encoder-decoder network · Soft Dice loss

1 Introduction

Glioma is a group of malignancies that arises from the glial cells in the brain. Currently, gliomas are the most common primary tumors of the central nervous system [2,6]. The symptoms of patients presenting with a glioma depend on the anatomical site of the glioma in the brain and can be too common (e.g headaches, nausea or vomiting, mood and personality alterations) to give an accurate diagnosis in early stages of the disease. The primary diagnosis is usually confirmed by magnetic resonance imaging (MRI) or computed tomography (CT) that provide additional structural information about the tumor.

Gliomas usually consist of heterogeneous sub-regions (edema, enhancing and non-enhancing tumor core, etc.) with variable histologic and genomic phenotypes [2]. Presently, multimodal MRI scans are used for non-invasive tumor evaluation and treatment planning, due to its ability to depict the tumor sub-regions with different intensities. However, segmentation of brain tumors in multimodal MRI scans is one of the most challenging tasks in medical imaging because of the high heterogenity in tumor appearances and shapes.

The brain tumor segmentation challenge (BraTS) [1] is aimed at development of automatic methods for the brain tumor segmentation. All participants of the

© Springer Nature Switzerland AG 2020
A. Crimi and S. Bakas (Eds.): BrainLes 2019, LNCS 11993, pp. 296–302, 2020.
https://doi.org/10.1007/978-3-030-46643-5_29

BraTS are provided with a clinically-acquired training dataset of pre-operative MRI scans (4 sequences per patient) and segmentation masks for three different tumor sub-regions, namely the GD-enhancing tumor, the peritumoral edema, and the necrotic and non-enhancing tumor core. The MRI scans were acquired with different clinical protocols and various scanners from multiple 19 institutions. Each scan was annotated manually by one to four raters and subsequently approved by expert raters.

The performance of proposed algorithms was evaluated by the Dice score, sensitivity, specificity and the 95th percentile of the Hausdorff distance.

2 Materials and Methods

2.1 Data Preprocessing

One of the main difficulties with applying automatic segmentation methods for MRI is that image intensities are not standardized and can exhibit a high variability in both intra- and inter-image domains. In order to make the intensities of MR images more homogeneous and use them as the input for our CNN, we applied Z-score normalization for each modality and each patient separately. The mean and the standard deviation were computed based on voxels with non-zero intensities corresponding to the brain region which helped in reducing the effect of the different brain sizes among patients. All background voxels remained unchanged after the normalization.

2.2 Network Architecture

The widely used 3D U-Net model [11] served as the basis to design our own neural network. Instead of using conventional convolutional blocks comprised of a $3 \times 3 \times 3$ convolution, a batch normalization layer (batch norm) and a ReLU activation function as a basic element of the network, we chose to rely upon a residual block with full pre-activation [13] supplemented by a concurrent spatial and channel squeeze & excitation (scSE) module [14]. In essence, this module is the version of the squeeze & excitation (SE) block [16], that was the key element of the SENetwork (the 1st prize in the object localization and classification task in the ImageNet 2017 challenge), modified for the image segmentation task. In order to include the scSE module in the residual block, we followed the same approach that was applied in SE-ResNet architectures [16]. Due to the high memory consumption working with 3D images, we switched from using batch norm layers to instance normalization (instance norm) [15] that was shown to work better in the small-batch regime [18].

We replaced max pooling operations in the encoder of the network by learnable downsampling blocks which consist of one $3 \times 3 \times 3$ strided convolutional layer, the instance norm, the ReLU activation and the scSE module [14]. Similarly, we implemented upsampling blocks in the decoder of the network using a $3 \times 3 \times 3$ transposed convolution instead.

To reduce memory consumption and increase the receptive field of the network without significant computational overhead, we implemented the first downsampling block with a kernel size of $7 \times 7 \times 7$ right after the input. The last convolutional layer followed by the softmax activation to produce the output of the model was applied with a kernel size $3 \times 3 \times 3$ in order to decrease the number of misclassified adjacent voxels and generate more smooth segmentation masks with less holes.

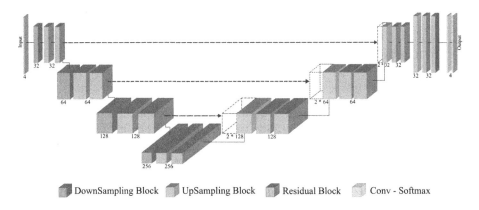

Fig. 1. Proposed encoder-decoder network with residual blocks.

2.3 Training Procedure

Due to the large size of provided MRIs, we trained the model on randomly extracted patches of the size $128 \times 160 \times 128$ voxels with a batch size of 1. First, the chosen patch size was large enough to cover the significant part of the brain region for all patients. Second, since we did not apply any explicit regularization techniques for the model, such as dropout layers or a weight decay, training on random patches with a batch size of 1 (*online learning*) helped to reduce overfitting.

We trained the model for 300 epochs using Adam optimizer with $\beta_1 = 0.9$ and $\beta_2 = 0.99$ for exponential decay rates for moment estimates. We experimentally determined reasonable bounds for the learning rate and applied a cosine annealing schedule gradually reducing the learning rate from $lr_{max} = 10^{-4}$ to $lr_{min} = 10^{-6}$ for every 25 epochs and performing the adjustment at each epoch. The described procedure provided better results in our experiments compared with other techniques, e.g. an exponential learning rate decay or reducing the learning rate on a plateau.

We experimented with different data augmentation methods during training. Our best results were received when applying mirroring on the axial plane for each training patch, gamma correction and random rotations with the angle uniformly sampled from the range $[5, 15]°$ along the random set of axes for each input MRI sequence before extracting the patch.

2.4   Loss Function

Considering the fact that the Dice score was one of the metrics used for the evaluation in the BraTS 2019 segmentation challenge, we trained our model with the Soft Dice Loss Function. Based on [12], the loss function for one training example can be written as

$$L(y, \hat{y}) = 1 - \frac{1}{\mathcal{C}} \sum_{c=1}^{\mathcal{C}} \frac{2 \sum_i^{\mathcal{N}} y_i^c \hat{y}_i^c + 1}{\sum_i^{\mathcal{N}} y_i^c + \sum_i^{\mathcal{N}} \hat{y}_i^c + 1} \tag{1}$$

where $y_i = \left[y_i^1, y_i^2, \ldots, y_i^C\right]^{\top}$ - the one-hot encoded label for the i-th voxel, $\hat{y}_i = \left[\hat{y}_i^1, \hat{y}_i^2, \ldots, \hat{y}_i^C\right]^{\top}$ - predicted probabilities for the i-th voxel. $\mathcal{N}$ and $\mathcal{C}$ are the total numbers of voxels and classes for the given example, respectively. Additionally we applied Laplacian smoothing by adding $+1$ to the numerator and the denominator in the loss function to avoid the zero division in cases when one or several labels were not represented in the training example.

It is important to notice that the training data in the BraTS challenge had segmentation labels for three tumor sub-regions, namely the necrotic and non-enhancing tumor core (NCR&NET), the peritumoral edema (ED) and the GD-enhancing tumor (ET). However, for the evaluation organizers used the GD-enhancing tumor (ET), the tumor core (TC), which was comprised of NCR&NET along with ET, and the whole tumor (WT) that combines all provided sub-regions. Hence, during the model training we minimized the loss directly on these nested tumor sub-regions.

2.5   Ensembling

For our experiments on the training dataset we applied stratified 10-folds cross-validation, splitting available images based on the source of the data which was extracted from patient IDs.

We combined all models trained on the different splits of the training data into an ensemble by averaging their predicted probability distributions.

The other challenging aspect of the competition was to correctly classify patients without the ET subregion. In order to reach this goal, we provided each model in the ensemble with *the right of veto*. If at least one model in the ensemble predicted the absence of the ET tumor subregion for the patient, we used the second highest probable label for all voxels that were classified as the ET by other models in the ensemble for this particular patient.

3   Results and Discussion

The results of the BraTS 2019 segmentation challenge are presented in Table 1 and Table 2. The Dice and Hausdorff distance scores were utilized for the evaluation. The results were obtained with the use of the online platform on the

Table 1. BraTS 2019 validation set results. The mean and the standard deviation of results for the single model were computed based on the performance of each individual model in the ensemble.

Metrics	Dice score			Hausdorff distance (95%)		
Class	ET	WT	TC	ET	WT	TC
Single model	0.67 ± 0.02	0.87 ± 0.01	0.79 ± 0.01	7.82 ± 1.07	8.35 ± 0.94	9.58 ± 1.35
Ensemble	0.72	0.89	0.81	4.86	6.19	7.68

Table 2. BraTS 2019 test set results. All metrics were computed for predictions of the ensemble.

Metrics	Dice score			Hausdorff distance (95%)		
Class	ET	WT	TC	ET	WT	TC
Mean	0.76	0.87	0.80	16.45	7.45	22.95
StdDev	0.22	0.14	0.27	69.57	29.25	79.59
Median	0.83	0.92	0.91	2	3.08	3
25% quantile	0.75	0.86	0.82	1.41	2	1.73
75% quantile	0.88	0.94	0.94	3	5.39	5.83

validation set with 125 patients (see Table 1) and the test set with 166 patients (see Table 2) without publicly available segmentation masks.

The high discrepancy in the mean Hausdorff distance between the validation set and the test set might be due to the fact that the model failed to make a prediction for one testing case (*BraTS19_Testing_123*).

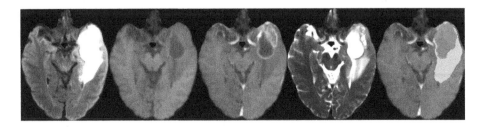

Fig. 2. MRI sequences and predicted segmentation for the patient *BraTS19_MDA_907_1* from the validation set with the Dice score equal to 0.74, 0.89 and 0.81 for enhanced tumor, whole tumor and tumor core sub-regions. From left: FLAIR, T1, T1ce, T2, T1 & segmentation mask.

Among all classes, the lowest average Dice score was obtained for the ET sub-regions. This can be partially explained by the relatively small size of the ET class. The other reason is that the Dice score metrics takes the lowest possible value 0 for all patients without the ET sub-region, if there is at least one false positive. Incorporating the right of veto in the ensemble helped to alleviate this problem but did not solved it completely. Another possible way to address this problem might be the use of some post-processing techniques on resulting segmentation masks.

We believe that one of the biggest drawbacks of our approach, which is directly related to the Dice score, is the soft Dice loss function which was used for training. Based on the Dice score, this loss function heavily penalizes false positives for all cases without the ET class which makes the optimization more difficult and unstable. Moreover, in these circumstances, the model becomes biased towards predicting the background class. In our opinion, the use of the soft Dice loss function in the combination with the other loss function, e.g the Cross-Entropy, might be a better alternative.

References

1. Menze, B.H., et al.: The multimodal brain tumor image segmentation benchmark (BRATS). IEEE TMI **34**(10), 1993–2024 (2015)
2. Bakas, S., et al.: Advancing the Cancer Genome Atlas glioma MRI collections with expert segmentation labels and radiomic features. Nat. Sci. Data (2017, in Press)
3. Bakas, S., et al.: Segmentation labels and radiomic features for the pre-operative scans of the TCGA-GBM collection. TCIA (2017)
4. Bakas, S., et al.: Segmentation labels and radiomic features for the pre-operative scans of the TCGA-LGG collection. TCIA (2017)
5. Bakas, S., Reyes, M., Jakab, A., Bauer, S., Rempfler, M., Crimi, A., et al.: Identifying the best machine learning algorithms for brain tumor segmentation, progression assessment, and overall survival prediction in the BRATS challenge. arXiv preprint arXiv:1811.02629 (2018)
6. Upadhyay, N., Waldman, A.D.: Conventional MRI evaluation of gliomas. Br. J. Radiol. **84**(Spec Iss 2), 107–111 (2011). Spec No. 2
7. Kamnitsas, K., et al.: Efficient multi-scale 3D CNN with fully connected CRF for accurate brain lesion segmentation. MIA **36**, 61–78 (2017)
8. Kamnitsas, K., et al.: Ensembles of multiple models and architectures for robust brain tumour segmentation. In: Crimi, A., Bakas, S., Kuijf, H., Menze, B., Reyes, M. (eds.) BrainLes 2017. LNCS, vol. 10670, pp. 450–462. Springer, Cham (2018). https://doi.org/10.1007/978-3-319-75238-9_38
9. Long, J., Shelhamer, E., Darrell, T.: Fully convolutional networks for semantic segmentation. In: Proceedings of the IEEE Conference on Computer Vision and Pattern Recognition, pp. 3431–3440 (2015)
10. Ronneberger, O., Fischer, P., Brox, T.: U-Net: convolutional networks for biomedical image segmentation. In: Navab, N., Hornegger, J., Wells, W.M., Frangi, A.F. (eds.) MICCAI 2015. LNCS, vol. 9351, pp. 234–241. Springer, Cham (2015). https://doi.org/10.1007/978-3-319-24574-4_28
11. Çiçek, Ö., Abdulkadir, A., Lienkamp, S.S., Brox, T., Ronneberger, O.: 3D U-Net: learning dense volumetric segmentation from sparse annotation. In: Ourselin, S., Joskowicz, L., Sabuncu, M.R., Unal, G., Wells, W. (eds.) MICCAI 2016. LNCS, vol. 9901, pp. 424–432. Springer, Cham (2016). https://doi.org/10.1007/978-3-319-46723-8_49

12. Milletari, F., Navab, N., Ahmadi, S.-A.: V-net: fully convolutional neural networks for volumetric medical image segmentation. In: International Conference on 3D Vision, pp. 565–571. IEEE (2016)
13. He, K., Zhang, X., Ren, S., Sun, J.: Identity mappings in deep residual networks. In: Leibe, B., Matas, J., Sebe, N., Welling, M. (eds.) ECCV 2016. LNCS, vol. 9908, pp. 630–645. Springer, Cham (2016). https://doi.org/10.1007/978-3-319-46493-0_38
14. Roy, A.G., Navab, N., Wachinger, C.: Concurrent spatial and channel squeeze & excitation in fully convolutional networks. arXiv preprint arXiv:1803.02579 (2018)
15. Ulyanov, D., Vedaldi, A., Lempitsky, V.: Instance normalization: the missing ingredient for fast stylization. arXiv preprint arXiv:1607.08022 (2016)
16. Hu, J., Shen, L., Sun, G.: Squeeze-and-excitation networks. CoRR, vol. abs/1709.01507. http://arxiv.org/abs/1709.01507 (2017)
17. He, K., Zhang, X., Ren, S., Sun, J.: Delving deep into rectifiers: surpassing human-level performance on imagenet classification. In: The IEEE International Conference on Computer Vision (ICCV), December 2015
18. Wu, Y., He, K.: Group normalization. In: Ferrari, V., Hebert, M., Sminchisescu, C., Weiss, Y. (eds.) ECCV 2018. LNCS, vol. 11217, pp. 3–19. Springer, Cham (2018). https://doi.org/10.1007/978-3-030-01261-8_1

Deep Convolutional Neural Networks for Brain Tumor Segmentation: Boosting Performance Using Deep Transfer Learning: Preliminary Results

Mostefa Ben Naceur[1,2(✉)], Mohamed Akil[1], Rachida Saouli[2], and Rostom Kachouri[1]

[1] Gaspard Monge Computer Science Laboratory, A3SI, ESIEE Paris, CNRS, University Paris-Est, Marne-la-Vallée, France
{mostefa.bennaceur,mohamed.akil,rostom.kachouri}@esiee.fr
[2] Smart Computer Sciences Laboratory, Computer Sciences Department, Exact.Sc, and SNL, University of Biskra, Biskra, Algeria
rachida.saouli@esiee.fr

Abstract. Brain tumor segmentation through MRI images analysis is one of the most challenging issues in medical field. Among these issues, Glioblastomas (GBM) invade the surrounding tissue rather than displacing it, causing unclear boundaries, furthermore, GBM in MRI scans have the same appearance as Gliosis, stroke, inflammation and blood spots. Also, fully automatic brain tumor segmentation methods face other issues such as false positive and false negative regions. In this paper, we present new pipelines to boost the prediction of GBM tumoral regions. These pipelines are based on 3 stages, first stage, we developed Deep Convolutional Neural Networks (DCNNs), then in second stage we extract multi-dimensional features from higher-resolution representation of DCNNs, in third stage we developed machine learning algorithms, where we feed the extracted features from DCNNs into different algorithms such as Random forest (RF) and Logistic regression (LR), and principal component analysis with support vector machine (PCA-SVM). Our experiment results are reported on BRATS-2019 dataset where we achieved through our proposed pipelines the state-of-the-art performance. The average Dice score of our best proposed brain tumor segmentation pipeline is 0.85, 0.76, 0.74 for whole tumor, tumor core, and enhancing tumor, respectively. Finally, our proposed pipeline provides an accurate segmentation performance in addition to the computational efficiency in terms of inference time makes it practical for day-to-day use in clinical centers and for research.

Keywords: Brain tumor segmentation · Convolutional Neural Networks · Support vector machine · Glioblastomas · Transfer learning · Principal component analysis

© Springer Nature Switzerland AG 2020
A. Crimi and S. Bakas (Eds.): BrainLes 2019, LNCS 11993, pp. 303–315, 2020.
https://doi.org/10.1007/978-3-030-46643-5_30

1   Introduction

Brain tumor is a growing abnormal cell in the brain or central spin canal [1]. Usually, a radiologist uses MRI scans as the most effective [2] technique to generate Multi-modal images and to identify different tumor regions in the soft tissues of the brain. In general, a radiologist generates four standard MRI images modalities for Gliomas tumors diagnosis [3]: T2-weighted fluid attenuated inversion recovery (Flair), T1-weighted (T1), T1-weighted contrast-enhanced (T1c), and T2-weighted (T2) for each patient. Furthermore, what makes the diagnosis hard for radiologist is that each patient has a different health condition, age, gender, in addition to Glioblastoma tumors is unexpected, in other words, these tumors could appear anywhere in the brain.

Current state-of-the-art methods in the field of computer vision are based on a deep learning, especially Convolutional Neural Networks (CNNs). Where in CNNs [4], we find a feature extractor with a bank of convolution layers, then pooling layers to make the images less sensitive and invariant to small translations, then the last step in CNNs is a classifier (in general Softmax layer) that classifies each pixel into one of a set of classes. After the breakthrough in 2012 of AlexNet [5] model that outperformed the state-of-the-art methods in the field of object recognition, many methods obtained high results in many fields especially in medical field such as [6–11]. In general, these methods are trained on 4 types of MRI images: Flair, T1, T1c, and T2.

Our ongoing work is based on our previous work [9,15]. In this paper, we are focusing on two major issues: (1) false positive regions – where the model predicts non-tumor regions as tumor regions but in fact they are not-, (2) false negative regions – where the model classifies some regions as non-tumor regions but in fact they are. In [9,15] we addressed the problem of false positive regions by two steps: we used a global threshold for each slice to remove small non-tumoral regions based on connected-components, then in second step, to enhance the post-processing step more we used a morphological opening operator. Despite the success of these two post-processing steps, further steps are required to improve the segmentation results. The main reason of these two issues (i.e. false positive regions and false negative regions) is the classifier of DCNNs, where in our case the classifier is the Softmax function. Softmax function gives an estimated vector at the end after each forward propagation of DCNNs, by normalizing the outputs to stay between 0 and 1, i.e., the outputs become as probabilities. Then, we pick the result of the forward pass based on the maximum probability among all probabilities, and this maximum probability represent a class, in our case, one class out of the 4 predefined classes (i.e., Necrotic and Non-Enhancing tumor, Peritumoral Edema, Enhancing tumor and healthy tissue). The Softmax function is a simple and an accurate function for training phase, but it is not adequate for the prediction or test phase for the problem of instance segmentation. The extracted features from MRI scans are complicated and in this case of DCNNs are hierarchical, so, classifying these features among a set of classes is not simple and intuitive for Softmax function. To overcome this issue, we developed two brain tumor segmentation pipelines, firstly, we extract feature maps from DCNNs, secondly, these features maps become the dataset of training and testing for another machine learning algorithms. The first pipeline is based on two algorithms: RF and LR, where the second pipeline is based on PCA-SVM.

The aim of this paper is to propose and develop new pipelines for brain tumor segmentation. Where we use the technique of transfer learning to extract features from DCNNs architecture then we feed these features into another machine learning algorithms (i.e., RF, LR and PCA-SVM), then for the first pipeline we combine the results of DCNNs, RF and LR into one 3D image, while for the second pipeline we train PCA-SVM on the extracted features maps. The proposed pipelines are used to segment the brain tumors of GBM with both high- and low-grade.

2   Proposed Method

In this paper, we proposed two pipelines to boost the segmentation performance of GBM brain tumor. One of the main issues for segmentation performance degradation is false positive and false negative regions, and the main reason of these issues is Softmax function, where this simple function does not provide an accurate results at the prediction phase (testing phase). Thus, to improve the accuracy of Softmax in the first proposed pipeline, we combine the results of Softmax with the results of two algorithms (i.e., RF, LR), while for the second pipeline, we feed the feature maps of the last layer of DCNNs to PCA-SVM. So for both pipelines, the first step, we extract the higher-resolution feature maps of the last layer before output layer (Softmax layer). The second step, we prepare these feature maps to become in a better representation for machine learning algorithms.

2.1   Features Visualization

In this paper, we trained a DCNNs model from scratch where this model is trained for brain tumor segmentation of GBM. DCNNs work by learning to extract hierarchical features from images. The algorithm of CNNs [4] is originally inspired by the visual system. In 1962, Hubel and Wiesel [16] discovered that each type of neurons in the visual system, responds to one specific feature: vertical lines, horizontal lines, shapes, etc. From this discovery, the algorithm of CNNs was developed. Thus, CNNs detect low-level features such as lines, edges, etc. and high-level features such as shapes and objects. These hierarchical features help CNNs algorithm to better locate the boundaries of the tumor regions (see Fig. 1). The Fig. 1 shows different features of a subject from our training set, as you can see, we can clearly distinguish the boundaries of each region.

2.2   Fusion Solution-Based Predictions Aggregation

The flowchart for detecting GBM brain tumors (see Fig. 2) is composed of two parts: the first part is composed of the DCNNs model and the second part is

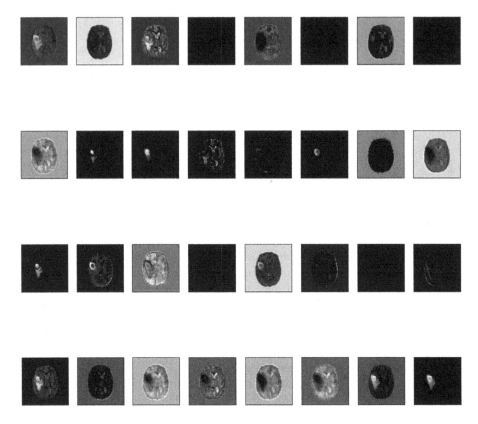

Fig. 1. Features visualization of the layer 16 of the CNNs model. Layer 16 of CNNs has 32 features maps, each has a filter to search for a specific feature the most representative in a MRI image

composed of machine learning algorithms such as RF and LR. To integrate the LR and RF in DCNNs, we first extracted the feature maps of the last layer before the output, then we replaced the Softmax layer with the aforementioned algorithms (i.e., LR and RF). Then we fused the results of these three algorithms (LR, RF and Softmax) into one result to diagnose the presence of tumor in each extracted feature maps and to diagnose the class of this tumor in each pixel.

To develop a DCNNs architecture, we have either pixel-wise approach or patch-wise approach. The first approach deals with pixels, while second approach deals with patches. In this paper, we used patch-wise classification, car it provides a good segmentation results [19–21] compared to pixel-wise classification, in addition it is less prone to overfitting, and these advantages are due to the parameters sharing between neurons in the network. Patch-wise approach takes as an input patches with limited size, and after extensive experiments to get the best patch's size (e.g., $32 \times 32 \times 4$, $64 \times 64 \times 4$) for our approach, we observed that patches with size ($64 \times 64 \times 4$) provide an accurate segmentation

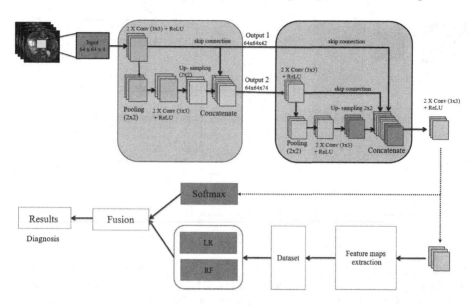

Fig. 2. Flowchart fusion solution for predicting GBM brain tumors. This flowchart has two parts: the first part (two green boxes) represent the DCNNs model, the second part (three blue boxes) represent the aggregation results of Sofmtax, LR and RF in a single prediction result (Color figure online)

performance in terms of the evaluation metrics (see Sect. 2.6). The optimization of DCNNs is done using stochastic gradient descent with mini-batch size equals to 8 and learning rate computed as follows:

$$LER_i = 10^{-3} \times 0.99^{LER_{i-1}} \tag{1}$$

Where the initial learning rate (LER) was $LER_0 = 0.001$, LER_i ($i \in \mathbb{N}^+$) is the new learning rate, LER_{i-1} is the learning rate of the last epoch, 0.99 is a decreasing factor. The DCNNs model was implemented on Keras which is a high-level open source Deep Learning library, and Theano as a Back-end, where Theano exploits GPUs to optimize Deep Learning architectures (i.e., to minimize the error). In this work, all our results are obtained using Python environment on windows 64-bit, Intel Xeon processor 2.10 GHz with 24 GB RAM, and the training is done on Nvidia Quadro GPU K1200 with 4 GB RAM.

After training DCNNs model, we extracted feature maps of the last layer before applying the Softmax classifier to train two other classifiers: LR, RF. These two classifiers are powerful to avoid the overfitting problem, in addition, RF is considered as an ensemble learning, where it is used to improve the system's performance in many applications.

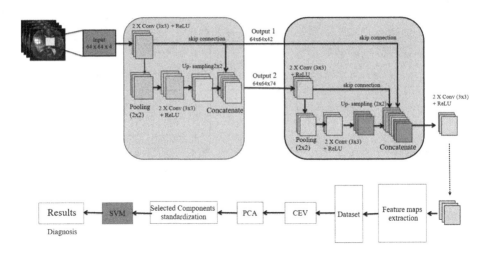

Fig. 3. Flowchart of PCA-SVM solution for predicting GBM brain tumors. This flowchart starts by a DCNNs model (two green boxes), then the extracted feature maps from DCNNs will be reduced by applying cumulative explained variance (CEV), and the reduced selected components are fed into SVM to predict the class of each pixel in GBM brain tumor. (Color figure online)

2.3   Semi Automatic-Based Support Vector Machine

In the flowchart in Fig. 3, we used an SVM method to segment the MRI images of patients with GBM. In this flowchart, we first trained DCNNs as we did in the first step (see Sect. 2.1). To improve the segmentation performance, we extracted and collected the feature maps of the last layer before the classifier (Softmax) into one dataset for training and testing (60% and 40%, respectively). Secondly, to reduce the huge dimensionality of the features maps, we computed the cumulative explained variance (CEV) in order to obtain the number of components that cover all variance (most useful information). By using these techniques, we can reduce the number of redundant features in addition to noise, because noise do not have a high variance allow it to be extracted among the first components. After applying CEV method on the extracted features, we observed that 99% of variances are concentrated on only 139 components out of 3200 components; which means that 3061 of components hold redundant features and noise (see Fig. 4).

As you can see from Fig. 4, the first figure (Fig. 4.a) represents the dimension of all features, where (Fig. 4.b) represents only 139 dimensions among 3200 dimensions, thus from these two figures we can conclude that 139 components represent 99% of variances (information). After computing CEV, we applied principal components analysis (PCA) to reduce the dimensionality of features until 139 components; for each patch, we use only 139 components instead of 3200 components. Reducing the dimensionality using PCA helps to get new

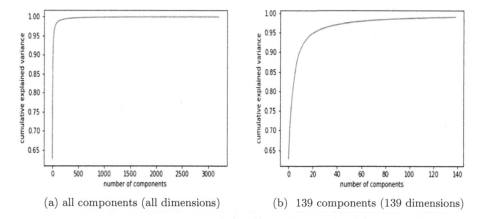

(a) all components (all dimensions)   (b) 139 components (139 dimensions)

Fig. 4. Cumulative explained variance curve for (a) all components and for (b) 139 components. It can be seen that we only need 139 components out of 3200 components to represent 99% of the features. The remaining components (i.e., 3061) represent redundant features and noise.

features more representative, without getting redundant features and noise which need a lot of preprocessing to remove them.

The last step in this flowchart (see Fig. 3) consists of applying SVM which is one of the most powerful methods for classification as it can deal with many forms of data and classification problems (binary, multi, linear, non-linear). To develop a SVM method: firstly, we need a lot of data, here we use the computed components by using PCA. Secondly, we need to specify the type of problem, in our case, it is a multi-classification issue (4 classes). Usually, multi-classification problem is a non linearly separable issue, so to verify the type of problem (linear, non-linear), we draw the first three components in 3D space (see Fig. 5):

As you can see in (Fig. 5.a and 5.b), we drew only 3 components from 139 components, and as expected the issue of multi-classification in our case is non linearly separable. From Fig. 5, we conclude that the issue that we are dealing with, is multi-classification and non linearly separable issue, thus this conclusion helps us to determine the hyperparameters of SVM especially the kernel. In general, in SVM, there are three types of kernels: linear, polynomial and radial basis function (RBF). Firstly we can eliminate linear kernel because it is used for linearly separable issues, secondly polynomial kernel is computationally expensive and needs a lot of memory, thus, in this paper we use the RBF kernel. Moreover, for the other hyperparameters: coefficient gamma γ and C slack-penalty: because SVM is sensitive to outliers and feature scaling and as we mentioned earlier the problem of multi-classification of brain tumor segmentation is non linearly separable issue which means in this case C slack-penalty is great than zero (soft-margin classification); some instances could be on the street of the decision boundary (margin violations). Furthermore, the value of gamma γ controls the influence of each feature in its search space, thus in this case and because of

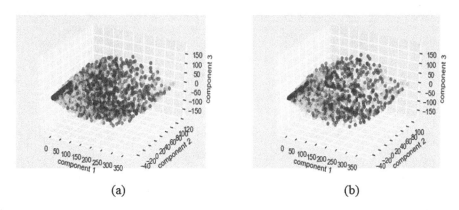

(a)   (b)

Fig. 5. Scatter plot for the first three components in 3D space in (a) training and (b) testing datasets. It can be seen that in the training and testing dataset, the issue of brain tumor segmentation is a nonlinear multi classification issue. Best viewed in color. (Color figure online)

outliers, gamma γ should be a little bit high. Because there is no straightforward method to select gamma γ and C, we used cross-validation with random search and grid search techniques on a small dataset of 30 brain images from the training set. The result of cross-validation as the following: gamma $\gamma = 0.001$ and $C = 5$. Moreover, SVM is sensitive to feature scaling because the kernel RBF is used to compute the objective function of SVM, thus, RBF assumes that all features are centered around zero and the variance's magnitude for all features is the same, so, we standardized the computed components of PCA, to be centered with 0 mean and a standard deviation of 1 as the following (see Eq. 2):

$$Standardization(X) = \frac{X - U}{\sigma} \tag{2}$$

Where X is a sample, U is the Mean of the training set and σ is the Standard Deviation.

Brain tumor segmentation is primarily used for diagnosis, treatment, and follow-up. The developed pipeline in this paper is applied to GBM which are brain tumors and life-threatening diseases. These tumors have four classes: Necrotic and Non-Enhancing tumor, Peritumoral Edema, Enhancing tumor and healthy tissue. To interpret MRI images, a radiologist employs a manual segmentation. Furthermore, it is known that the manual segmentation in MRI images is a time-consuming and a tedious procedure. In general, there are three methods to obtain a brain tumor segmentation image: manual, semi-automatic and fully automatic. In this paper, we investigate the segmentation performance using extracted features from DCNNs architecture and a SVM method which is a supervised learning method. Our approach is semi-automatic, where this approach needs a user interaction to reduce the misclassified regions by the SVM method, that are in most cases false positives.

2.4   Dataset

We have used a publicly available dataset called BraTS 2019 dataset [12–14,17, 18], where the training set has 259 patient's brain images with high-grade (HGG) and 76 patient's brain images with low-grade (LGG). Each patient's brain image comes with 4 MRI sequences (i.e., Flair, T1, T1c and T2) and the ground truth of 4 segmentation labels which are obtained manually by radiologists: Healthy tissue, Necrotic and Non-Enhancing tumor, Peritumoral Edema, Enhancing core. BRATS 2019 validation and testing sets contain 125 and 166 images respectively of patients with unknown grade, i.e. the validation and testing sets do not have the Ground Truth labels. Our DCNNs model is built upon 2D image patches (Slice), where this model predicts the pixel's class which is the center of the 2D patch.

2.5   Pre-processing

To enhance the quality of the MRI scans and to remove some noise, we applied 3 steps:

1. Removing 1% highest and lowest intensities: this technique helps to remove some noise at the tail of the histogram, where this step has provided good results in many research [11].
2. Subtracting the mean and dividing by the standard deviation of non-zero values in all channels: this technique is used to center and to put the data in the same scale, i.e. bringing the mean intensity value and the variance between one and minus one.
3. In this step, we try to isolate the background from the tumoral regions by assigning the minimum values to -9, where it has been observed that using integer numbers between -5 to -15, fit our DCNNs model. The application of the second pre-processing step, led to bringing the mean value in the range $[-1, 1]$, in other words, the intensities of all regions in addition to healthy and background became between -1 and 1. As we know, the intensity of background pixels of MRI images in BRATS data equals to 0, thus to isolate the zero pixels (background) from the other regions, we normalized the histogram of the MRI images by shifting the zero pixels to another bin outside the range $[-1, 1]$. We found that the bin -9 in many experiments, gives good results in the training and testing phases.

2.6   Evaluation

To evaluate the performance of the proposed flowchart (see Fig. 2), we used BRATS online evaluation system[1]. This system evaluates the uploaded images using four metrics: Dice score, Sensitivity, Specificity, and Hausdorff distance:

$$\text{Dice (P,T)} = \frac{|P_1 \wedge T_1|}{(|P_1| + |T_1|)/2}, \quad \text{Sensitivity (P,T)} = \frac{|P_1 \wedge T_1|}{|T_1|}, \quad \text{Specificity (P,T)} = \frac{|P_0 \wedge T_0|}{|T_0|},$$

$$\text{Hausdorff (P,T)} = max \left\{ \sup_{p \in \partial P_1} \inf_{t \in \partial T_1} d(p , t) , \sup_{t \in \partial T_1} \inf_{p \in \partial P_1} d(t , p) \right\}$$

---

[1] https://ipp.cbica.upenn.edu/.

3   Results

In this section, we evaluate our proposed brain tumor segmentation pipelines on a public BRATS 2019 dataset using the online evaluation system.

Table 1. Evaluation results of fusion solution-based predictions aggregation pipeline on BRATS 2019 validation set. WT, TC, ET denote whole tumor, tumor core, enhancing tumor, respectively.

	Dice score			Sensitivity			Specificity			Hausdorff		
	WT	TC	ET	WT	TC	ET	WT	TC	ET	WT	TC	ET
Mean	0.84	0.70	0.61	0.84	0.71	0.69	0.99	0.99	1.0	22.64	20.40	13.76
Standard deviation	0.13	0.23	0.33	0.16	0.26	0.28	0.01	0.01	0.01	26.10	24.63	24.90
Median	0.88	0.77	0.77	0.89	0.80	0.81	0.99	1.0	1.0	8.37	10.68	3.16
25 quantile	0.84	0.58	0.43	0.82	0.61	0.60	0.99	0.99	1.0	3.61	6.40	2.0
75 quantile	0.91	0.87	0.86	0.94	0.92	0.89	1.0	1.0	1.0	37.29	22.20	11.0

Table 2. Evaluation results of fusion solution-based predictions aggregation pipeline on BRATS 2019 testing set. WT, TC, ET denote whole tumor, tumor core, enhancing tumor, respectively.

	Dice score			Hausdorff		
	WT	TC	ET	WT	TC	ET
Mean	0.84709	0.75889	0.73703	12.99701	15.4957	6.03933
Standard deviation	0.15312	0.25993	0.23841	23.97851	25.62727	16.45033
Median	0.89588	0.85913	0.8148	4.30077	8.09315	2.23607
25 quantile	0.83621	0.74323	0.70943	3	4	1.41421
75 quantile	0.92368	0.9143	0.87902	7.95064	14.65374	3.74166

Table 1 and Table 2 show the segmentation results of our proposed pipeline (see Fig. 2) for fully automatic brain tumor segmentation. The prediction of tumoral regions is performed using 2D patches with size equals to $64 \times 64 \times 4$ (4 corresponds to using different modalities such as T1, post-contrast T1, T2 and FLAIR). Then, we extract the feature maps of the last layer before Softmax function, then we feed these features into different machine learning algorithms such as RF and LR. Last step, we combine the results of DCNNs, RF and LR into one 3D image using voting technique; where the most predicted label among the predictions (e.g., 1, 1 and 0) of these classifiers (algorithms) become the result (in this case the label becomes 1). Table 1 and Table 2 shows the validation and testing scores, and as you can see, we are able to achieve segmentation results comparable to the top performing methods in state-of-the-art such as the work of [10]. Moreover, the achieved median score is high: 0.88, 0.89 for whole tumor on dice score (validation and testing, respectively), this high values is due to achieving a good segmentation performance for most MRI images.

Table 3. Evaluation results of semi-automatic-based support vector machine pipeline on some subjects of BRATS 2019 training set. WT, TC, ET denote whole tumor, tumor core, enhancing tumor, respectively.

	Dice score			Sensitivity			Specificity			Hausdorff		
	WT	TC	ET	WT	TC	ET	WT	TC	ET	WT	TC	ET
"BraTS19_2013_10_1"	1	0.58996	0.24228	1	0.74294	0.19889	1	0.96818	0.99043	0	29.03446	26.41969
"BraTS19_2013_11_1"	1	0.60334	0.14959	1	0.80945	0.25156	1	0.96026	0.97204	0	13	13
"BraTS19_2013_12_1"	1	0.46689	0.18173	1	0.79719	0.28464	1	0.89892	0.93791	0	27.09243	30.09983

Table 3 shows the segmentation results of our semi-automatic method that is based on PCA and SVM. In this table, we show different metrics (see Sect. 2.6). As you can see our method provides good segmentation results especially on tumor core and enhancing tumor regions and that is due to the extracted features from the layer 16 of the DCNNs model and the selected components of PCA. Please note that this method is semi automatic, where it needs a user interaction to select the tumoral regions, that's why the whole tumor is 1 for Dice score, Sensitivity and Specificity, and 0 for Hausdorff distance. Please note also that the validation and testing sets do not have the ground truth labels for the four regions. In the future, we will study the impact of each layer in the DCNNs model in addition to the standardized components of PCA on the segmentation results.

4   Discussion and Conclusions

In this paper, we developed two brain tumor segmentation pipelines for GBM brain tumors, these pipelines are based on DCNNs and learned features maps. The proposed DCNNs model uses skip connections and up-sampling filters to maximize the features representation inside the model. Also, using short skip connections helps to complete the missing information during the pooling layers and convolution striding, and long skip connections encourage the feature reuse which assists the model to combine the low-level and the high-level features and to better locate the tumor regions. Moreover, to overcome the issues of false positive and false negative regions, we extracted the feature maps to train another two machine learning algorithms: random forest, logistic regression, and SVM. These algorithms showed a high impact on the segmentation performance.

Our experimental results show that our proposed brain tumor segmentation pipelines improved the evaluation metrics (.i.e., Dice score, Sensitivity, Specificity, Hausdorff). The Mean Dice score of our best proposed fully automatic brain tumor segmentation pipeline (see Fig. 2) is 0.85, 0.76, 0.74 for whole tumor,

tumor core, and enhancing tumor, respectively. The second pipeline (see Fig. 3) is a semi automatic method based on PCA, SVM and learned feature maps of DCNNs. In this study, we used cumulative explained variance with PCA to reduce the dimension of features to 139 components that are enough to provide 99% of variances for each patient image. Then, we applied a SVM method to predict the class of each pixel. The showing segmentation results are promising and give a high segmentation performance, in which we can enhance it in the future with more investigation in the different phases from feature extraction to prediction using machine learning algorithms. Moreover, the proposed pipeline is suitable for adopting in research and as a part of different clinical settings.

As a perspective of this research, we intend to investigate principal component analysis (PCA) to explore and reduce the features dimensionality, where with this technique we can improve the results by using only the features that have a huge impact on the segmentation results. After this study, we intend to integrate support vector machine (SVM), PCA and DCNNs into an end-to-end supervised learning algorithm.

References

1. Young, R.J., Knopp, E.A.: Brain MRI: tumor evaluation. J. Magn. Reson. Imaging: Official J. Int. Soc. Magn. Reson. Med. **24**(4), 709–724 (2006)
2. Akram, M.U., Usman, A.: Computer aided system for brain tumor detection and segmentation. In: 2011 International Conference on Computer Networks and Information Technology (ICCNIT), pp. 299–302. IEEE (2011)
3. Işın, A., Direkoglu, C., Şah, M.: Review of MRI-based brain tumor image segmentation using deep learning methods. Proc. Comput. Sci. **102**, 317–324 (2016)
4. LeCun, Y., Bottou, L., Bengio, Y., Haffner, P., et al.: Gradient-based learning applied to document recognition. Proc. IEEE **86**(11), 2278–2324 (1998)
5. Krizhevsky, A., Sutskever, I., Hinton, G.E.: ImageNet classification with deep convolutional neural networks. In: Advances in Neural Information Processing Systems, pp. 1097–1105 (2012)
6. Davy, A., et al.: Brain tumor segmentation with deep neural networks. In: Proceedings of the MICCAI Workshop on Multimodal Brain Tumor Segmentation Challenge BRATS, pp. 01–05 (2014)
7. Pereira, S., Pinto, A., Alves, V., Silva, C.A.: Deep convolutional neural networks for the segmentation of gliomas in multi-sequence MRI. In: Proceedings of the MICCAI Workshop on Multimodal Brain Tumor Segmentation Challenge BRATS, pp. 52–55 (2015)
8. Chang, P.D., et al.: Fully convolutional neural networks with hyperlocal features for brain tumor segmentation. In: Proceedings MICCAI-BRATS Workshop, pp. 4–9 (2016)
9. Ben Naceur, M., Saouli, R., Akil, M., Kachouri, R.: Fully automatic brain tumor segmentation using end-to-end incremental deep neural networks in MRI images. Comput. Methods Programs Biomed. **166**, 39–49 (2018)
10. Zhao, X., Wu, Y., Song, G., Li, Z., Zhang, Y., Fan, Y.: A deep learning model integrating FCNNs and CRFs for brain tumor segmentation. Medical Image Anal. **43**, 98–111 (2018)

11. Havaei, M., et al.: Brain tumor segmentation with deep neural networks. Med. Image Anal. **35**, 18–31 (2017)
12. Menze, B.H., et al.: The multi-modal brain tumor image segmentation benchmark (BRATS). IEEE Trans. Med. Imaging **34**(10), 1993–2024 (2015). https://doi.org/10.1109/TMI.2014.2377694
13. Bakas, S., et al.: Advancing The Cancer Genome Atlas glioma MRI collections with expert segmentation labels and radiomic features. Nat. Sci. Data **4**, 170117 (2017). https://doi.org/10.1038/sdata.2017.117
14. Bakas, S., et al.: Identifying the Best machine learning algorithms for brain tumor segmentation, progression assessment, and overall survival prediction in the BRATS challenge. arXiv preprint arXiv:1811.02629 (2018)
15. Ben Naceur, M., Kachouri, R., Akil, M., Saouli, R.: A new online class-weighting approach with deep neural networks for image segmentation of highly unbalanced glioblastoma tumors. In: Rojas, I., Joya, G., Catala, A. (eds.) IWANN 2019. LNCS, vol. 11507, pp. 555–567. Springer, Cham (2019). https://doi.org/10.1007/978-3-030-20518-8_46
16. Hubel, D.H., Wiesel, T.N.: Receptive fields, binocular interaction and functional architecture in the cat's visual cortex. J. Physiol. **160**(1), 106–154 (1962)
17. Bakas, S., et al.: Segmentation labels and radiomic features for the pre-operative scans of the TCGA-GBM collection. The Cancer Imaging Archive (2017). https://doi.org/10.7937/K9/TCIA.2017.KLXWJJ1Q
18. Bakas, S., et al.: Segmentation labels and radiomic features for the pre-operative scans of the TCGA-LGG collection. The Cancer Imaging Archive (2017). https://doi.org/10.7937/K9/TCIA.2017.GJQ7R0EF
19. Ronneberger, O., Fischer, P., Brox, T.: U-Net: convolutional networks for biomedical image segmentation. In: Navab, N., Hornegger, J., Wells, W.M., Frangi, A.F. (eds.) MICCAI 2015. LNCS, vol. 9351, pp. 234–241. Springer, Cham (2015). https://doi.org/10.1007/978-3-319-24574-4_28
20. Çiçek, Ö., Abdulkadir, A., Lienkamp, S.S., Brox, T., Ronneberger, O.: 3D U-Net: learning dense volumetric segmentation from sparse annotation. In: Ourselin, S., Joskowicz, L., Sabuncu, M.R., Unal, G., Wells, W. (eds.) MICCAI 2016. LNCS, vol. 9901, pp. 424–432. Springer, Cham (2016). https://doi.org/10.1007/978-3-319-46723-8_49
21. Milletari, F., Navab, N., Ahmadi, S.A.: V-Net: fully convolutional neural networks for volumetric medical image segmentation. In: 2016 Fourth International Conference on 3D Vision (3DV), pp. 565–571. IEEE, October 2016

Multimodal Brain Tumor Segmentation with Normal Appearance Autoencoder

Mehdi Astaraki[1,2(✉)], Chunliang Wang[1], Gabriel Carrizo[1],
Iuliana Toma-Dasu[2], and Örjan Smedby[1]

[1] Department of Biomedical Engineering and Health Systems,
KTH Royal Institute of Technology, Hälsovägen 11C, 14157 Huddinge, Sweden
{mehast, chunwan}@kth.se
[2] Karolinska Institutet, Department of Oncology-Pathology,
Karolinska Universitetssjukhuset, 17176 Solna, Stockholm, Sweden

Abstract. We propose a hybrid segmentation pipeline based on the autoencoders' capability of anomaly detection. To this end, we, first, introduce a new augmentation technique to generate synthetic paired images. Gaining advantage from the paired images, we propose a Normal Appearance Autoencoder (NAA) that is able to remove tumors and thus reconstruct realistic-looking, tumor-free images. After estimating the regions where the abnormalities potentially exist, a segmentation network is guided toward the candidate region. We tested the proposed pipeline on the BraTS 2019 database. The preliminary results indicate that the proposed model improved the segmentation accuracy of brain tumor subregions compared to the U-Net model.

Keywords: Brain tumor segmentation · Variational Autoencoder · Paired-image generation

1 Introduction

Glioma is one of the most common types of malignant tumors originating within the brain. Due to different characteristics of the tumor cells, gliomas can be categorized into two major classes. Regardless of whether the tumors are classified as Low-Grade Gliomas (LGGs) with less aggressive behavior or High-Grade Gliomas (HGGs), more aggressive ones, the treatment outcome will most likely be poor [1]. Recent evidence suggests that the existence of intratumor heterogeneity is likely a key factor in treatment failures [2]. This heterogeneity can be observed in multimodal Magnetic Resonance (MR) scans where the tumor subregions, i.e. peritumoral edema, necrotic core, enhancing and non-enhancing tumor core are portrayed with varying shape, appearance and intensity patterns. In addition to this highly heterogenous appearance, gliomas can appear anywhere in the brain with the same appearance as gliosis, and more importantly they invade the surrounding tissues instead of displacing them. Therefore, segmentation of brain tumors in multi modal MR scans is one of the most challenging tasks in medical image analysis [3].

Existing brain tumor segmentation methods can be categorized as either generative or discriminative models. The generative approaches are based on using prior information

© Springer Nature Switzerland AG 2020
A. Crimi and S. Bakas (Eds.): BrainLes 2019, LNCS 11993, pp. 316–323, 2020.
https://doi.org/10.1007/978-3-030-46643-5_31

such as probabilistic anatomical atlases [4]. On the other hand, discriminative models are aimed at classifying the image voxels as healthy or abnormal tissue. Although encouraging results have been achieved by extracting well-defined handcrafted features such as textures and local histogram, and then employing classification algorithms such as Support Vector Machines (SVMs) [5], accurate segmentation of non-trivial cases such as LGGs without enhancing part remains challenging. More recently, developing Convolutional Neural Networks (CNNs) for brain tumor segmentation tasks have resulted in significant improvement of the segmentation accuracy [6, 7]. While 3D CNNs are theoretically capable of taking full advantageous from volumetric MR images, the computational cost may make this approach impractical, and therefore volumetric segmentation of gliomas with 2.5D CNNs has been investigated [8].

In this paper, we propose a novel segmentation pipeline based on the generation of synthetic paired images. By adopting a blending algorithm, first, we generate synthetic tumor images from healthy images. Then, generated paired images are employed to learn a Variational Autoencoder (VAE) [9] in order to automatically remove the tumors from real tumor images. Finally, the simulated healthy images are post-processed to be added to a segmentation network.

2   Methods

2.1   Data and Pre-processing

In this study, we used the dataset of 2019 Brain Tumor Segmentation (BraTS) challenge which consists of the training set, validation set and a testing set [10–14]. The training set includes 259 HGG and 76 LGG subjects with the corresponding annotations. The manual segmentation masks contain four labels: 1 for necrotic (NCR) and Non-Enhancing Tumor (NET), 2 for Peritumoral Edema (ED), 4 for Enhancing Tumor (ET) and 0 for the rest of the image. The validation set includes 125 subjects without the grades and segmentation masks. Each subject contains four MR scans, namely T1-weighted (T1), post-contrast T1-weighted (T1Gd), T2-weighted (T2), and T2 Fluid Attenuated Inversion Recovery (FLAIR). The provided data set were already pre-processed, i.e., registered to a template and skull-stripped. Apart from the mentioned steps, we applied a bias correction field algorithm to fix the intensity inhomogeneities in MR images [15]. The BraTS challenge aims at the segmentation of three subregions: enhancing tumor, tumor core, and whole tumor.

2.2   Paired-Image Generation

Anomaly detection is a conventional approach to learn the distribution and appearance of healthy anatomical structures which is often achieved via unsupervised learning [16]. However, the latent representation of healthy brain structures cannot be easily discriminated from challenging unhealthy structures such as LGGs. Accordingly, we performed the anomaly detection task in a supervised fashion to be able to distinguish the subtle differences. This approach, however, requires the availability of both healthy and tumor images, paired for each subject, which is not clinically realistic. To tackle this limitation, we introduce a novel augmentation technique to generate synthetic paired images. By employing a 2 dimensional Poisson blending algorithm [17], real tumoral image slices were blended with healthy brain images. In fact, real tumors were extracted from the BraTS training set, therefore since both the tumors and healthy slices were already aligned in a same coordinate system, no registration step was required. By applying geometric and intensity-based transformations on the tumor images, a wide variety of synthetic tumor images were produced (Fig. 1).

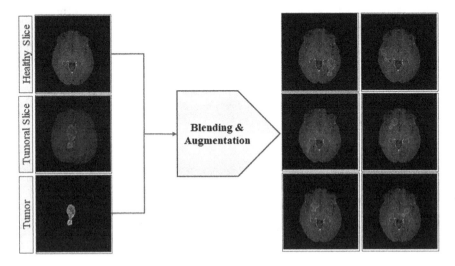

Fig. 1. Blending the tumors over healthy normal image slices.

2.3   Automatic Tumor Removal with Normal Appearance Autoencoder

Detecting abnormalities that deviate from the training data has been a main field of research in machine learning. By imposing a constraint on the latent variables, autoencoders turn into a probabilistic framework (VAE) that would be able to generate new images by sampling from the learned latent distribution. In this study, having the advantages of paired images, we modified the training of the VAE and turned it into a supervised generative problem. By showing the synthetic tumor images to the network and requiring the model to reconstruct corresponding healthy images, the latent variables are constrained to learn the distribution of only the healthy anatomical part of the brain. Therefore, during the training phase, only synthetic paired images were used, but in the inference phase real tumor images were shown to the model. We named the proposed model, Normal Appearance Autoencoder (NAA) [18].

2.4   Tumor Segmentation

The difference between the input tumor images and the reconstructed tumor-free images were calculated through intensity thresholding and morphological operators.

The difference images were then considered as a prior source of information by adding them as an additional channel to the segmentation network in order to guide the attention of the model toward the candidate regions. To implement this, we developed a 5-channel U-Net [19] like network (4 channels for image slices and 1 channel for the prior slice) to segment the tumor subregions with 3 independent 2D networks for 3 orthogonal views (axial, coronal, and sagittal). A majority voting algorithm was then applied to the segmentation results in order to reconstruct the volumetric masks. To perform a fair comparison, the segmentation experiments were repeated with the same model but only with 4 input channels as image slices (base model) (Fig. 2).

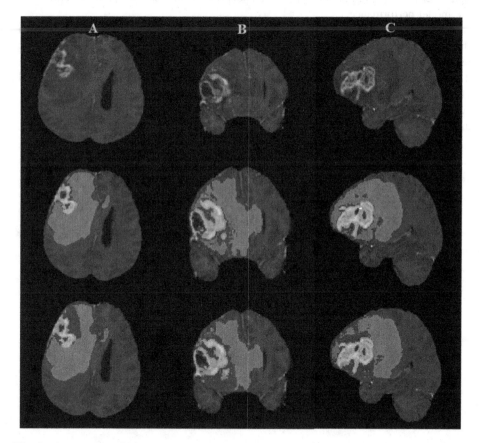

Fig. 2. An example of improved segmentation accuracy by adding the prior channel. First row: the original image slices in 3 orthogonal views as Axial (A), Coronal (B), and Sagittal (C). Second row: segmentation results of the base model. Third row: segmentation results of the prior-based model. The yellow color represents the enhancing part, red color represents the non-enhancing and necrotic subregions, and green indicates the edema subregion. (Color figure online)

3   Experiments and Results

The NAA contains 4 convolutional blocks as the encoder, one convolutional block as latent space, and 4 up-convolutional blocks as the decoder. A new regularization term was added to the NAA loss function to minimize the inconsistency between the latent distribution of input images and reconstructed images [20]. Accordingly, the optimization procedure of the NAA was performed through a multi-loss function. The segmentation network contains the same number of convolutional/up-convolutional blocks in encoder and decoder. Each block includes two convolutional layers, one max-pooling and one batch normalization layer. Both models were trained with the Adam optimizer for a maximum number of 200 epochs with early stopping criteria defined on the model performance.

The segmentation models were trained with a 5-fold cross-validation approach on the training data set, and then all the training data were employed to train a model to be used for the validation and test data sets. The segmentation results were evaluated with 4 metrics including Dice Coefficient (DC), sensitivity (SEN), specificity (SPC), and Hausdorff Distance (HD) directly on the online system[1]. It should be mentioned that we submitted the segmentation masks under the name of "KTH-MTH2019" (See Tables 1 and 2).

Table 1. Mean values of the four metrics on BraTS 2019 training set.

Sub-region	Metric	Method	
		Base model	Priori model
Enhanced	DC	0.76 ± 0.21	0.80 ± 0.18
	SEN	0.72 ± 0.23	0.77 ± 0.20
	SPC	0.99 ± 0.00	0.99 ± 0.00
	HD	3.44 ± 4.36	2.13 ± 2.60
Core	DC	0.74 ± 0.25	0.85 ± 0.10
	SEN	0.70 ± 0.27	0.80 ± 0.15
	SPC	0.99 ± 0.00	0.99 ± 0.00
	HD	7.38 ± 7.12	4.79 ± 3.41
Whole	DC	0.84 ± 0.13	0.90 ± 0.06
	SEN	0.80 ± 0.17	0.87 ± 0.10
	SPC	0.99 ± 0.00	0.99 ± 0.00
	HD	6.39 ± 5.50	3.82 ± 3.64

---

[1] https://ipp.cbica.upenn.edu.

Table 2. Mean values of the four metrics on BraTS 2019 validation set.

Sub-region	Metric	Method	
		Base model	Priori model
Enhanced	DC	0.70 ± 0.29	0.71 ± 0.28
	SEN	0.73 ± 0.26	0.80 ± 0.17
	SPC	0.99 ± 0.00	0.99 ± 0.00
	HD	5.61 ± 10.28	6.02 ± 11.95
Core	DC	0.78 ± 0.24	0.81 ± 0.14
	SEN	0.74 ± 0.27	0.85 ± 0.14
	SPC	0.99 ± 0.00	0.99 ± 0.00
	HD	7.17 ± 9.51	7.16 ± 10.85
Whole	DC	0.87 ± 0.10	0.87 ± 0.09
	SEN	0.85 ± 0.14	0.88 ± 0.11
	SPC	0.99 ± 0.00	0.99 ± 0.01
	HD	5.73 ± 7.72	5.90 ± 8.76

4 Discussion and Conclusion

In this study, we proposed a new pipeline for brain tumor segmentation which includes three steps: synthetic paired image generation, reconstruction of tumor-free images, and a prior-based segmentation network. While the introduced blending step is beneficial for supervised training algorithms, it can be also considered as an image augmentation technique that can be helpful to tackle the unavailability of the large number of tumoral images. Moreover, by modifying the VAE into NAA, we showed that the performance of the model in learning the distribution of healthy anatomical structures can be improved. In addition, instead of directly segmenting the tumor by post-processing the output of the NAA, we add it as a prior source of information to the segmentation model. The results indicate that adding the prior information to the U-Net could successfully improve the segmentation accuracy. In fact, the higher values of Dice and sensitivity metrics achieved from the proposed priori model, convey the message that adding representative information to even simple models (simple U-Net with 4 levels of depth) can potentially improve the model performance. It is worth mentioning that applying the bias field correction would lead to, slightly, reduce the contrast in Flair images, therefore it would cause some effects on ED segmentation accuracy. Although the ED segmentation is not evaluated independently, such effect should be quantified in future studies.

In conclusion, we have proposed a hybrid brain tumor segmentation model based on the capability of autoencoders in abnormality detection. The proposed method delivered promising results and further development will include an extension of the model to 3D.

Acknowledgement. This study was supported by the Swedish Childhood Cancer Foundation (grant no. MT2016-0016), the Swedish innovation agency Vinnova (grant no. 2017-01247) and the Swedish Research Council (VR) (grant no. 2018-04375).

References

1. Havaei, M., et al.: Brain tumor segmentation with deep neural networks. Med. Image Anal. **35**, 18–31 (2017). https://doi.org/10.1016/J.MEDIA.2016.05.004

2. Sottoriva, A., et al.: Intratumor heterogeneity in human glioblastoma reflects cancer evolutionary dynamics. Proc. Natl. Acad. Sci. U.S.A. **110**, 4009–4014 (2013). https://doi.org/10.1073/pnas.1219747110

3. Zhao, X., Wu, Y., Song, G., Li, Z., Zhang, Y., Fan, Y.: A deep learning model integrating FCNNs and CRFs for brain tumor segmentation. Med. Image Anal. **43**, 98–111 (2018). https://doi.org/10.1016/J.MEDIA.2017.10.002

4. Menze, B.H., van Leemput, K., Lashkari, D., Weber, M.-A., Ayache, N., Golland, P.: A generative model for brain tumor segmentation in multi-modal images. In: Jiang, T., Navab, N., Pluim, J.P.W., Viergever, M.A. (eds.) MICCAI 2010. LNCS, vol. 6362, pp. 151–159. Springer, Heidelberg (2010). https://doi.org/10.1007/978-3-642-15745-5_19

5. Li, H., Fan, Y.: Label propagation with robust initialization for brain tumor segmentation. In: 2012 9th IEEE International Symposium on Biomedical Imaging (ISBI), pp. 1715–1718. IEEE (2012)

6. Isensee, F., Kickingereder, P., Wick, W., Bendszus, M., Maier-Hein, K.H.: No new-net. In: Crimi, A., Bakas, S., Kuijf, H., Keyvan, F., Reyes, M., van Walsum, T. (eds.) BrainLes 2018. LNCS, vol. 11384, pp. 234–244. Springer, Cham (2019). https://doi.org/10.1007/978-3-030-11726-9_21

7. Myronenko, A.: 3D MRI brain tumor segmentation using autoencoder regularization. In: Crimi, A., Bakas, S., Kuijf, H., Keyvan, F., Reyes, M., van Walsum, T. (eds.) BrainLes 2018. LNCS, vol. 11384, pp. 311–320. Springer, Cham (2019). https://doi.org/10.1007/978-3-030-11726-9_28

8. Wang, C., Smedby, Ö.: Automatic brain tumor segmentation using 2.5D U-Nets. In: MICCAI Workshop on Multimodal Brain Tumor Segmentation (BRATS) Challenge, Québec, pp. 292–296 (2017)

9. Kingma, D.P., Welling, M.: Auto-encoding variational Bayes (2013)

10. Menze, B.H., Jakab, A., Bauer, S., Kalpathy-Cramer, J., Farahani, K., et al.: The multimodal brain tumor image segmentation benchmark (BRATS). IEEE Trans. Med. Imaging **34**, 1993–2024 (2015). https://doi.org/10.1109/tmi.2014.2377694

11. Bakas, S., Akbari, H., Sotiras, A., Bilello, M., Rozycki, M., et al.: Advancing the cancer genome atlas glioma MRI collections with expert segmentation labels and radiomic features. Sci. Data. **4** (2017). https://doi.org/10.1038/sdata.2017.117. Article no. 170117

12. Bakas, S., Reyes, M., Jakab, A., Bauer, S., Rempfler, M., et al.: Identifying the Best Machine Learning Algorithms for Brain Tumor Segmentation, Progression Assessment, and Overall Survival Prediction in the BRATS Challenge. (2018)

13. Bakas, S., Akbari, H., Sotiras, A., Bilello, M., Rozycki, M., et al.: Segmentation labels and radiomic features for the pre-operative scans of the TCGA-GBM collection. Cancer Imaging Arch. (2017). https://doi.org/10.7937/k9/tcia.2017.klxwjj1q

14. Bakas, S., Akbari, H., Sotiras, A., Bilello, M., Rozycki, M., et al.: Segmentation labels and radiomic features for the pre-operative scans of the TCGA-LGG collection. Cancer Imaging Arch. (2017). https://doi.org/10.7937/k9/tcia.2017.gjq7r0ef

15. Tustison, N.J., et al.: N4ITK: improved N3 bias correction. IEEE Trans. Med. Imaging **29**, 1310–1320 (2010). https://doi.org/10.1109/tmi.2010.2046908

16. Baur, C., Wiestler, B., Albarqouni, S., Navab, N.: Deep autoencoding models for unsupervised anomaly segmentation in brain MR images. In: Crimi, A., Bakas, S., Kuijf, H., Keyvan, F.,

Reyes, M., van Walsum, T. (eds.) BrainLes 2018. LNCS, vol. 11383, pp. 161–169. Springer, Cham (2019). https://doi.org/10.1007/978-3-030-11723-8_16

17. Pérez, P., Gangnet, M., Blake, A., Pérez, P., Gangnet, M., Blake, A.: Poisson image editing. In: ACM SIGGRAPH 2003 Papers on - SIGGRAPH 2003, p. 313. ACM Press, New York (2003)

18. Astaraki, M., Toma-Dasu, I., Smedby, Ö., Wang, C.: Normal appearance autoencoder for lung cancer detection and segmentation. In: Shen, D., et al. (eds.) MICCAI 2019. LNCS, vol. 11769, pp. 249–256. Springer, Cham (2019). https://doi.org/10.1007/978-3-030-32226-7_28

19. Ronneberger, O., Fischer, P., Brox, T.: U-Net: convolutional networks for biomedical image segmentation. In: Navab, N., Hornegger, J., Wells, W.M., Frangi, A.F. (eds.) MICCAI 2015. LNCS, vol. 9351, pp. 234–241. Springer, Cham (2015). https://doi.org/10.1007/978-3-319-24574-4_28

20. Chen, X., Konukoglu, E.: Unsupervised detection of lesions in brain MRI using constrained adversarial auto-encoders (2018)

Knowledge Distillation for Brain Tumor Segmentation

Dmitrii Lachinov[1]([✉]) [iD], Elena Shipunova[2], and Vadim Turlapov[3] [iD]

[1] Department of Ophthalmology and Optometry, Medical University of Vienna,
Vienna, Austria
`dmitrii.lachinov@meduniwien.ac.at`
[2] Intel, Moscow, Russian Federation
`elena.shipunova@intel.com`
[3] Lobachevsky State University, Nizhny Novgorod, Russia
`vadim.turlapov@itmm.unn.ru`

Abstract. The segmentation of brain tumors in multimodal MRIs is one of the most challenging tasks in medical image analysis. The recent state of the art algorithms solving this task are based on machine learning approaches and deep learning in particular. The amount of data used for training such models and its variability is a keystone for building an algorithm with high representation power.

In this paper, we study the relationship between the performance of the model and the amount of data employed during the training process. On the example of brain tumor segmentation challenge, we compare the model trained with labeled data provided by challenge organizers, and the same model trained in omni-supervised manner using additional unlabeled data annotated with the ensemble of heterogeneous models.

As a result, a single model trained with additional data achieves performance close to the ensemble of multiple models and outperforms individual methods.

Keywords: BraTS · Segmentation · Knowledge distillation · Deep learning

1 Introduction

Brain tumor segmentation is a reliable instrument for disease monitoring. Moreover, it is a central and informative tool for planning further treatment and assessing the way the disease progresses. However, manual segmentation is a tedious and time-consuming procedure that requires a lot of attention from the grader.

To simplify clinicians workload, many automatic segmentation algorithms have been proposed recently. The majority of them utilize machine learning techniques and deep learning in particular. The downside of these approaches

---

https://github.com/lachinov/brats2019.

ⓒ Springer Nature Switzerland AG 2020
A. Crimi and S. Bakas (Eds.): BrainLes 2019, LNCS 11993, pp. 324–332, 2020.
https://doi.org/10.1007/978-3-030-46643-5_32

is the amount of data required to successfully train a deep learning model. To capture all possible variations of biological shapes, it is desired to have them present in the manually labeled dataset.

The dataset provided in the scope of Brain Tumor Segmentation Challenge (BraTS 2019) [1,4,11] is the largest publicly available dataset [2,3] with MRI scans of brain tumors. This year, it contains 259 High-Grade Gliomas (HGG) and 76 Low-Grade Gliomas (LGG) in the training set. Each MRI scan describes native (T1), post-contrast T1-weighted (T1Gd), T2-weighted (T2) and T2 Fluid Attenuated Inversion Recovery (T2-FLAIR) volumes. These images were obtained using different scanner and protocols from multiple institutions. Each scan is registered to the same anatomical template, skull stripped and resampled to the isotopic resolution. All the images were manually segmented by multiple graders who followed the same annotation protocol. The GD-enhancing tumor, the peritumoral edema, the necrotic and non-enhancing tumor core were annotated on the scans.

The BraTS dataset is considered to be one of the largest publicly available medical dataset with 3D data. Despite the dataset size, it is still considered small compared to the natural images datasets, that may contain millions of samples. Datasets of such scale cover broad number of training examples, allowing algorithms to catch details that are not present in smaller size datasets. We think that the limited amount of tumor shapes and locations is the main reason why regularization techniques [12] work especially well in the provided scope. In this study, we are trying to solve the problem of limited dataset size from a different angle. We propose to utilize all available unlabeled data in the training process using the knowledge distillation [7,14].

Originally the knowledge distillation was proposed by Hinton et al. [7] for transferring the knowledge of an ensemble to the single neural network. Authors introduced the new type of ensemble consisting of several big models and many specialist models which learned to distinguish fine-grained classes that the 'big' models were misclassifying. The soft labels could be utilized as a regularization during the training of the final model. The effectiveness of the proposed approach was demonstrated on MNIST dataset as well as on the speech recognition task.

The variation of knowledge distillation called data distillation was introduced by Radosavovic et al. [14], where authors investigated omni-supervised learning, a special case of semi-supervised learning with available labeled data as well as unlabeled internet-scale data sources. Authors proposed to annotate unlabeled data with a single model by averaging predictions produced by differently transformed input data. The automatically annotated data was then used for training of a student model on the combined dataset. The data distillation was demonstrated on the example of human keypoint detection and general object detection, where student models outperform models trained solely on labeled data.

Inspired by previous papers, in this study we employ the knowledge distillation method to train student model with labeled data from BraTS 2019 challenge, automatically labeled data from BRaTS 2016 and unlabeled data from

BraTS 2019 and 2018. To provide annotation for unlabeled datasets we train an ensemble of models.

2   Building an Ensemble

The key idea of our approach is to enhance the generalization power of the student model by training it on the larger scale dataset. We achieve this by utilizing the unlabeled data available in the scope of the BraTS challenge.

Starting from 2017, training dataset of the challenge includes manually annotated data, as well as data from 2012 and 2013 challenges, that was also graded manually. However, data from 2014–2016 challenges was segmented by the fusion of best-performing methods from the challenges of previous years. Later, that data was discarded. Even though the quality of annotation grew significantly, the number of available samples is still incomparable to the scale of natural images datasets. In our opinion, the quantity of training samples is as important as the quality of annotation. And at the current moment, the data we have doesn't represent all possible variations of tumor shapes and structures.

During the last year's challenge, Isensee et al. [9] demonstrated that the accuracy of the model benefits from adding data from 2014–2016 challenges. We believe that this is one of the reasons why the mentioned method was ranked high during the 2018 competition. To further investigate the effectiveness of adding additional data into the training, we propose to utilize data from 2014–2016 and 2018 challenges. More specifically, we employ BraTS 2016 dataset, that is segmented automatically by the organizers. We also utilize this year's validation dataset, as well as 2018 testing dataset. Since the ground truth is unavailable, we first train an ensemble of the models. In the ensemble, we include multiple architectures [9,10,12] that demonstrated their performance during previous challenges. Then we automatically annotate unlabeled data with this ensemble trained solely on the BraTS 2019 training dataset; and we train a single model on the extended dataset. We describe the baseline methods that form our ensemble in the following section.

2.1   No New Net

The first method we employ in our ensemble is UNet [15] with 3D convolutions [5] and minor modifications employed by Isensee et al. [9] for participation in BraTS 2018 challenge, where this method was ranked second.

UNet is a fully convolutional decoder-encoder network with skip connections, that allows to segment fine structures well. It is especially effective in segmentation of biomedical imaging data. With slight modification its capable of showing state of the art results. For instance, Isensee et al. replaced ReLU activations with its leaky variant with negative slope equals to 10^{-2}. At the same time, trilinear upsampling was used in the decoder with prior filter number reduction. Instance Normalization [16] was used instead of Batch Normalization [8] due to

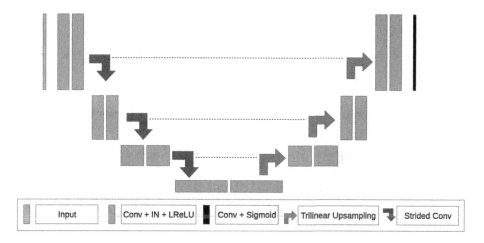

Fig. 1. The architecture of UNet that we use in our experiments. Dotted lines represent skip-connections. Instead of max pooling we use strided convolutions.

inaccurate statistics estimation with small batch sizes. The network architecture is illustrated on Fig. 1.

We train this network with a patch size of $128 \times 128 \times 128$ voxels and batch size of 2 for 160k iterations of weight updates. The initial learning rate was set to 10^{-4} and then dropped by a factor of 10 at 120k iterations. As in the paper [9] we employed a mixture of Soft Dice Loss and Binary Cross-Entropy for training. Similarly, instead of predicting 4 target classes, we segment three overlapping regions: Whole Tumor, Tumor Core and Enhancing Tumor.

$$L^{dice}(g,p) = \frac{1}{K} \cdot \sum_{k=1}^{K} \frac{2 \sum p_k \cdot g_k}{\sum p_k^2 + g_k^2}$$

$$L^{bce}(g,p) = -\frac{1}{N} \cdot \sum_{k=1}^{K} \sum (g_k \cdot log(p_k) + (1 - g_k) \cdot log(1 - p_k))$$

where N - is a number of voxels in the output, K - number of classes, g_k and p_k is ground truth and predicted probabilities of class k respectively. The resulting loss function is calculated as a sum of Soft Dice loss and BCE loss:

$$L^{final}(g,p) = L^{dice}(g,p) + L^{bce}(g,p)$$

2.2 UNet with Residual Connections

The second method we use in the ensemble is a UNet [5,15] with residual connections [6]. This model coupled with the autoencoder branch that imposes additional regularization on the encoder was employed by A. Myronenko [12], who took the first place at previous year's challenge.

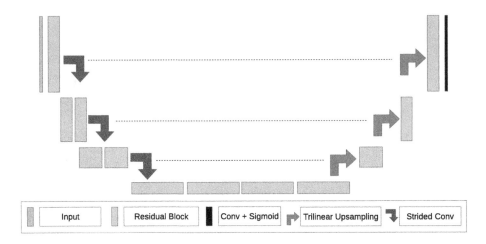

Fig. 2. UNet with residual blocks. Dotted arrows represent skip-connections.

Opposite to standard UNet, authors propose to use asymmetrically large encoder compared to the relatively thin decoder. ReLU activations are used as nonlinearities in this model. As a normalization layer, Group Normalization [17] is employed due to the small batch size used during training. Group Normalization computations are independent of batch size, thus GN provides similar performance on small and large batch sizes. In the encoder part, feature maps are progressively downsampled by 2 and increased in number by the same factor. The number of residual blocks at each level equals to 1, 2, 2 and 4. In the decoder, however, the number of residual blocks remains equal to 1 across all levels. Authors propose to use additional Variational Auto-Encoder (VAE) branch for regularization. We follow this choice for training the ensemble model, but we decided against using it in the student model. The network architecture is illustrated in Fig. 2.

We train this model with Adam optimizer using the same learning rate and learning schedule as for the previous baseline method. The loss function we used is the same as in the previous case. The patch size of $144 \times 144 \times 128$ gave us better results compared to the smaller ones.

2.3 Cascaded UNet

The third baseline method we use was previously introduced by the corresponding authors for participating in BraTS 2018. There we used a cascade of UNets [10], each one has multiple encoders that correspond to input modalities. In this method, we employed ReLU as the activation function. The network was built of basic pre-activation residual blocks that consist of two instance normalization layers, two ReLU activation layers and two convolutions with a kernel size of 3. We build a cascade of the models with the same topology that operate on the different scales of the input volume. Thus, we achieve an extremely large receptive field that can capture the global context. The architecture of base network in the cascade is illustrated in Fig. 3.

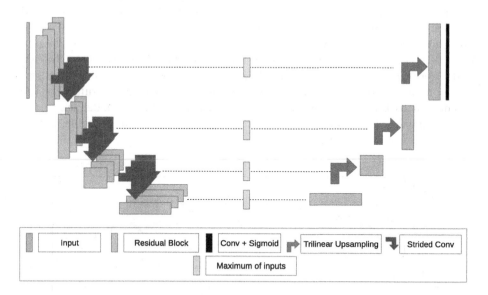

Fig. 3. Schematic representation of base network in the cascade. Each input modality has separate encoder. The encoder outputs are then merged together with elementwise maximum.

The model is trained with SGD with initial learning rate of 0.1, exponential learning rate decay with rate 0.99 for every epoch, the momentum of 0.9 and minibatch size equal to 4. All input samples were resampled to $128 \times 128 \times 128$ resolution. The training was performed for 500 epochs.

3 Data Preprocessing and Augmentation

As a preprocessing step, we normalize each input volume (modality) to have zero mean and unit variance for non-zero foreground voxels. This normalization is done independently for each input image. For training Cascaded UNet we also resample input images to the resolution of $128 \times 128 \times 128$.

To enhance generalization capabilities of the networks we perform a large set of data augmentations during training. First, we crop random regions of the images in the way, that at least one non background pixel is present in the cropped fragment. Then we randomly scale, rotate and mirror these images across X and Y axes. We decided to keep the original orientation along Z axis. Finally, we apply intensity shift and contrast augmentations for each modality independently.

4 Knowledge Distillation

We build the ensemble of above-named models by averaging the outputs. Next, we annotate all unlabeled data we have with this ensemble and use a combined set of manually and automatically labeled data as a training dataset. We pick architecture described in Sect. 2.2 without VAE branch as a student model and train it in the same way. Our experiments demonstrated that model with VAE under performs compared to the model without it. Due to increased dataset size we are no longer required to use regularization to get plausible results. For the submission to the evaluation system we trained the student model on all the data we have, with manual annotations and annotations from ensemble, except the dataset we are evaluating on.

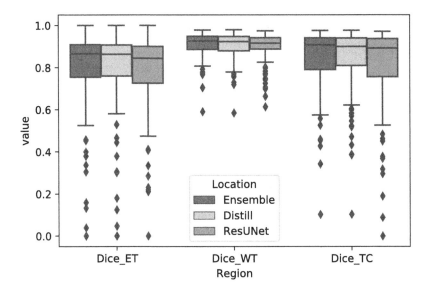

Fig. 4. The boxplot of the evaluation results on validation dataset.

Table 1. Performance of the methods. Local validation scores are reported.

Method	Dice ET	Dice WT	Dice TC
UNet	0.7836	0.9152	0.8743
Res UNet	0.7392	0.9204	0.8754
Casc UNet	0.9235	0.8925	0.8719
Distilled	0.7440	0.9218	0.8835

Table 2. Performance of the methods. The scores were evaluated on validation dataset.

Method	Dice ET	Dice WT	Dice TC
UNet	0.7402	0.8974	0.8349
Res UNet	0.7424	0.9018	0.8278
Casc UNet	0.7307	0.8997	0.8335
Ensemble	0.7562	0.9072	0.8435
Distilled	0.7563	0.9045	0.8420

Table 3. Performance of the student model on testing dataset.

Method	Dice ET	Dice WT	Dice TC
Distilled	0.7949	0.8874	0.8292

5 Results

The training was performed with PyTorch Framework [13] and single Nvidia 2080Ti. Evaluation was performed locally using stratified train-test split, as well as using validation dataset and online evaluation platform. Due to sensitivity of validation metric to borderline cases (ex. empty mask), Cascaded UNet demonstrated high Dice score segmenting Enhancing Tumor by correctly predicting these few borderline cases in the validation split.

On the validation dataset the ensemble of the models got the highest score. At the same time, the model trained with knowledge distillation scored almost the same as ensemble. As can be seen on Fig. 4 and Table 2, that represent results on validation dataset, student (distilled) model has lower Dice score variance of Enhancing tumor and Tumor core regions. On the local validation dataset (Table 1) the performance of the student model is comparable to the performance of the ensemble, except for ET class, where many empty masks are present. The final submission results of the distilled model on the test dataset are present in the Table 3. Surprisingly, the score for Enhancing Tumor got slightly increased compared to the validation results. That fact may indicate lower number of empty ET regions in test set compared to validation set.

Performance of the student model across all the evaluation is comparable to the performance of the ensemble. Initially, we expected it to even surpass ensemble performance as it was demonstrated in [14] due to larger number of samples to train on.

References

1. Bakas, S., et al.: Advancing the cancer genome atlas glioma MRI collections with expert segmentation labels and radiomic features. Sci Data 4 (2017). https://doi.org/10.1038/sdata.2017.117, http://www.ncbi.nlm.nih.gov/pmc/articles/PMC5685212/. 28872634[pmid], article no. 170117

2. Bakas, S., et al.: Segmentation labels and radiomic features for the pre-operative scans of the TCGA-GBM collection. Cancer Imaging Arch. (2017). https://doi.org/10.7937/K9/TCIA.2017.KLXWJJ1Q

3. Bakas, S., et al.: Segmentation labels and radiomic features for the pre-operative scans of the TCGA-LGG collection. Cancer Imaging Arch. (2017). https://doi.org/10.7937/K9/TCIA.2017.GJQ7R0EF

4. Bakas, S., et al.: Identifying the best machine learning algorithms for brain tumor segmentation, progression assessment, and overall survival prediction in the brats challenge. ArXiv abs/1811.02629 (2018)

5. Çiçek, Ö., Abdulkadir, A., Lienkamp, S.S., Brox, T., Ronneberger, O.: 3D U-Net: learning dense volumetric segmentation from sparse annotation. In: Ourselin, S., Joskowicz, L., Sabuncu, M.R., Unal, G., Wells, W. (eds.) MICCAI 2016. LNCS, vol. 9901, pp. 424–432. Springer, Cham (2016). https://doi.org/10.1007/978-3-319-46723-8_49

6. He, K., Zhang, X., Ren, S., Sun, J.: Deep residual learning for image recognition. In: 2016 IEEE Conference on Computer Vision and Pattern Recognition (CVPR), pp. 770–778, June 2016. https://doi.org/10.1109/CVPR.2016.90

7. Hinton, G., Vinyals, O., Dean, J.: Distilling the knowledge in a neural network. arXiv e-prints arXiv:1503.02531, March 2015

8. Ioffe, S., Szegedy, C.: Batch normalization: accelerating deep network training by reducing internal covariate shift. CoRR abs/1502.03167 (2015). http://arxiv.org/abs/1502.03167

9. Isensee, F., Kickingereder, P., Wick, W., Bendszus, M., Maier-Hein, K.H.: No new-net. In: Crimi, A., Bakas, S., Kuijf, H., Keyvan, F., Reyes, M., van Walsum, T. (eds.) BrainLes 2018. LNCS, vol. 11384, pp. 234–244. Springer, Cham (2019). https://doi.org/10.1007/978-3-030-11726-9_21

10. Lachinov, D., Vasiliev, E., Turlapov, V.: Glioma segmentation with cascaded UNet. In: Crimi, A., Bakas, S., Kuijf, H., Keyvan, F., Reyes, M., van Walsum, T. (eds.) BrainLes 2018. LNCS, vol. 11384, pp. 189–198. Springer, Cham (2019). https://doi.org/10.1007/978-3-030-11726-9_17

11. Menze, B.H., et al.: The multimodal brain tumor image segmentation benchmark (BRATS). IEEE Trans. Med. Imaging **34**(10), 1993–2024 (2015). https://doi.org/10.1109/TMI.2014.2377694

12. Myronenko, A.: 3D MRI brain tumor segmentation using autoencoder regularization. CoRR abs/1810.11654 (2018). http://arxiv.org/abs/1810.11654

13. Paszke, A., et al.: Automatic differentiation in PyTorch. In: NIPS Autodiff Workshop (2017)

14. Radosavovic, I., Dollar, P., Girshick, R., Gkioxari, G., He, K.: Data distillation: towards omni-supervised learning. arXiv e-prints arXiv:1712.04440, December 2017

15. Ronneberger, O., Fischer, P., Brox, T.: U-Net: convolutional networks for biomedical image segmentation. In: Navab, N., Hornegger, J., Wells, W.M., Frangi, A.F. (eds.) MICCAI 2015. LNCS, vol. 9351, pp. 234–241. Springer, Cham (2015). https://doi.org/10.1007/978-3-319-24574-4_28

16. Ulyanov, D., Vedaldi, A., Lempitsky, V.S.: Instance normalization: the missing ingredient for fast stylization. CoRR abs/1607.08022 (2016). http://arxiv.org/abs/1607.08022

17. Wu, Y., He, K.: Group normalization. CoRR abs/1803.08494 (2018). http://arxiv.org/abs/1803.08494

Combined MRI and Pathology Brain Tumor Classification

Brain Tumor Classification Using 3D Convolutional Neural Network

Linmin Pei[1], Lasitha Vidyaratne[1], Wei-Wen Hsu[2],
Md Monibor Rahman[1], and Khan M. Iftekharuddin[1(✉)]

[1] Vision Lab, Electrical and Computer Engineering, Old Dominion University,
Norfolk, VA 23529, USA
{lxpei001,lvidy001,mrahm006,kiftekha}@odu.edu
[2] Electrical and Computer Engineering, Old Dominion University,
Norfolk, VA 23529, USA
wxhsu001@odu.edu

Abstract. In this paper, we propose a deep learning-based method for brain tumor classification. It is composed of two parts. The first part is brain tumor segmentation on the multimodal magnetic resonance image (mMRI), and the second part performs tumor classification using tumor segmentation results. A 3D deep neural network is implemented to differentiate tumor from normal tissues, subsequentially, a second 3D deep neural network is developed for tumor classification. We evaluate the proposed method using pateint dataset from Computational Precision Medicine: Radiology-Pathology Challenge (CPM: Rad-Path) on Brain Tumor Classification 2019. The result offers 0.749 for dice score and 0.764 for F1 score for validation data, while 0.596 for dice score and of 0.603 for F1 score for testing data, respectively. Our team was ranked second in the CPM:Rad-Path challenge on Brain Tumor Classification 2019 based on overall testing performance.

Keywords: Deep neural network · Brain tumor classification · Magnetic resonance image · Histopathology image

1 Introduction

Brain tumor is a fatal and complex disease. The overall average annual age-adjusted incidence rate for all primary brain and other CNS tumors was 23.03 per 100,000 population [1]. The estimated five- and ten- year relative survival rates are 35.0% and 29.3% for patients with malignant brain tumor, respectively, according to a report from 2011–2015 [1]. Even with substantial improvements in the treatment process, the median survival period of patients with glioblastoma (GBM), a progressive primary brain tumor, is still 12–16 months [2]. In general, patient's survival period highly depends on the grade of the tumor. Therefore, accurate tumor classification is imperative for proper prognosis. Diagnosis and grading of brain tumor are conventionally done by pathologists, who examine tissue sections fixed on glass slides under a light microscope. However, this manual process is time-consuming, tedious, and susceptible to human errors. Therefore, a computer-aided automatic brain tumor classification method is

© Springer Nature Switzerland AG 2020
A. Crimi and S. Bakas (Eds.): BrainLes 2019, LNCS 11993, pp. 335–342, 2020.
https://doi.org/10.1007/978-3-030-46643-5_33

highly desirable. Some traditional machine learning methods, such as Support Vector Machine (SVM) [3], Random Forest (RF) [4], etc. are widely used for brain tumor classification. However, these methods require hand-crafted features extraction.

The World Health Organization (WHO) revised the central nervous system (CNS) tumor classification criteria by including both pathology image and molecular information in 2016 [5]. However, non-invasive magnetic resonance image (MRI) is regarded as an alternative source to classify tumor. Hence, many works have proposed the use of MRI in designing automated tumor classification methods [3, 6–9]. In recent years, deep learning has shown great success in many areas, and specifically in brain tumor classification [10–12].

The whole slide histopathology image size may be usually up to 1–2 GB and computation burden is a challenge because of its massive size. A related challenge is to effectively select the region-of-interest (ROI). This work attempts to analyze both pathology image and MRI as the sources for the brain tumor classification task. The available training data includes multimodal MRI (mMRI) sequences such as T1-weighted MRI (T1), T1-weighted MRI with contrast enhancement (T1ce), T2-weighted MRI (T2), and T2-weighted MRI with fluid-attenuated inversion recovery (T2-FLAIR).

In this paper, we investigate deep learning-based tumor classification method using mMRI and histopathology data. Our results suggest that analysis of mMRI image alone may be sufficient for the desired tumor classification task. It is composed of two parts, brain tumor segmentation and tumor classification. For brain tumor segmentation, we take advantage of the Multimodal Brain Tumor Segmentation Challenge 2019 (BraTS 2019) training data set [13–17], and use a 3D UNet-like CNN [18]. Subsequently, we use a regular 3D CNN with the tumor segmentation output of the previous part to accomplish the final classification task.

2   Method

Brain tumor segmentation is utilized as a prerequisite for tumor classification. Accordingly, the accuracy of brain tumor segmentation impacts the final tumor classification performance. The framework of brain tumor segmentation is shown in Fig. 1. We use a UNet-like architecture for brain tumor segmentation, and then apply a regular 3D CNN architecture for the tumor classification task. The input brain tumors are categorized into one of three sub-types: glioblastoma (G), oligodendroglioma (O), and astrocytoma (A).

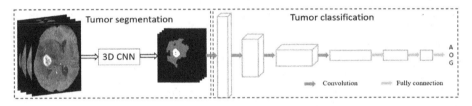

Fig. 1. Pipeline of the proposed method.

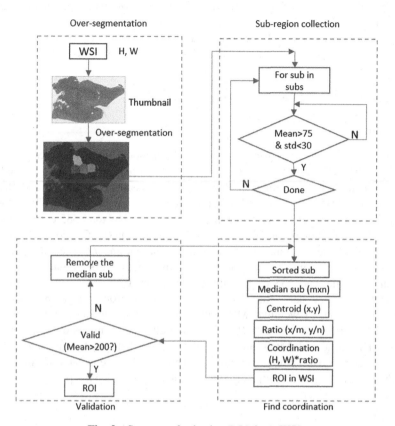

Fig. 2. Strategy of selecting ROI from WSI.

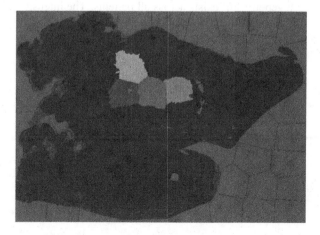

Fig. 3. A case example of selecting ROI from WSI.

We assume that the cancerous cells have lower intensity on WSI. We select ROI from WSI following the steps as shown Fig. 2. We perform over-segmentation [20] on the thumbnail WSI images. From the super-pixels, we select the region with mean intensity at 10-percentile of all regions. Finally, we map the location to original WSI, and take the corresponding region as the ROI. Figure 3 shows an example.

3   Materials and Pre-processing

3.1   Data

The training data is obtained from two sources: Multimodal Brain Tumor Segmentation Challenge 2019 (BraTS 2019) [13–17] and the Computational Precision Medicine: Radiology-Pathology Challenge on Brain Tumor Classification 2019 (CPM-RadPath 2019). Data of BraTS 2019 is used for tumor segmentation model, and CPM-RadPath 2019 is for tumor classification. For BraTS 2019 data, it has total 335 cases consisting of 259 high-grade glioma (HGG) and 76 low-grade glioma (LGG). For CPM-RadPath data, it has 221 cases in training, 35 cases in validation, and 73 cases in testing data, respectively. The training dataset contains 54 cases with astrocytoma, 34 cases with oligodendroglioma, and 133 cases with glioblastoma. Each case contains both radiology and pathology data. Radiology data consists of four image modalities: T1-weighted MRI (T1), T1-weighted MRI with contrast enhancement (T1ce), T2-weighted MRI (T2), and T2-weighted MRI with fluid-attenuated inversion recovery (T2-FLAIR). The structural MRIs are co-registered, skull-stripped, denoised, and bias corrected. Each case has paired pathology as well, and the size is varied from 50 Mb to 1 Gb. Note that the ground truths of validation and testing data are privately owned by the challenge organizer.

3.2   Pre-processing of Radiology Images

Even though the mMRIs are co-registered, skull-stripped, denoised, and bias corrected, the intensity inhomogeneity still exists across all patients. The intensity inhomogeneity may result in tumor misclassification. To reduce the impact of intensity inhomogeneity, z-score normalization is employed. Figure 4 shows an example with z-score normalization.

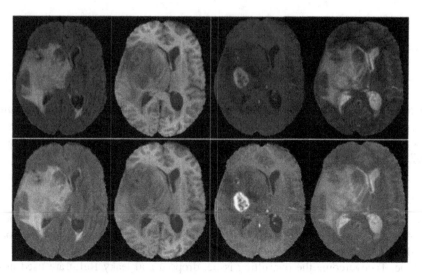

Fig. 4. An example of applying z-score normalization. Top row: raw images. Bottom row: z-score normalized images. From left to right: T2-FLAIR, T1, T1ce, and T2 image.

4   Experiments and Results

The dice score obtained using pathology image analysis is 0.716, which is less than that of using only radiology MRI. Hence, the remaining methods in this work use mMRI for analysis.

4.1   Hyper-parameter Setting

We empirically crop size of all mMRIs to $160 \times 192 \times 128$ to reduce the computational burden on the graphics processing unit (GPU). However, we ensure that all appropriate brain information is retained in the cropped version. The corresponding cross-entropy loss function is computed as follows:

$$LL_{ce} = -\sum\nolimits_{c=1}^{M} y_{o,c} log(p_{o,c}), \qquad (1)$$

where M is the number of classes, y is a binary indictor (0 or 1) if class label c is the correct classification for observation o. p is predicted probability observation o belonging to class c. We use Adam [19] optimizer with initial learning rate of $lr_0 = 0.001$ in training phase, and the learning rate (lr_i) is gradually reduced as follows:

$$lr_i = lr_0 * \left(1 - \frac{i}{N}\right)^{0.9}, \qquad (2)$$

where i is epoch counter, and N is total number of epochs in training.

4.2   Brain Tumor Segmentation

We randomly split the training dataset to have 80% for training, and the rest 20% used as validation based on high-grade glioma (HGG) and low-grade glioma (LGG). We then save the best model according to the validation performance. The saved model is then used for our final tumor segmentation on radiology data of CPM-RadPath 2019. The result offers dice score of 0.73659, 0.89612, and 0.81043 for enhancing tumor (ET), whole tumor (WT), and tumor core (TC), respectively. After post-processing steps (removing small objects and filling hole), the performance reaches dice of 0.76758, 0.90456, and 0.81657 for ET, WT, and TC, respectively.

4.3   Brain Tumor Classification

With the segmented tumor, we use a regular 3D CNN for classification. The tumor is categorized as glioblastoma, oligodendroglioma, or astrocytoma. We also randomly split data as training data and validation data with a ratio of 4:1 according to the tumor sub-type. To overcome the overfitting issue, dropout and leaky ReLu are used in the network. We achieve a training accuracy of 81%.

4.4   Online Evaluation

We applied the proposed method to CPM-RadPath 2019 validation data set, which includes 35 cases with radiology and pathology images. We achieve a best validation accuracy of 0.749. All the performance parameters are reported in Table 1.

Table 1. Online evaluation result by using our proposed method.

Phase	Dice	Average	Kappa	Balance_acc	F1_micro
Validation	0.749	0.764	0.715	0.749	0.829
Testing	0.596	NA	0.39	0.596	0.603

4.5   Discussion

In this work, we also try the experiment by integrating radiology images with pathology image for tumor classification. The histopathology images are traditionally used for tumor classification in clinical settings. However, there are some challenges for processing the pathology image. First, the size of whole slide image (WSI) has very large variation within the dataset. Second, the region-of-interest (ROI) selection is very challenging. The quality of selected ROI impacts the final tumor classification. Finally, ROI selection in WSI is quite computationally expensive for this task.

The dice score shows large difference in the Validation and Testing phase. This may be due to small data sample size and data imbalance. The training data consists of only 221 cases, which may be insufficient for deep learning-based method. Of this, 133 cases are glioblastoma, and rest 88 cases are divided between oligodendroglioma and astrocytoma, creating a data imbalance.

5 Conclusion

This work proposes a deep learning-based method for brain tumor classification. We first segment the brain tumor using a context encoding-based 3D deep neural network. We then apply a regular 3D convolutional neural network on the segmented tumor outputs to obtain the tumor classification. The online evaluation results show a promising performance. In future, we plan to utilize the digital pathology image along with mMRI for categorizing tumor sub-type, especially for distinguishing oligodendroglioma and astrocytoma.

Acknowledgements. This work was partially funded through NIH/NIBIB grant under award number R01EB020683.

References

1. Ostrom, Q.T., Gittleman, H., Truitt, G., Boscia, A., Kruchko, C., Barnholtz-Sloan, J.S.: CBTRUS statistical report: primary brain and other central nervous system tumors diagnosed in the United States in 2011–2015. Neuro-oncol. **20**(suppl_4), iv1–iv86 (2018)
2. Chen, J., McKay, R.M., Parada, L.F.: Malignant glioma: lessons from genomics, mouse models, and stem cells. Cell **149**(1), 36–47 (2012)
3. Zacharaki, E.I., et al.: Classification of brain tumor type and grade using MRI texture and shape in a machine learning scheme. Magn. Reson. Med. Official J. Int. Soc. Magn. Reson. Med. **62**(6), 1609–1618 (2009)
4. Reza, S.M., Samad, M.D., Shboul, Z.A., Jones, K.A., Iftekharuddin, K.M.: Glioma grading using structural magnetic resonance imaging and molecular data. J. Med. Imaging **6**(2), 024501 (2019)
5. Louis, D.N., et al.: The 2016 World Health Organization classification of tumors of the central nervous system: a summary. Acta Neuropathol. **131**(6), 803–820 (2016)
6. Pei, L., Reza, S.M., Li, W., Davatzikos, C., Iftekharuddin, K.M.: Improved brain tumor segmentation by utilizing tumor growth model in longitudinal brain MRI. In: Medical Imaging 2017: Computer-Aided Diagnosis, 2017, vol. 10134, p. 101342L. International Society for Optics and Photonics (2017)
7. Machhale, K., Nandpuru, H.B., Kapur, V., Kosta, L.: MRI brain cancer classification using hybrid classifier (SVM-KNN). In: 2015 International Conference on Industrial Instrumentation and Control (ICIC), pp. 60–65. IEEE (2015)
8. Alfonse, M., Salem, A.-B.M.: An automatic classification of brain tumors through MRI using support vector machine. Egypt. Comput. Sci. J. **40**(03), 11–21 (2016). (ISSN: 1110–2586)
9. Zulpe, N., Pawar, V.: GLCM textural features for brain tumor classification. Int. J. Comput. Sci. Issues (IJCSI) **9**(3), 354 (2012)
10. Sajjad, M., Khan, S., Muhammad, K., Wu, W., Ullah, A., Baik, S.W.: Multi-grade brain tumor classification using deep CNN with extensive data augmentation. J. Comput. Sci. **30**, 174–182 (2019)
11. Barker, J., Hoogi, A., Depeursinge, A., Rubin, D.L.: Automated classification of brain tumor type in whole-slide digital pathology images using local representative tiles. Med. Image Anal. **30**, 60–71 (2016)

12. Sultan, H.H., Salem, N.M., Al-Atabany, W.: Multi-classification of brain tumor images using deep neural network. IEEE Access **7**, 69215–69225 (2019)
13. Bakas, S. et al.: Identifying the best machine learning algorithms for brain tumor segmentation, progression assessment, and overall survival prediction in the BRATS challenge, arXiv preprint arXiv:1811.02629 (2018)
14. Bakas, S., et al.: Segmentation labels and radiomic features for the pre-operative scans of the TCGA-LGG collection. Cancer Imaging Arch. **286** (2017)
15. Menze, B.H., et al.: The multimodal brain tumor image segmentation benchmark (BRATS). IEEE Trans. Med. Imaging **34**(10), 1993–2024 (2014)
16. Bakas, S., et al.: Segmentation labels and radiomic features for the pre-operative scans of the TCGA-GBM collection. The cancer imaging archive (2017). ed 2017
17. Bakas, S., et al.: Advancing the cancer genome atlas glioma MRI collections with expert segmentation labels and radiomic features. Sci. Data **4**, 170117 (2017)
18. Myronenko, A.: 3D MRI brain tumor segmentation using autoencoder regularization. In: Crimi, A., Bakas, S., Kuijf, H., Keyvan, F., Reyes, M., van Walsum, T. (eds.) BrainLes 2018. LNCS, vol. 11384, pp. 311–320. Springer, Cham (2019). https://doi.org/10.1007/978-3-030-11726-9_28
19. Kingma, D.P., Ba, J.: Adam: A method for stochastic optimization, arXiv preprint arXiv: 1412.6980 (2014)
20. Achanta, R., Shaji, A., Smith, K., Lucchi, A., Fua, P., Süsstrunk, S.: SLIC superpixels compared to state-of-the-art superpixel methods. IEEE Trans. Pattern Anal. Mach. Intell. **34** (11), 2274–2282 (2012)

Brain Tumor Classification with Multimodal MR and Pathology Images

Xiao Ma[1,2] and Fucang Jia[1,2(✉)]

[1] Shenzhen Institutes of Advanced Technology, Chinese Academy of Sciences,
Shenzhen, China
fc.jia@siat.ac.cn

[2] Shenzhen College of Advanced Technology, University of Chinese Academy
of Sciences, Shenzhen, China

Abstract. Gliomas are the most common primary malignant tumors of
the brain caused by glial cell canceration of the brain and spinal cord.
Its incidence accounts for the vast majority of intracranial tumors and
has the characteristics of high incidence, high recurrence rate, high mortality, and low cure rate. Gliomas are graded into I to IV by the World
Health Organization (WHO) and the treatment is highly dependent on
the grade. Diagnosis and classification of brain tumors are traditionally
done by pathologists, who examine tissue sections fixed on glass slides
under a light microscope. This process is time-consuming and labor-
intensive and does not necessarily lead to perfectly accurate results. The
computer-aided method has the potential to improve tumor classification
process. In this paper, we proposed two convolutional neural networks
based models to predict the grade of gliomas from both radiology and
pathology data. (1) 2D ResNet-based model for pathology whole slide
image classification. (2) 3D DenseNet-based model for multimodal MRI
images classification. Finally, we achieve first place in CPM-RadPath-
2019 [1] challenge using these methods for the tasks of classifying lower
grade astrocytoma (grade II or III), oligodendroglioma (grade II or III)
and glioblastoma (grade IV).

Keywords: Glioma classification · Machine learning · Pathology ·
Multimodal MRI

1 Introduction

Brain tumors are one of the common diseases in the central nervous system,
which are classified into primary and secondary tumors. Primary brain tumors
originate from brain cells, while secondary tumors metastasize from other organs
to the brain. Glioma is the first of all types of tumors in the brain and has the
highest incidence of intracranial malignant tumors. The World Health Organization (WHO) grades I-IV for gliomas based on malignant behavior for clinical
purposes. Preoperative glioma grading is critical for prognosis prediction and

© Springer Nature Switzerland AG 2020
A. Crimi and S. Bakas (Eds.): BrainLes 2019, LNCS 11993, pp. 343–352, 2020.
https://doi.org/10.1007/978-3-030-46643-5_34

treatment planning. Grade I has the highest survival rate and Grade IV has the lowest.

The current standard diagnosis and grading of brain tumors are done by pathologists who test Hematoxylin and Eosin (H&E) staining tissue sections fixed on glass slides under an optical microscope after resection or biopsy. The cell of these subtypes has distinct features that pathologists use for grade confirmation. Some of the gliomas have the mixed above-mentioned features so it is difficult to distinguish accurately. The whole confirmation process is usually time-consuming, invasive, prone to sampling error and user interpretation. The previous study has shown that CNN can be helpful for glioma classification using pathology images [7]. CNN has shown its ability to work well with large labeled datasets of computer vision tasks, such as ImageNet. But it is not easy for medical imaging fields because of the limitations of data acquisition. Besides, especially for WSI images, it is a difficult point because the size is too large, such as 40k * 40k pixels.

Magnetic Resonance Imaging (MRI) is the standard medical imaging technique for brain tumor diagnosis in clinical practice. Usually, several complimentary 3D MRI modalities are obtained - such as Fluid Attenuation Inversion Recover (FLAIR), T1, T1 with contrast agent (T1c), and T2. Many works have been done on trying to grade gliomas using MRI images by radiomics feature-based machine learning methods [4] and CNN based deep learning approaches [5,6,13,14] since MRI is relatively safe and non-invasive. However, since the golden standard for glioma classification is based on pathological information, it is difficult to predict the subtype of the glioma using only the MRI data.

With the development of machine learning, especially deep learning and computing ability, the automatic diagnosis technology based on computer vision has been applied in many fields, and has made great success in judging the types of diseases and the segmentation of lesions. It is natural to use deep learning to combine pathological and MRI images to predict gliomas subtypes.

In this work, we present a deep learning-based method for the glioma classification of (1) lower grade astrocytoma, IDH-mutant (Grade II or III), (2) oligodendroglioma, IDH-mutant, 1p/19q codeleted (Grade II or III), and (3) glioblastoma and diffuse astrocytic glioma with molecular features of glioblastoma, IDH-wildtype (Grade IV). We try to utilize both pathology and MRI data to train the model for a better classification result. Our model won first place in the MICCAI 2019 Computational Precision Medicine: Radiology-Pathology Challenge.

2   Dataset and Method

In this section, we describe the CPM-RadPath-2019 dataset and our method for preprocessing and prediction.

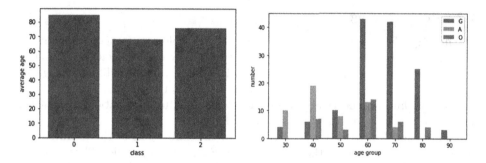

Fig. 1. Statistical information on the age of the patients from the dataset.

2.1 Dataset

The CPM-RadPath-2019 training dataset consists of a total of 221 paired radiology scans and digitized histopathology images. Three subtypes which need to be classified are glioblastoma, oligodendroglioma, and astrocytoma with the number of 133, 54 and 34. The data was provided in two formats: *.tiff for pathology images and *.nii.gz for MRI images with modalities of FLAIR, T1, T1c, and T2. In addition, we also get age information of the patient in the format of days (Fig. 1). The size of the validation and testing set is 35 and 73 respectively.

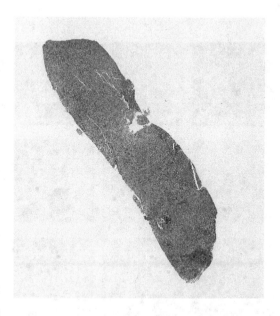

Fig. 2. Visualization of pathological whole slide image.

2.2   Preprocessing

Pathology. For all patients, each one has a pathological image, respectively (Fig. 2). In each picture, only a small part of it is where the cells are stained, and a large part of it is a white background. Besides, there are different degrees of damage in some pictures, such as some strange color areas (blue or green). So we need to find an effective way to extract meaningful information. Pathology images of this dataset are compressed tiff format files which are very big after decompression (up to about 30 GB). Due to the limitations of the resources, it is not feasible to process the whole image. So we use openslide [8] to extract patches of 512 * 512 pixel with a stride of 512 in both directions from the whole slide images (see Fig. 3). During the extraction, we set several constraints to prevent sampling to data that we are not expecting like backgrounds and damaged images.

1) $100 <$ mean of the patch < 220.
2) Standard deviation of the patch > 100.
3) Convert the patch from RGB to HSV and the mean of channel 0 of it > 135 with the standard deviation > 25.

The iteration process will be stopped if two thousand patches are extracted or exceeded the limits of maximum time. The number of patches of each label is 108744, 49067 and 30318 with the order of G, A, and O. Although it may not be very accurate, we set the labels of all extracted patches to the labels of the entire WSI image.

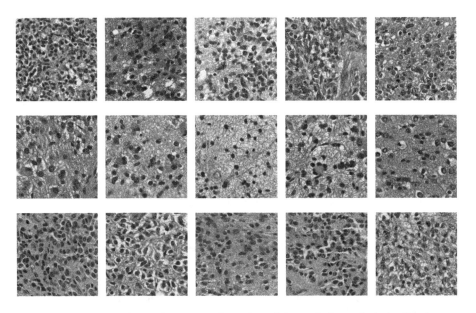

Fig. 3. Visualization of sample extracted patches of the pathological whole slide images. (First row : glioblastoma. Second row: astrocytoma. Third row: oligodendroglioma).

Because of the different conditions of staining slice, using color normalization can make the result of the image better and extract more effective information. Now there are many robust color normalization methods for preprocessing the image [12]. In our model, for the sake of simplicity, we transform the original RGB images into grayscale images. The entire preprocessing process of pathological images can be seen in Fig. 4. All these extracted patches will be used as the input of the neural network for the tumor classification.

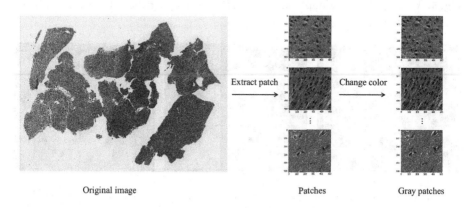

Fig. 4. Pathological image preprocessing process.

Magnetic Resonance Images. In general, for magnetic resonance images, extracting valid features requires information on the tumor mask, and then models are established by these radiomics features to predict the desired outcome. However, since the data set does not provide information about the mask, we do not intend to use additional information. Although the MR image of this dataset is very similar to the BraTS dataset [2], we have tried to train a segmentation network using data from BraTS. Four raw modalities data are used for the convolutional neural network to extract the 3D texture information. We use SimpleITK to convert the MRI images to numpy arrays and normalize them by subtracting the mean and scaling it to unit variance independently by each modality. The volume size of the input data for the convolutional neural network is $4 * 155 * 240 * 240$. The sample MRI images can be seen in Figs. 5 and 6.

2.3 Convolutional Neural Network

Pathology. The extracted patches of the original WSI are grayscale *.png files. We use well-known ResNet34 and ResNet50 [9] deep learning network architectures for this classification task. The results of patch based classification on cross-validation set can be seen in Table 1. The size of the training and cross-validation set is 150504 and 37625 respectively.

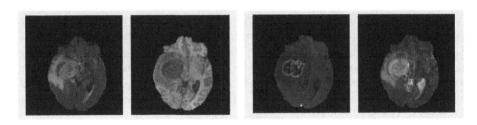

Fig. 5. Visualization of sample MRI data. (From left to right: FLAIR, T1, T1c, T2 respectively)

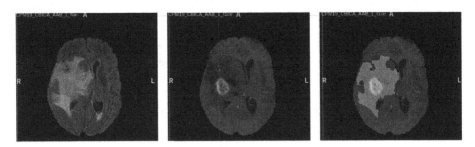

Fig. 6. An example of the segmentation results trained using BraTS18 dataset. From left to right are FLAIR, T1ce, and segmentation mask.

ImageNet statistics are used for the normalization of the input images. The batch size is 24 due to the limitation of the GPU memory. The cross-validation set is obtained by randomly splitting the training set by 0.2. Adam optimizer is employed with the initial learning rate $1e-4$ and weight decay $1e-5$. The loss function is cross entropy loss. Extensive data augmentations are used including random crop, rotation, zoom, translation and color change.

Table 1. Performance of patch classification accuracy.

	Accuracy on cross-validation set
ResNet50	96.7%
ResNet34	95.6%

Magnetic Resonance Images. 3D DenseNet is employed to explore the MRI volumes' capabilities of the classification of gliomas because two-dimensional networks may lose some spatial contextual information and dense connections work better with small datasets [10]. The input and output layers are modified to meet our needs. Five-fold cross validation is employed to get a more robust model. The average results obtained could be found in Table 2. Besides, the confusion matrix can be seen in Fig. 7.

Table 2. Performance of MRI classification accuracy.

	Accuracy on cross-validation set
DenseNet-8	81.82%
DenseNet-121	90.9%

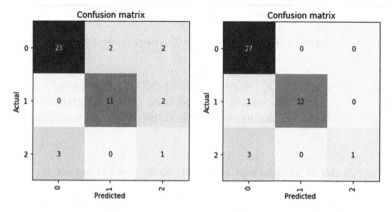

Fig. 7. Confusion matrix of DenseNet-8 and DenseNet-101 on cross-validation set respectively.

The batch size is 2 because of GPU memory efficiency. Mirroring, rotations, scaling, cropping, and color augmentation are used for data augmentation. The learning rate, loss function, and optimizer are set to be the same as the ResNet model described above. All models are implemented on PyTorch platform [11] and use NVIDIA Tesla V100 16GB GPU.

In the training stage, we train two models separately. In the inference stage, we make a simple regression model get the final result according to the outputs of the two models. Test time augmentation is performed at both the convolutional neural network models for a more stable prediction. In general, the results produced by the combined use of the outputs of the two models are more robust and accurate. The whole process can be seen in Fig. 8.

3   Results and Discussion

In this section, we will briefly present some of the best evaluation results and discuss the methods we have experimented with.

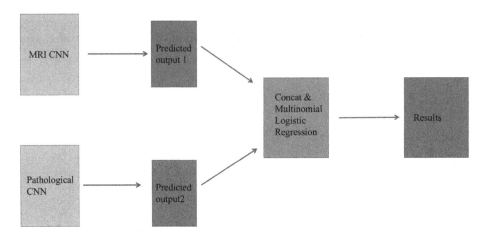

Fig. 8. The whole process of the prediction stage.

As you can see in Table 3, the results are obtained from the online evaluation tool. Above mentioned approaches are evaluated on the validation set. Several metrics are available from online evaluation to estimate the performance of our models.

Table 3. Results on CPM-RadPath-2019 validation set.

	Balanced accuracy	Kappa	F1_micro
Pathology only ResNet50	0.833	0.866	0.914
Mri only DenseNet121	0.711	0.748	0.829
Ensemble	0.889	0.903	0.943

Among the metrics, balanced accuracy is defined as the average recall obtained in each class. Kappa coefficients represent the proportion of error reduction between classifications and completely random classifications. p_o is the sum of the number of samples for each correct classification divided by the total number of samples. p_e is the expected agreement when both annotators assign labels randomly. The F1 score can be interpreted as a weighted average of the precision and recall, where an F1 score reaches its best value at 1 and the worst score at 0. [3] TP, TN, FN, FP stand for true positive, true negative, false negative, and false positive respectively.

$$precision = \frac{TP}{(TP + FP)}$$

$$recall = \frac{TP}{(TP + FN)}$$

$$Balanced\ accuracy = \sum^{Classes} recall/Classes \tag{1}$$

$$\kappa = \frac{(p_o - p_e)}{(1 - p_e)} \tag{2}$$

$$F1 = \frac{2*(precision*recall)}{(precision + recall)} \tag{3}$$

For pathological data, we have used directly extracted pictures and binarized and denoised pictures after extraction. The first method can obtain above 95% accuracy easily by using a convolutional neural network, while the second approach is stuck on about 75% accuracy. As for the patch extraction method, I believe there could be some better constraints other than what we have mentioned above. The speed and the rate of successful sampling of pre-set value need to be further improved. Besides, the further annotation of pathological whole slide images could be a help for our method.

For MRI data, in addition to the methods mentioned above, we also try to use the segmentation tumors regions as the input of the neural network. The segmentation network is trained by BraTS2018 challenge. However, we do not use the tumor segmentation information because we do not want to use external datasets. 2D network has also experimented and the results are worse than the 3D network as we can infer.

Overall, our experiments show the power of the convolutional neural network on the task of glioma classification. Through pathological and MRI data, our model could support the diagnosis and treatment planning of glioma for the pathologists and radiologist. Effective combination of the computer-aided and manual method can improve efficiency of the remedy process.

4   Conclusion

In this paper, we explore the potential of computer-aided method's ability to the diagnosis and grading of glioma by developing a CNN based ensemble model from pathology and radiology data. Our method achieves first place for astrocytoma, oligodendroglioma, and glioblastoma classification problem from CPM-RadPath-2019. The results suggest that our model could be useful for improving the accuracy of glioma grading. The proposed model could be further developed in the future.

Acknowledgements. This work was supported in part by Shenzhen Key Basic Science Program (JCYJ20170413162213765 and JCYJ20180507182437217), the Shenzhen Key Laboratory Program (ZDSYS201707271637577), the NSFC-Shenzhen Union Program (U1613221), and the National Key Research and Development Program (2017YFC0110903).

References

1. Computational precision medicine: Radiology-pathology challenge on brain tumor classification 2019, CBICA. https://www.med.upenn.edu/cbica/cpm2019.html
2. Multimodal brain tumor segmentation challenge 2018. CBICA. https://www.med.upenn.edu/sbia/brats2018/evaluation.html
3. Buitinck, L., et al.: API design for machine learning software: experiences from the SCIKIT-learn project. In: ECML PKDD Workshop: Languages for Data Mining and Machine Learning, pp. 108–122 (2013)
4. Chen, Q., Wang, L., Wang, L., Deng, Z., Zhang, J., Zhu, Y.: Glioma grade predictions using scattering wavelet transform-based radiomics (2019)
5. Citak-Er, F., Firat, Z., Kovanlikaya, I., Ture, U., Ozturk-Isik, E.: Machine-learning in grading of gliomas based on multi-parametric magnetic resonance imaging at 3T. Comput. Biol. Med. **99**, 154–160 (2018)
6. Decuyper, M., Van Holen, R.: Fully automatic binary glioma grading based on pre-therapy MRI using 3D convolutional neural networks (2019)
7. Ertosun, M.G., Rubin, D.L.: Automated grading of gliomas using deep learning in digital pathology images: a modular approach with ensemble of convolutional neural networks. In: AMIA Annual Symposium Proceedings, vol. 2015, p. 1899. American Medical Informatics Association (2015)
8. Goode, A., Gilbert, B., Harkes, J., Jukic, D., Satyanarayanan, M.: Openslide: a vendor-neutral software foundation for digital pathology. J. Pathol. Inform. **4**, 27 (2013)
9. He, K., Zhang, X., Ren, S., Sun, J.: Deep residual learning for image recognition. CoRR abs/1512.03385 (2015). http://arxiv.org/abs/1512.03385
10. Huang, G., Liu, Z., Weinberger, K.Q.: Densely connected convolutional networks. CoRR abs/1608.06993 (2016). http://arxiv.org/abs/1608.06993
11. Paszke, A., et al.: Automatic differentiation in pytorch (2017)
12. Tellez, D., et al.: Quantifying the effects of data augmentation and stain color normalization in convolutional neural networks for computational pathology. CoRR abs/1902.06543 (2019). http://arxiv.org/abs/1902.06543
13. Wang, X., et al.: Machine learning models for multiparametric glioma grading with quantitative result interpretations. Front. Neurosci. **12**, 1046 (2018)
14. Yang, Y., et al.: Glioma grading on conventional MR images: a deep learning study with transfer learning. Front. Neurosci. **12**, 804 (2018)

Automatic Classification of Brain Tumor Types with the MRI Scans and Histopathology Images

Hsiang-Wei Chan, Yan-Ting Weng, and Teng-Yi Huang[✉]

Department of Electrical Engineering, National Taiwan University
of Science and Technology, Taipei, Taiwan
tyhuang@mail.ntust.edu.tw

Abstract. In the study, we used two neural networks, including VGG16 and Resnet50, to process the whole slide images with feature extracting. To classify the three types of brain tumors (i.e., glioblastoma, oligodendroglioma, and astrocytoma), we tried several clustering methods include k-means and random forest classification methods. In the prediction stage, we compared the prediction results with and without MRI features. The results support that the classification method performed with image features extracted by VGG16 has the highest prediction accuracy. Moreover, we found that combining with radiomics generated from MR images slightly improved the accuracy of the classification.

1 Introduction

Classification of brain tumors could be done by examining tissue sections fixed on glass slides using a microscope. Recent advances of MRI methods could also provide information about the brain-tumor classification. It is potentially useful to combine MRI and histopathology images to diagnose the brain tumor. and In 2019, the Computational Precision Medicine: Radiology-Pathology Challenge on Brain Tumor Classification 2019 (CPM-RadPath) is going to held in MICCAI 2019 Brain Lesions (BrainLes) Workshop [1–4]. We attempt to use the neural networks to process the data and compare the performance in each method.

2 Method

2.1 Data

The organizer of CPM-RadPath 2019 provided radiology scans and digitized histopathology images of the same patient. The radiology scans are MRI images with four contrasts, and those are T1-weighted, T1-weighted with contrast enhancement, T2-weighted, and FLAIR. For each subject, the MRI dataset consists of four 3D Nifty files (matrix: $240 \times 240 \times 155$). The digitized histopathology images are provided as the whole-slide TIFF format. The diagnostic classification of each patient is also provided as a spreadsheet file. They were classified as three sub-types. Type A is lower-grade astrocytoma and IDH-mutant (Grade II or III). The type O is oligodendroglioma,

© Springer Nature Switzerland AG 2020
A. Crimi and S. Bakas (Eds.): BrainLes 2019, LNCS 11993, pp. 353–359, 2020.
https://doi.org/10.1007/978-3-030-46643-5_35

IDH-mutant, and 1p/19q codeleted (Grade II or III). The type G is glioblastoma, diffuse astrocytic glioma with molecular features of glioblastoma, and IDH-wildtype (Grade IV). The CPM-RadPath-2019 database consists of 221, 35, 73 subjects for training, validation, and testing stages, respectively. For local validation, we randomly selected 30 subjects from the training set. Thus, in this article, we report the performance metrics based on 191 and 30 subjects for training and validation, respectively.

2.2   Preprocessing and Feature Extraction of the Whole-Slide Images

First, we processed the whole-slide TIFF files. Because the matrix size of the TIFF file is too large to be loaded into our GPU hardware, we split the whole-slide image into batches with a 1024 × 1024 matrix, resized them into 224 × 224 batches, and converted them into a three-channel RGB format. Then, we calculated the sum of standard deviations of three channels of each image and excluded the batches with the values less than a threshold (e.g., 100 or 50) to reduce the number of batches. We use two deep learning models to extract the feature from images. One is VGG16, and the other is ResNet50 [5]. The architecture of our VGG16 implementation is shown in Fig. 1. It has thirteen convolution layers and three fully connected layers. The shape of the input image for VGG16 is 224 × 224. The architecture of ResNet50 implementation is shown in Fig. 2. The ResNet50 is in-depth with more layers, and it used residual learning to solve the problem of degradation in deeper network. The shape of the input image for ResNet50 is also 224 × 224. The Keras framework (https://keras.io) provided the model weights based on the ImageNet database. In VGG16, we extracted 4096 features from the layer of fc2 for each image. In ResNet50, we extracted the 2048 features from the layer of avg-pool. Then, we stacked all the extracted features to create an npz file (a python data format) for each TIFF file. The file is termed TIFF-npz hereafter. In summary, the pre-processing and feature extraction steps convert a TIFF file of each patient into a feature matrix with a size of N × 4096 or N × 2048, where N is the amount of obtained batches. We totally generated 3 TIFF-npz files, VGG16_100_1024.npz, VGG16_50_1024.npz and Resnet50_50_1024.npz. Using VGG16_100_1024.npz as an example, the naming conventions of the TIFF-npz files are"VGG16" for the deep learning model, "100" for the threshold of batch exclusion, and "1024" for the original matrix size of each batch.

VGG-16 Architecture

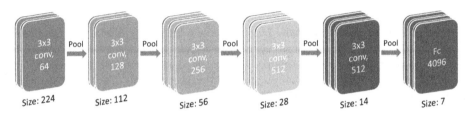

Fig. 1. The VGG16 network has thirteen convolution layers and three fully connected layers. We used the fully connected layer, which named fc2, to extract the 4096 features.

Resnet50 Architecture

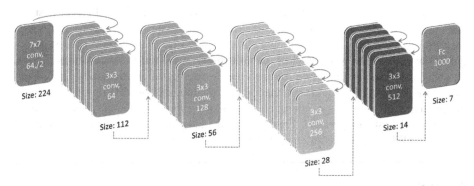

Fig. 2. The shape of the input image for ResNet50 needs is 224 × 224. We used the layer before fully connected layer named avg-pool to extract 2048 features.

2.3 Preprocessing and Feature Extraction of the MRI Images

For each patient, we have four 3D brain volumes with the Nifty (nii.gz) format, and the matrix size of each data is 240 × 240 × 155. We stacked the four brain volumes (i.e., T1, T1ce, T2, and FLAIR) and padded zeros of the volumes to convert the MRI dataset into a 4D tensor (256 × 256 × 256 × 4). Then, we used the brain-tumor segmentation method that we developed in the BraTs 2018 challenge to obtain the core region of the tumor. The algorithm used Tensorflow as the framework, and the deep learning network was based on SegNet [6]. The procedure generated a mask file for the predicted region of the tumor core. Then, we used the pyradiomics module (https://www.radiomics.io) [7] provided by Computational Imaging and Bioinformatics Lab in Harvard Medical School to generate radiomic features (107 × 3 = 321) from the original three MRI files (T1ce, T2 and FLAIR) and the mask of the predicted tumor core. The 428 radiomics features of each subject were stored in a npy file (a python file format) for further processing. The file is termed MRI-npy hereafter.

2.4 Clustering of Batches and Building Classification Models

The batches extracted from the TIFF file are not all belong to the tumor region. Including all batches for machine learning may hamper the performance of classification. It is desirable to collect batches that provide information to distinguish tumor types of patients. However, the label of the regions of tumors is not provided in this challenge. Thus, we came out with an unsupervised clustering method to distribute the batches into several groups and evaluate the classification performance of each group to determine the "right" group of the batches. First, we load npz files of all 191 patients and used a k-means algorithm to cluster all the batches into 1 to 8 groups. Finally, we trained a random forest model [8] (Scikit-learn 0.20) for each group of batches and evaluated the prediction accuracy of all the models. For MRI features, we trained a random forest model to distinguish tumor types based on 321 MRI radiomic features.

2.5   Prediction

During the prediction stage, we first pre-processed the whole-slide image and the MRI files to obtain the TIFF-npz and MRI-npy files for each subject. For TIFF-npz, first, we used the k-means clustering model to distribute the batches into several groups. For each group, we had random forest models to classify the batches into different tumor types. For each batch, we calculated the probability of the three classifications (i.e., A, O, and G). For each patient, we summed the probability of all the batches and determined the tumor class from the summation of probabilities. For MRI-npy, we produced the predicted probabilities of the three classes using the MRI features and the corresponding random forest model. Finally, we calculated the sum of the probabilities of MRI and TIFF to produce an ensemble prediction.

3   Result

Table 1 lists the macro average of each split model we obtained in the VGG16_100_1024.npz. The macro average is the average of f1-score in each model, and every model had the same npz files used in prediction. The macro average of split6 model with cluster label 3 had the best average. Table 2 lists the macro average of each split model we obtained in VGG16_50_1024.npz. We found that the best average appeared in split2 model with cluster label 2, it means that the more splits did not help the network to cluster in the right label. For both VGG16 cases, we found that the more images did not improve the performance of the network. Table 3 lists the macro average of each split model we obtained in the Resnet50_50_1024.npz. Table 4 lists the f1-score for each tumor type and macro average. With the features extracted from MR images, the f1-score for each class is slightly higher.

Table 1. The classification performance (f1-score) of VGG16_100_1024.npz

| | **VGG16_100_1024** | | | | |
| | split2 | split3 | split4 | split5 | split6 |
Cluster labels	Macro avg	Macro avg	Macro avg	Macro avg	Macro avg
0	0.53	0.47	0.40	0.60	0.36
1	0.64	0.39	0.36	0.49	0.27
2		0.45	0.33	0.40	0.14
3			0.49	0.14	0.72
4				0.19	0.35
5					0.49

Table 2. The classification performance (f1-score) of VGG16_50_1024.npz

| | VGG16_50_1024 | | | | |
| | split2 | split3 | split4 | split5 | split6 |
Cluster labels	Macro avg	Macro avg	Macro avg	Macro avg	Macro avg
0	0.41	0.38	0.55	0.38	0.38
1	0.61	0.33	0.33	0.46	0.14
2		0.51	0.19	0.23	0.46
3			0.43	0.50	0.19
4				0.14	0.37
5					0.33

Table 3. The classification performance (f1-score) of Resnet50_50_1024.npz

| | Resnet50_50_1024 | | | | |
| | split2 | split3 | split4 | split5 | split6 |
Cluster labels	Macro avg	Macro avg	Macro avg	Macro avg	Macro avg
0	0.48	0.22	0.14	0.14	0.34
1	0.14	0.22	0.33	0.22	0.28
2		0.34	0.19	0.17	0.13
3			0.22	0.32	0.41
4				0.38	0.14
5					0.14

Table 4. K-means split6 model with cluster label '3' in VGG16_100_1024.npz

VGG16_100_1024
Kmeans split6 cluster label '3'

	original	with MRI
	f1-score	f1-score
G	0.67	0.78
A	0.81	0.86
O	0.67	0.71
Marco avg	0.72	0.78

4   Discussions and Conclusions

In the study, we attempt to classify the brain-tumor types by using the whole-slide and MRI images. The major challenge of applying AI on the whole-slide images is the large matrix. In our GPU hardware, it is not possible to build a deep-learning network for the whole matrix of the whole-slide image. Thus, we developed a method to unsupervisedly distribute batches of the whole-slide image into several groups. In our result, the k-means method with a cluster size of 6 outperformed the other cluster sizes. For MRI images, the major challenge is how to extract features from a 3D volume. We used the segmentation method that we developed the previous BraTs challenge and extracted the corresponding radiomic features. In our preliminary result, the classification models based on MRI radiomic features did not perform as accurate as the whole-slide features. During this challenge, we keep investigating the optimal method to combine the whole-slide and MRI features for improving the accuracy of the tumor classification.

References

1. Bakas, S., et al.: Segmentation labels and radiomic features for the pre-operative scans of the TCGA-GBM collection. The Cancer Imaging Archive (2017). https://doi.org/10.7937/K9/TCIA.2017.KLXWJJ1Q
2. Bakas, S., et al.: Segmentation labels and radiomic features for the pre-operative scans of the TCGA-LGG collection. The Cancer Imaging Archive (2017). https://doi.org/10.7937/K9/TCIA.2017.GJQ7R0EF
3. Menze, B.H., et al.: The multimodal brain tumor image segmentation benchmark (BRATS). IEEE Trans. Med. Imaging **34**(10), 1993–2024 (2015)

4. Bakas, S., et al.: Advancing the cancer genome atlas glioma MRI collections with expert segmentation labels and radiomic features. Nat. Sci. Data **4**, 170117 (2017). https://doi.org/10.1038/sdata.2017.117
5. He, K., Zhang, X., Ren, S., Sun, J.: Deep residual learning for image recognition. In: CVPR (2016)
6. Badrinarayanan, V., Kendall, A., Cipolla, R.: SegNet: a deep convolutional encoder-decoder architecture for image segmentation. CoRR (2015). http://arxiv.org/abs/1511.00561
7. Griethuysen, J.J.M., et al.: Computational radiomics system to decode the radiographic phenotype. Cancer Res. **77**(21), e104–e107 (2017). https://doi.org/10.1158/0008-5472.CAN-17-0339
8. Breiman, L.: Manual on setting up, using, and understanding. Random Forests v4.0 (2003)

Brain Tumor Classification with Tumor Segmentations and a Dual Path Residual Convolutional Neural Network from MRI and Pathology Images

Yunzhe Xue[1], Yanan Yang[1], Fadi G. Farhat[1], Frank Y. Shih[1], Olga Boukrina[2],
A. M. Barrett[3], Jeffrey R. Binder[4], William W. Graves[5],
and Usman W. Roshan[1(✉)]

[1] Department of Computer Science, New Jersey Institute of Technology, Newark, NJ,
USA
usman@njit.edu
[2] Stroke Rehabilitation Research, Kessler Foundation, West Orange, NJ, USA
[3] Emory University and Atlanta VA Medical Center, Atlanta, GA, USA
[4] Department of Neurology, Medical College of Wisconsin, Milwaukee, WI, USA
[5] Department of Psychology, Rutgers University – Newark, Newark, NJ, USA

Abstract. Brain tumor classification plays an important role in brain
cancer diagnosis and treatment. Pathologists typically have to work
through numerous pathology images that can be in the order of hundreds or thousands which takes time and is prone to manual error. Here
we investigate automating this task given pathology images as well as
3D MRI volumes without lesion maps. We use data provided by the
CPM-RadPath 2019 MICCAI challenge. We first evaluate accuracy on
the validation dataset with MRI and pathology images separately. We
predict the 3D tumor mask with our custom developed tumor segmentation model that we used for the BraTS 2019 challenge. We show that
the predicted tumor segmentations give a higher validation accuracy of
77.1% vs. 69.8% with MRI images when trained by a 3D residual convolutional neural network. For pathology images we train a 2D residual
network and obtain a 66.2% validation accuracy. In both cases we find
high training accuracies above 95% which suggests overfitting. We propose a dual path residual convolutional neural network model that trains
simultaneously from both MRI and pathology images and we use a simple
method to prevent overfitting. One path of our network is fully 3D and
considers 3D tumor segmentations as input while the other path considers pathology images. To prevent overfitting we stop training after 90%
training accuracy at the epoch number where our network loss increases
in the following one. With this approach we achieve a validation accuracy
of 84.9% showing that indeed combining the two image sources yields a
better overall accuracy.

Keywords: Convolutional neural networks · Residual networks ·
Pathology · Brain MRI

© Springer Nature Switzerland AG 2020
A. Crimi and S. Bakas (Eds.): BrainLes 2019, LNCS 11993, pp. 360–367, 2020.
https://doi.org/10.1007/978-3-030-46643-5_36

1   Introduction

Brain cancer tumors fall into different categories as given by the World Health Organization [1–3]. The correct prediction of tumor type plays a key role in diagnosis and treatment. However, pathologists typically have to browse numerous images to determine the tumor type which requires considerable training, is time intensive, and is prone to manual errors. The automated classification of tumor type can greatly speed up physician diagnosis and lead to better care and treatment.

The CPM-RadPath 2019 MICCAI challenge is to automatically predict three tumor types given below.

- Lower grade astrocytoma, IDH-mutant (Grade II or III)
- Oligodendroglioma, IDH-mutant, 1p/19q codeleted (Grade II or III)
- Glioblastoma and Diffuse astrocytic glioma with molecular features of glioblastoma, IDH-wildtype (Grade IV).

The contest provides MRI and pathology images from 221 patients as training data and 35 as validation. For each patient we have 3D MRI images in four modalities: native (T1), post-contrast T1-weighted (T1Gd), T2-weighted (T2), and T2 Fluid Attenuated Inversion Recovery (T2-FLAIR). All brain scans were obtained with different clinical protocols and from various scanners from different institutions. The images were all co-registered to the same anatomical template, interpolated to the same resolution (1 mm^3) and skull-stripped.

We are also given varying number of pathology images for each patient. These are digitized whole slide tissue images captured from Hematoxylin and Eosin (H&E) stained tissue specimens. The tissue specimens were scanned at 20x or 40x magnifications. In Fig. 1 we see a cropped pathology image with a Grade IV tumor (class G) from this dataset.

Inspired by the success of convolutional neural networks in image recognition tasks, we present a dual-path residual convolutional neural network solution to this problem. We find that using predicted tumor segmentations of each MRI image leads to higher overall validation accuracy than if we used the original MRI images. We also see that our model achieves above 95% training accuracy which suggests overfitting. With careful training we achieve a validation accuracy of 84.9% with both datasets which is higher than the accuracy with predicted tumor segmentations or pathology images alone.

2   Methods

2.1   Custom Designed U-Network for Predicting Tumor Segmentations

In Fig. 2 we show our custom designed U-Network to predict tumor segmentations from MRI images [4]. We trained our network on data from the Brain Tumor Segmentation (BraTS) 2019 MICCAI challenge [5,6]. We see our network takes images in four modalities and predicts segmentations of three regions of the tumor.

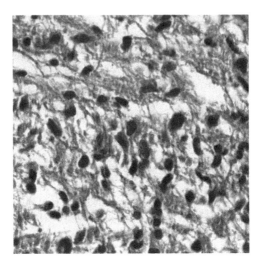

Fig. 1. A typical cropped pathology image taken from the CPM-RadPath dataset with a Grade IV tumor (class G)

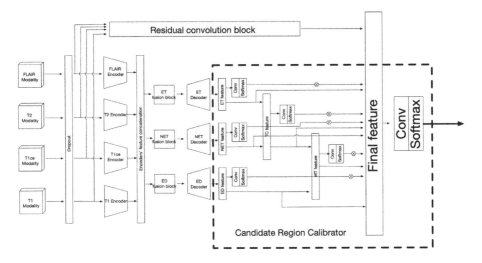

Fig. 2. Our custom designed multi-modal tumor segmentation network

2.2 Dual-Path Residual Convolutional Neural Network

The ResNet18 architecture [7] uses residual connections between layers to prevent gradient vanishing problems and is a highly successful approach. In Fig. 3(a) and 3(b) we show the ResNet18 convolutional neural network architectures that we use separately on MRI and pathology images respectively. We combine them in a dual-path model as shown in Fig. 4.

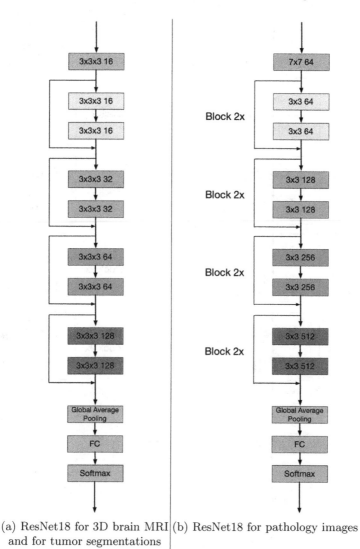

(a) ResNet18 for 3D brain MRI (b) ResNet18 for pathology images
and for tumor segmentations

Fig. 3. Our ResNet18 networks for 3D tumor and pathology images. In each block is shown the size and number of convolutional kernels all with stride 1 except for the first convolutional block that has stride 2.

2.3   Model Training and Parameters

We use the standard cross-entropy loss function [8] to predict the three tumor classes. We implement our network using the Pytorch library [9].

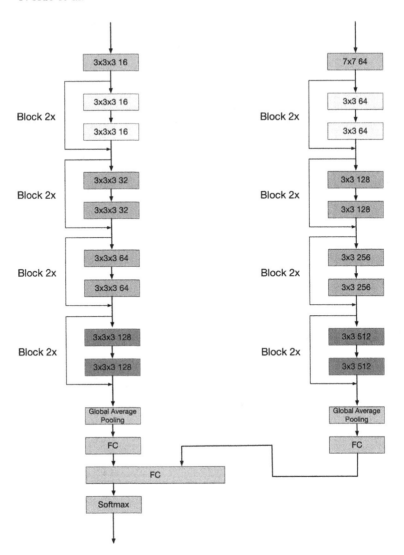

Fig. 4. Our combined network model for both tumor segmentations and pathology images. In each block is shown the size and number of convolutional kernels all with stride 1 except for the first convolutional block that has stride 2.

3D ResNet18 Training. We train our network for 60 epochs, learning rate of 0.01, stochastic gradient descent with Nesterov, a batch size of 8, and no weight decay.

2D ResNet18 Training. We train our network for 100 epochs, learning rate of 0.01, stochastic gradient descent with Nesterov, a batch size of 128, and no weight decay.

Combined Model Training. Our combined model takes in both tumor segmentations and pathology images as input for each patient. For each tumor segmentation we randomly pick 8 pathology images of the patient that go into the same batch during training. If a patient has less than 8 pathology images (which occur in some cases) we simply select randomly with replacement. At the end of the 2D part of our combined model is an average operation that averages the features of the 8 images into one layer that is then concatenated into the 3D part (see Fig. 4).

We train our network for 50 epochs, learning rate of 0.01, stochastic gradient descent with Nesterov, a batch size of 8, and no weight decay.

Early Stopping to Prevent Overfitting. To prevent overfitting we train our model until it reaches a 90% training accuracy. After that we will stop at the epoch if loss increases in the following one.

2.4   Data Preprocessing and Augmentation

3D ResNet18 Data Preprocessing. We normalize the data by subtracting the mean and dividing by standard deviation to give 0 mean and unit variance. We crop and pad original images from dimensions $240 \times 240 \times 155$ to $160 \times 192 \times 160$.

2D ResNet18 Preprocessing. We randomly crop each image from dimensions 512×512 to 224×224. We also study a center crop variant. We perform random horizontal flip on images during both the training and inference processes.

Combined Model Preprocessing. Here we preprocess the MRI images and pathology ones using the same methods described above in the individual networks.

3   Results

We use our custom designed 3D network [4] for the Brain Tumor Segmentation (BraTS) 2019 MICCAI challenge [5,6] to predict tumor segmentations for each MRI image. In Fig. 5 we show tumor segmentations by our BraTS model for each of the three different axial planes of a given slice of an MRI image. We see that the predicted tumor is highly accurate when compared with the true tumor segmentation across all four image modalities. We conjecture that the tumor position and size play a bigger role in determining the tumor type than the entire MRI image. Thus we consider these as inputs to our models vs. the original MRI images.

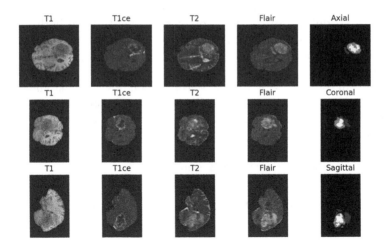

Fig. 5. Tumor segmentations given by our BraTS model in all three axial planes for a given slice across four image modalities. We use the predicted tumor segmentations (that we see are highly accurate in this example) as input to our model to classify the tumor type.

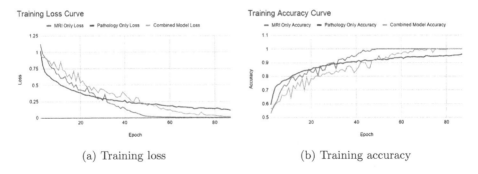

(a) Training loss   (b) Training accuracy

Fig. 6. Our individual and combined model training loss and accuracy.

We first examine our model training lose and accuracy. In Fig. 6 we see the training loss and accuracy of our models on the predicted tumor segmentations, pathology images, and combined images model. In all three cases we see a high training accuracy suggesting that we may be overfitting. In order to avoid this we perform early stopping as described above.

We now proceed to the validation accuracies with different training datasets in Table 1. First we see that indeed the tumor segmentations give a higher validation accuracy of 77.1% than using MRI images alone which give 69.8%. We also see that validation accuracy on pathology images alone is lower than that of MRI and tumor images. In the case of random crops on pathology images it varies between 66.2% and 69.2%.

Combining the MRI images with pathology images under random crops gives us 78.7% validation accuracy whereas combining with tumor segmentations gives

Table 1. Validation accuracy from different training datasets

Brain MRI images	69.8
Predicted tumor segmentations	77.1
Pathology (center crop)	66.2
Pathology (random crop)	66.2–69.2
Combined MRI + pathology (random crop)	78.7
Combined MRI + pathology (center crop)	78.7
Combined tumor + pathology (random crop)	81.6
Combined tumor + pathology (center crop)	84.9

us 81.6%. Finally combing MRI images with pathology under center crop also gives 78.7% while combining tumor segmentations with pathology images under center crop gives us the best validation accuracy of 84.9%.

4 Conclusion

We show that with predicted tumor segmentations we can achieve a higher accuracy for predicting tumor category than if we used the original MRI images. We present a dual path residual convolutional neural network trained on both tumor segmentations and pathology images simultaneously. We show that the combined model achieves a higher accuracy of 84.9% than if we used the tumor or pathology images alone which achieve 77.1% and 66.2% respectively.

References

1. Louis, D.N., et al.: The 2016 world health organization classification of tumors of the central nervous system: a summary. Acta Neuropathol. **131**(6), 803–820 (2016)
2. Kleihues, P., et al.: The WHO classification of tumors of the nervous system. J. Neuropathol. Exp. Neurol. **61**(3), 215–225 (2002)
3. Brat, D.J., et al.: cIMPACT-NOW update 3: recommended diagnostic criteria for diffuse astrocytic glioma, IDH-wildtype, with molecular features of glioblastoma, WHO grade IV. Acta Neuropathol. **136**(5), 805–810 (2018)
4. Xue, Y., et al.: A multi-path decoder network for brain tumor segmentation. In: Proceedings of MICCAI BraTS 2019 Challenge (2019)
5. Menze, B.H., et al.: The multimodal brain tumor image segmentation benchmark (BRATS). IEEE Trans. Med. Imaging **34**(10), 1993–2024 (2014)
6. Bakas, S., et al.: Identifying the best machine learning algorithms for brain tumor segmentation, progression assessment, and overall survival prediction in the BRATS challenge. arXiv preprint arXiv:1811.02629 (2018)
7. He, K., Zhang, X., Ren, S., Sun, J.: Deep residual learning for image recognition. In: Proceedings of the IEEE Conference on Computer Vision and Pattern Recognition, pp. 770–778 (2016)
8. Goodfellow, I., Bengio, Y., Courville, A.: Deep Learning. MIT press, Cambridge (2016)
9. Paszke, A., et al.: Automatic differentiation in PyTorch. In: NIPS-W, Alban Desmaison (2017)

Tools AllowingClinical Translation of Image Computing Algorithms

From Whole Slide Tissues to Knowledge: Mapping Sub-cellular Morphology of Cancer

Tahsin Kurc[1]([⊠]), Ashish Sharma[2], Rajarsi Gupta[1], Le Hou[3], Han Le[3], Shahira Abousamra[3], Erich Bremer[1], Ryan Birmingham[2], Tammy DiPrima[1], Nan Li[2], Feiqiao Wang[1], Joseph Balsamo[1], Whitney Bremer[2], Dimitris Samaras[3], and Joel Saltz[1]

[1] Biomedical Informatics Department, Stony Brook University, Stony Brook, NY, USA
tahsin.kurc@stonybrook.edu
[2] Biomedical Informatics Department, Emory University, Atlanta, GA, USA
[3] Computer Science Department, Stony Brook University, Stony Brook, NY, USA

Abstract. Digital pathology has made great strides in the past decade to create the ability to computationally extract rich information about cancer morphology with traditional image analysis and deep learning. High-resolution whole slide images of cancer tissue samples can be analyzed to quantitatively extract and characterize cellular and sub-cellular phenotypic imaging features. These features combined with genomics and clinical data can be used to advance our understanding of cancer and provide opportunities to the discovery, design, and evaluation of new treatment strategies. Researchers need reliable and efficient image analysis algorithms and software tools that can support indexing, query, and exploration of vast quantities of image analysis data in order to maximize the full potential of digital pathology in cancer research. In this paper we present a brief overview of recent work done by our group, as well as others, in tissue image analysis and digital pathology software systems.

Keywords: Digital pathology · Image analysis · Databases and software tools · Information technology · Cancer research

1 Introduction

In the clinical setting, pathologists microscopically examine tissue samples from patients in order to diagnose and grade the severity of cancer by interpreting changes in tissue, cellular, and sub-cellular morphology. These evaluations are typically qualitative and often subject to inter-observer variability. Moreover, this type of specialized manual assessment of tissue samples is time and cost prohibitive and not scalable for studies with thousands of tissue samples. Advanced tissue imaging technologies that capture high-resolution images of tissue samples have made it possible to perform *quantitative analyses* of phenotypic features of cancer histomorphology. Image-based phenotypic data (Pathomics data) generated from quantitative analysis can be utilized to characterize tumor biology. Pathomics refers to the process of generating large

© Springer Nature Switzerland AG 2020
A. Crimi and S. Bakas (Eds.): BrainLes 2019, LNCS 11993, pp. 371–379, 2020.
https://doi.org/10.1007/978-3-030-46643-5_37

volumes of imaging features from digital pathology image data. Pathomics data can be combined with genomics, clinical outcomes, and therapeutics data to further refine our collective understanding of cancer in terms of onset, progression, survival, and treatment response. Thus, computational analyses of pathology images are being increasingly viewed as an essential component of future clinical applications in precision medicine [1, 2].

Image analyses of tissue samples in pathology studies have been demonstrated to provide valuable information in a variety of correlative and prognostic studies (e.g., [3–7]). Tissue images, however, present many computational and methodology challenges. State-of-the-art whole slide tissue imaging technology can typically scan tissue specimens at 50,000 × 50,000 pixels to over 100,000 × 100,000 pixels in resolution. A typical whole slide tissue image (WSI) of a cancer tissue sample will contain hundreds to thousands of tissue-specific structures (e.g. lymphovascular vessels and other stromal connective tissue elements) and several hundreds of thousands to more than a million cells and nuclei. The complexity arising from data density is further compounded by the diverse nature of tissue specimens across the wide spectrum of different types and subtypes of cancer. A tissue specimen may contain heterogeneous microanatomic regions of cancer and normal tissue with obvious and subtle distinctions and variations in appearances throughout and across these regions. There will be transitional areas where shape and texture characteristics that represent one type of region will be different from the features of tissue in adjacent regions since morphologic appearance and phenotypic features are highly variable in different types and stages of cancer. Tissue samples are routinely stained with Hematoxylin and Eosin (H&E) in order to identify tissue structures and objects, where nuclei are typically purple and cytoplasmic and extracellular material appear pink in color. H&E staining can vary within and across labs, which leads to variations in color intensity, distribution, and noise in pathology images. This is further complicated by variations in imaging instrumentation that create challenges for computational algorithms to delineate boundaries of different regions and objects during image segmentation and extraction of salient image features for classification. Challenges arising from tissue preparation and image acquisition processes also point to a need for improved standardization in these processes and for efficient methods for normalization of image data.

In this paper we present a brief overview of recent work, done by our group and other research teams, in developing tissue image analysis pipelines for segmentation and classification of image data and software systems to support the management, processing, and querying of large collections of WSIs, as well as analytic results. There has been a rapid proliferation of new methodologies to extract and interpret detailed information from tissue images due to advances in tissue scanning instruments and increased computational and storage capacity of modern computer systems. These methods are augmented by an increasing array of open source software systems that facilitate desktop and web-based access to large volumes of image data and analyses.

2 Methods and Software to Transform Tissue Data into Knowledge

This section is organized into two subsections. In the first subsection, we review examples of recent work on tissue image analysis techniques, including analysis pipelines developed by our group. In the second subsection, we briefly describe examples of projects that develop software infrastructure and tools to support tissue image management, analysis and visualization. All of our methods and software tools described in this paper have been implemented and released as containers, more specifically as Docker containers [8], for ease of deployment and use of distributed and Cloud computing environments. We provide links to the source repositories of our codes so that interested parties can download, build and deploy the analysis pipelines and the data management and visualization services and web-based applications.

2.1 Analysis of Tissue Images

There is a rapidly growing body of work on methods for segmentation and classification of histopathology images. Overviews of image analysis work can be found in several review papers (e.g., [9–11]). Here we describe examples of recent work.

Wang et al. designed a multi-path dilated residual network based on the Mask RCNN model to detect and segment dense small nuclei [12]. The network consists of a feature extraction model, which is implemented by using a dense Resnet, a feature pyramid network, a candidate region generation network, and a model for detection and segmentation of candidate regions. The Contour-aware Information Aggregation Network (CIA-Net) [13] implemented a multi-level information aggregation module. The information aggregation module was placed between two decoder networks that are designed for refinement by taking advantage of spatial and texture dependencies to provide a hierarchical refinement process. Experimental results show significant performance improvements over other traditional and deep learning-based methods. Graham et al. proposed a deep learning approach for simultaneous segmentation and classification of nuclei [14]. The algorithm takes into account the vertical and horizontal distances of nuclei to their centers of mass in order to separate touching or overlapping nuclei. After segmentation, the algorithm predicts the type of each segmented nucleus via an up-sampling branch in the deep learning network. The authors noted the need for good training datasets for accurate and robust deep learning-based segmentation networks and released a dataset of more than 24,000 manually segmented and labeled nuclei. A generative adversarial network (GAN) architecture is used by Mahmood et al. to generate and train deep learning segmentation networks by using real and synthetic data [15]. The authors implemented an unpaired GAN network to generate synthetic tissue images and the segmentation masks. These images and masks were then combined with real, manually generated segmentation data from six different tissue types to train a conditional GAN network for segmentation. The authors experimentally demonstrated that training with a large set of synthetic data and relatively small real dataset results in a segmentation model that performs significantly better than conventional methods.

Mobadersany et al. integrated image analysis results and genomics signatures to predict outcome in lower grade glioma and glioblastoma multiforme cases [16]. The authors designed a convolution deep learning network to classify image patterns related to patient survival. In their implementation, the convolutional layers were connected to a series of fully connected layers which transformed deep learning image features from the convolutional layers for survival analysis. A Cox proportional hazard module was used to carry out the survival analysis. Cruz-Roa et al., proposed and evaluated convolutional neural networks that were trained to detect and characterize breast cancer regions [17]. The proposed approach implemented an adaptive sampling method to select the most relevant patches and regions instead of processing all of the patches in a WSI. Their experimental evaluation showed that the sampling approach accelerated the prediction phase significantly with a relatively small reduction in prediction accuracy. Ren et al. [18] looked at the problems arising from variability in tissue staining and image acquisition when tissue specimens are handled at different laboratories and in different batches. The authors argued that such variations can significantly degrade the performance and reliability of a classification model, when the classification model is trained with one batch of images and applied to another batch of images. They investigated the use of adversarial training to address this problem. The adversarial training strategy employed convolutional neural networks and Siamese architecture as a mechanism for unsupervised domain adaptation by minimizing distribution difference between the source domain and the target domain. The experimental results showed significant classification improvements when this training strategy is executed.

Our group also developed a suite of algorithms for tissue image segmentation and classification. We developed a method to generate synthetic tissue images and corresponding segmentation masks [19]. This method uses a generative adversarial network (GAN) to synthesize tissue image tiles with different types of nuclei and distribution patterns. These images are refined in the GAN network and fed into a segmentation convolutional neural network on the fly for training. We have used this method to segment over 5,000 WSIs of cancer tissue samples in the Cancer Genome Atlas (TCGA) repository. The pipeline is implemented as a Docker container and it is available as an analysis component in our QuIP software distribution (see Sect. 2.2): https://github.com/SBU-BMI/quip_cnn_segmentation.

We developed a suite of classification methods that are aimed at generating and classifying maps of lymphocyte distributions, as well as detecting and characterizing cancer regions. We designed and implemented a deep learning workflow to generate spatial maps of tumor infiltrating lymphocytes (TILs) in whole slide tissue images [20]. This analysis workflow partitions each WSI in a dataset into 50-by-50 micron patches and classifies each patch as TIL-positive or TIL-negative. The deep learning network is trained in an iterative process, where predictions from the trained network are reviewed by pathologists and then corrected or refined. The corrected classifications are added to the training set in order to re-train and refine the classification model. A separate convolutional neural network is implemented to detect and classify regions of necrosis. This is done to reduce false positives because necrotic cells have condensed and fragmented nuclei that can make them appear similar to lymphocytes. In a more recent work, we trained a variety of existing deep learning methods using both manual ground truth data for TILs and weakly labeled data from our pipeline in the previous work [20].

The pipelines are available as Docker containers (https://github.com/SBU-BMI/u24_lymphocyte and https://github.com/SBU-BMI/quip_classification) in our QuIP software (see Sect. 2.2).

Lymphocyte distributions can be analyzed for an entire slide or with respect to tumor regions. Spatial relationships between lymphocytes and tumor regions can provide additional insights into cancer mechanisms and tumor immune interactions. We designed and implemented analysis pipelines to detect and characterize tumor regions in tissue images. Our breast tumor detection work [21] implements deep learning models based on the VGG16, Resnet34, and Inception-v4 networks. Our approach also executes a post-processing step on predictions from the deep learning models to incorporate spatial information about image patches. Our experimental results showed that Resnet34 with the post-processing step performs better than the other models and achieves an F1 score of 0.89. In another work, we extended the classification workflow to predict and characterize regions of pancreatic cancer [22]. The extended workflow uses noisy label data along with a small set of clean training samples to train a classification model. Training samples from noisy training dataset are assigned appropriate weights to reduce sample noise. This is done online during training so that the network loss can better approximate the loss due to clean samples. Our experiments demonstrated that the model trained with the noisy and clean data performs 2.9–3.7% better than a model trained with clean training samples only. These deep learning pipelines are implemented in Docker containers (https://github.com/SBU-BMI/quip_cancer_segmentation and https://github.com/SBU-BMI/quip_paad_cancer_detection) in the QuIP framework.

2.2 Software Systems for Handling Tissue Images and Analysis Results

Several research groups, including our team, have researched and developed tools and software infrastructure to address the challenges of managing, processing, and interacting with datasets of WSIs and analysis results [23–29]. The Open Microscopy Environment (OME) is a framework that includes components for data management, conversion of vendor file formats, and visualization of image and analysis data [23]. The project has developed a data model that is implemented in XML to store metadata about images and information about analysis results. The OME framework also implements a library called Bio-Formats that enables parsing and transformation of a variety of vendor specific image file formats. The Digital Slide Archive (DSA) software developed by Gutman et al. [25] is a framework of services and web-based applications. The software infrastructure provides support for management of large datasets of images and analysis results, as well as the ability to interact with image and analysis data. The backend database management system of DSA tracks image metadata, organizes images into collections, and handles user authentication and authorization. Martel et al. implemented a desktop application, the Pathology Image Informatics Platform, which uses freely available viewer software and extends its capabilities via plug-ins [27]. The implementation provides several types of image processing plugins out of the box, such as modules for detection of out-of-focus regions in images and immunohistochemistry tissue image analysis. The Cytomine project developed a web-based software system to support collaborative analysis of tissue images through sharing of histology and molecular imaging data [26]. It facilitates manual annotations and computer analyses of images.

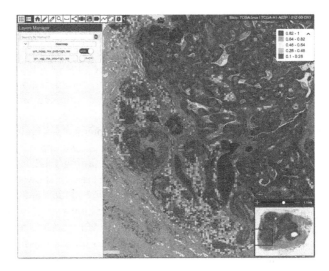

Fig. 1. Visualization of tumor infiltrating lymphocyte results as a heatmap overlaid on a whole slide tissue image in QuIP. Each colored square represents a 50x50-micron patch. The color value is assigned based on a mapping of the classification probability of said patch. The user can zoom and pan to look at different regions of the tissue and TIL map.

We have developed a web-based and containerized software system called QuIP, which provides services, tools, and analysis pipelines [29]. QuIP is implemented as a fully containerized software infrastructure. This architecture decision is driven by primarily two observations. First, most software tools for data management, visualization, and analysis in digital pathology use a wide range of third-party libraries and software modules. These libraries and modules may be implemented in different languages, may further depend on other libraries and service frameworks, and may require different deployment mechanisms. This significantly complicates the deployment of the software infrastructure and limits its portability across operating system platforms. Containerization provides portability and simplifies the deployment process by isolating the modules and services of a software system. Second, there is growing interest in the use of cloud computing for digital pathology that is driven by the rapid decline in costs, which facilitates the utilization of cloud computing as a cost-effective solution for large-scale computing. Containerization makes it easier to deploy and execute analysis pipelines, data management, and user-facing services in the cloud. In QuIP, all analysis pipelines and backend services for data management, image access, and user-facing components are implemented as Docker containers. Users interact with QuIP services through a suite of Web-based applications. These applications provide capabilities to view whole slide tissue images, visualize analysis results as overlays on image data, and manually annotate images. Figure 1 shows an example visualization of a TIL analysis results, generated from one of our recent classification models (published in [30]), via the QuIP web interface. The TIL heatmap represents the prediction

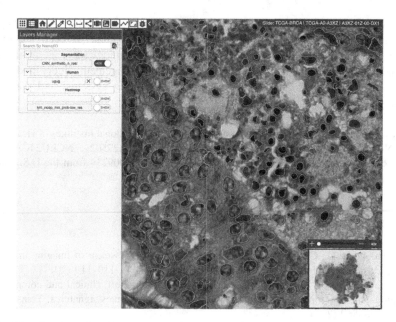

Fig. 2. Visualization of nucleus segmentation results as an overlay of polygons on the whole slide tissue image of an H&E stained tissue specimen. Each polygon represents the boundaries of a segmented nucleus.

probability map, i.e., each image patch is assigned a probability of being TIL-positive. It is overlaid on the whole slide image. A user can pan and zoom to look for regions of TILs. Figure 2 shows visualization of results from nucleus segmentation generated from our CNN based algorithm (published in [19]). The segmented nuclei are overlaid on the whole slide tissue image as polygons.

3   Conclusions

In the past decade, digital pathology has transformed into a highly active field. This is in part due to advances in tissue scanning technologies and in part because of the increased computational and storage capacity of computing systems. We expect that in the next 5–10 years, most institutions will routinely generate digital whole slide tissue images of traditional glass slides in anatomic pathology and clinical research. This will create immensely valuable local and national big data resources for biomedical research. Realizing the full potential of these resources requires continued investment in efficient and reliable image analysis algorithms and software systems that can handle vast amounts of image data and features – indeed, some of the databases managed in our QuIP software contain over a billion segmented nuclei and imaging features from thousands of WSIs. In this paper we have presented overviews of example projects in histopathology image analysis and open source digital pathology software systems. This work by no means provides a complete or exhaustive coverage of exiting work.

Research and development efforts in digital pathology are rapidly increasing and evolving. We are seeing a significant increase in application of deep learning in histopathology image analysis and a rapid shift towards micro services architectures (via containerization of analysis pipelines and data management, query and visualization services), Cloud and high-performance computing systems, and web-based technologies.

Acknowledgements. This work was supported in part by the National Institutes of Health under grants NCI:U24CA180924, NCI:U24CA215109, NCI:UG3CA225021, NCI:U24CA189523, NINDS:R01NS042645, as well as R01LM011119 and R01LM009239 from the U.S. National Library of Medicine.

References

1. Chennubhotla, C., Clarke, L.P., Fedorov, A., et al.: An assessment of imaging informatics for precision medicine in cancer. Yearb Med. Inform. **26**(1), 110–119 (2017)
2. Colen, R., Foster, I., Gatenby, R., et al.: NCI workshop report: clinical and computational requirements for correlating imaging phenotypes with genomics signatures. Transl. Oncol. **7**(5), 556–569 (2014)
3. Beck, A.H., Sangoi, A.R., Leung, S., et al.: Systematic analysis of breast cancer morphology uncovers stromal features associated with survival. Sci Transl. Med. **3**(108), 108ra113 (2011)
4. Cheng, J., Mo, X., Wang, X., et al.: Identification of topological features in renal tumor microenvironment associated with patient survival. Bioinformatics **1**, 7 (2017)
5. Luo, X., Zang, X., Yang, L., et al.: Comprehensive computational pathological image analysis predicts lung cancer prognosis. J. Thorac. Oncol. **12**(3), 501–509 (2017)
6. Wang, C., Pécot, T., Zynger, D.L., et al.: Identifying survival associated morphological features of triple negative breast cancer using multiple datasets. J. Am. Med. Inform. Assoc. **20**(4), 680–687 (2013)
7. Yu, K.-H., Zhang, C., Berry, G.J., et al.: Predicting non-small cell lung cancer prognosis by fully automated microscopic pathology image features. Nat. Commun. **7**, 12474 (2016)
8. Anderson, C.: Docker [software engineering]. IEEE Software **32**(3), 102-c3 (2015)
9. Janowczyk, A., Madabhushi, A.: Deep learning for digital pathology image analysis: a comprehensive tutorial with selected use cases. J. Pathol. Inform. **7** (2016)
10. Komura, D., Ishikawa, S.: Machine learning methods for histopathological image analysis. Comput. Struct. Biotechnol. J. **16**, 34–42 (2018)
11. Xing, F., Yang, L.: Robust nucleus/cell detection and segmentation in digital pathology and microscopy images: a comprehensive review. IEEE Rev. Biomed. Eng. **9**, 234–263 (2016)
12. Wang, E.K., Zhang, X., Pan, L., et al.: Multi-path dilated residual network for nuclei segmentation and detection. Cells **8**(5), 499 (2019)
13. Zhou, Y., Onder, O.F., Dou, Q., Tsougenis, E., Chen, H., Heng, P.-A.: CIA-Net: robust nuclei instance segmentation with contour-aware information aggregation. In: Chung, A.C. S., Gee, J.C., Yushkevich, P.A., Bao, S. (eds.) IPMI 2019. LNCS, vol. 11492, pp. 682–693. Springer, Cham (2019). https://doi.org/10.1007/978-3-030-20351-1_53
14. Graham, S., Vu, Q.D., Raza, S.E.A., et al.: Hover-Net: Simultaneous segmentation and classification of nuclei in multi-tissue histology images. Med. Image Anal. **58**, 101563 (2019)

15. Mahmood, F., Borders, D., Chen, R., et al.: Deep adversarial training for multi-organ nuclei segmentation in histopathology images. IEEE Trans. Med. Imaging (2019)
16. Mobadersany, P., Yousefi, S., Amgad, M., et al.: Predicting cancer outcomes from histology and genomics using convolutional networks. Proc. Natl. Acad. Sci. U.S.A. **115**(13), E2970–E2979 (2018)
17. Cruz-Roa, A., Gilmore, H., Basavanhally, A., et al.: High-throughput adaptive sampling for whole-slide histopathology image analysis (HASHI) via convolutional neural networks: application to invasive breast cancer detection. PLoS One **13**(5), e0196828 (2018)
18. Ren, J., Hacihaliloglu, I., Singer, E.A., Foran, D.J., Qi, X.: Unsupervised domain adaptation for classification of histopathology whole-slide images. Front. Bioeng. Biotechnol. **7** (2019)
19. Hou, L., Agarwal, A., Samaras, D., et al.: Robust histopathology image analysis: to label or to synthesize? In: Proceedings of the IEEE Conference on Computer Vision and Pattern Recognition, pp. 8533–8542 (2019)
20. Saltz, J., Gupta, R., Hou, L., et al.: Spatial organization and molecular correlation of tumor-infiltrating lymphocytes using deep learning on pathology images. Cell Rep. **23**(1), 181 (2018)
21. Le, H., Gupta, R., Hou, L., et al.: Utilizing automated breast cancer detection to identify spatial distributions of tumor infiltrating lymphocytes in invasive breast cancer. arXiv preprint arXiv:1905.10841 (2019)
22. Le, H., Samaras, D., Kurc, T., Gupta, R., Shroyer, K., Saltz, J.: Pancreatic cancer detection in whole slide images using noisy label annotations. In: Shen, D., Liu, T., Peters, T.M., Staib, L.H., Essert, C., Zhou, S., Yap, P.-T., Khan, A. (eds.) MICCAI 2019. LNCS, vol. 11764, pp. 541–549. Springer, Cham (2019). https://doi.org/10.1007/978-3-030-32239-7_60
23. Allan, C., Burel, J.-M., Moore, J., et al.: OMERO: flexible, model-driven data management for experimental biology. Nat. Methods **9**(3), 245–253 (2012)
24. Bankhead, P., Loughrey, M.B., Fernández, J.A., et al.: QuPath: Open source software for digital pathology image analysis. Sci. Rep. **7**(1), 16878 (2017)
25. Gutman, D.A., Khalilia, M., Lee, S., et al.: The digital slide archive: a software platform for management, integration, and analysis of histology for cancer research. Cancer Res. **77**(21), e75–e78 (2017)
26. Marée, R., Rollus, L., Stévens, B., et al.: Cytomine: an open-source software for collaborative analysis of whole-slide images. Diagn. Pathol. **1**(8), 2016 (2016)
27. Martel, A.L., Hosseinzadeh, D., Senaras, C., et al.: An image analysis resource for cancer research: PIIP—pathology image informatics platform for visualization, analysis, and management. Cancer Res. **77**(21), e83–e86 (2017)
28. Williams, E., Moore, J., Li, S.W., et al.: Image data resource: a bioimage data integration and publication platform. Nat. Methods **14**, 775 (2017)
29. Saltz, J., Sharma, A., Iyer, G., et al.: A containerized software system for generation, management, and exploration of features from whole slide tissue images. Cancer Res. **77**(21), e79–e82 (2017)
30. Abousamra, S., Hou, L., Gupta, R., et al.: Learning from thresholds: fully automated classification of tumor infiltrating lymphocytes for multiple cancer types. arXiv preprint arXiv:1907.03960 (2019)

The Cancer Imaging Phenomics Toolkit (CaPTk): Technical Overview

Sarthak Pati[1], Ashish Singh[1], Saima Rathore[1,2], Aimilia Gastounioti[1,2], Mark Bergman[1], Phuc Ngo[1,2], Sung Min Ha[1,2], Dimitrios Bounias[1], James Minock[1], Grayson Murphy[1], Hongming Li[1,2], Amit Bhattarai[1], Adam Wolf[1], Patmaa Sridaran[1], Ratheesh Kalarot[1], Hamed Akbari[1,2], Aristeidis Sotiras[1,3], Siddhesh P. Thakur[1], Ragini Verma[1,2], Russell T. Shinohara[1,4], Paul Yushkevich[1,2,5], Yong Fan[1,2], Despina Kontos[1,2], Christos Davatzikos[1,2], and Spyridon Bakas[1,2,6(✉)]

[1] Center for Biomedical Image Computing and Analytics (CBICA), University of Pennsylvania, Philadelphia, PA, USA.
sbakas@upenn.edu
[2] Department of Radiology, Perelman School of Medicine, University of Pennsylvania, Philadelphia, PA, USA
[3] Department of Radiology and Institute for Informatics, School of Medicine, Washington University in St. Louis, Saint Louis, MO, USA
[4] Penn Statistics in Imaging and Visualization Endeavor (PennSIVE), University of Pennsylvania, Philadelphia, PA, USA
[5] Penn Image Computing and Science Lab., University of Pennsylvania (PICSL), Philadelphia, PA, USA
[6] Department of Pathology and Laboratory Medicine, Perelman School of Medicine, University of Pennsylvania, Philadelphia, PA, USA

Abstract. The purpose of this manuscript is to provide an overview of the technical specifications and architecture of the **Ca**ncer imaging **P**henomics **T**oolkit (CaPTk www.cbica.upenn.edu/captk), a cross-platform, open-source, easy-to-use, and extensible software platform for analyzing 2D and 3D images, currently focusing on radiographic scans of brain, breast, and lung cancer. The primary aim of this platform is to enable swift and efficient translation of cutting-edge academic research into clinically useful tools relating to clinical quantification, analysis, predictive modeling, decision-making, and reporting workflow. CaPTk builds upon established open-source software toolkits, such as the Insight Toolkit (ITK) and OpenCV, to bring together advanced computational functionality. This functionality describes specialized, as well as general-purpose, image analysis algorithms developed during active multi-disciplinary collaborative research studies to address real clinical requirements. The target audience of CaPTk consists of both computational scientists and clinical experts. For the former it provides i) an efficient image viewer offering the ability of integrating new algorithms, and ii) a library of readily-available clinically-relevant algorithms, allowing batch-processing of multiple subjects. For the latter it facilitates

---

S. Pati and A. Singh—Equally contributing authors.

© Springer Nature Switzerland AG 2020
A. Crimi and S. Bakas (Eds.): BrainLes 2019, LNCS 11993, pp. 380–394, 2020.
https://doi.org/10.1007/978-3-030-46643-5_38

the use of complex algorithms for clinically-relevant studies through a user-friendly interface, eliminating the prerequisite of a substantial computational background. CaPTk's long-term goal is to provide widely-used technology to make use of advanced quantitative imaging analytics in cancer prediction, diagnosis and prognosis, leading toward a better understanding of the biological mechanisms of cancer development.

Keywords: CaPTk · Cancer · Imaging · Phenomics · Toolkit · Radiomics · Radiogenomics · Radiophenotype · Segmentation · Deep learning · Brain tumor · Glioma · Glioblastoma · Breast cancer · Lung cancer · ITCR

1   Introduction

The bane of computational medical imaging research has been its translation to the clinical setting. If novel research algorithms can be validated in ample and diverse data, then they have the potential to contribute to our mechanistic understanding of disease and hence substantially increase their value and impact in both the clinical and scientific community. However, to facilitate this potential translation a Graphical User Interface (GUI) is essential, supportive, and tangential to medical research. Designing and packaging such a GUI is a lengthy process and requires deep knowledge of technologies to which not all researchers are exposed. Providing a computational researcher an easy way to integrate their novel algorithms into a well-designed GUI would facilitate the use of such algorithms by clinical researchers, hence bringing the algorithm closer to clinical relevance.

Towards this effort, numerous open-source applications have been developed [1–4] with varying degrees of success in either the computational and/or clinical research communities. However, each of these had their own limitations. To address these limitations, we developed the **C**ancer imaging **P**henomics **T**oolkit (CaPTk[1]) [5,6], which introduces a mechanism to integrate algorithms written in any programming language, while maintaining a lightweight viewing pipeline. CaPTk is a cross-platform (Windows, Linux, macOS), open-source, and extensible software platform with an intuitive GUI for analyzing both 2D and 3D images, and is currently focusing on radiographic scans of brain, breast, and lung cancer. CaPTk builds upon the integration of established open-source toolkits to bring together advanced computational functionality of specialized, as well as general-purpose, image analysis algorithms developed during active multidisciplinary collaborative research studies.

2   Tools Preceding CaPTk

Prior to beginning the development of CaPTk in 2015, we conducted a literature review of the current open-source applications developed by the community

---

[1] www.cbica.upenn.edu/captk.

to address specific scientific needs such as Quantitative Image Phenomic (QIP) extraction and image annotation. While some of these tools (such as MIPAV [11] and MedINRIA [4]) could not be extended with customized applications, others (namely, 3D-Slicer [1] and MITK [2]) had complex software architecture that would be challenging for a limited team to extensively modify. Additional considerations in the decision to launch CaPTk as an independent toolkit included the substantial code base internal to the Center for Biomedical Image Computing and Analytics (CBICA) relating to existing software applications implemented in C++, existing research directions, software development expertise in C++, and licensing considerations. Furthermore, none of these applications could provide customized interactions for initializing seedpoints, as required by applications such as GLISTR [7] and GLISTRboost [8,9]. Below we refer to specific packages that preceded the development of CaPTk and we attempt to describe their advantages and disadvantages (at that time) compared to the requirements of CaPTk that led to its design and development.

2.1 Medical Imaging Interaction Toolkit (MITK)

One of the earliest works towards an open source extensible toolkit by the scientific community, MITK [2] had spearheaded the adoption of DICOM ingestion by the research community. Designed to be cross-platform, it had been extensively tested on various types of datasets, modalities and organ-systems and has arguably provided one of the best DICOM parsing engines in the community with a very nice mechanism for incorporating plugins that integrate natively into the application. Although MITK's native plugin integration was not as user-/developer-friendly and required expert knowledge of MITK, it made incorporation of other tools such as XNAT [10] possible. However, we ended up not choosing MITK as an option for our solution as integrating an application written in a different programming language could not be done easily and any modifications of the graphical capabilities, such as custom interaction (e.g., GLISTR/GLISTRboost initializations) or quick GUI modification, was not possible without extensive background knowledge of toolkit's internals.

2.2 Medical Image Processing, Analysis, and Visualization (MIPAV)

One of the earliest tools funded by the National Institutes of Health (NIH), the MIPAV [11] package was written to provide an easy-to-use user interface for image visualization and interaction. MIPAV is a cross-platform application, with a back-end written in Java. There were some performance issues with algorithms that required a high amount of computation capabilities and it was targeted for specific functionality and had no way to integrate third-party applications.

2.3 3D Slicer

Perhaps the most popular open source tool for the imaging community, 3D Slicer incorporated a lot of functionality via its extension management system.

It was relatively easy for a developer to write a Slicer extension since it provided out-of-the-box support for Python, which potentially resulted in performance issues. It was cross-platform and very well tested by the community and had extensive DICOM support. The user interface, however, was not very friendly for clinical researchers and any changes to either the interactive capabilities or the user interface could not be done easily. In addition, integrating an application written in a language that was not in C++ or Python was also not trivial.

2.4 MeVisLab

MeVisLab [3] was a semi-commercial tool made available by Fraunhofer MEVIS[2] which was well tested by their industrial partners, had extensive DICOM support and was well-received by clinicians as well. However, it had limited capability of handling the number (5 in total) and size (2 kB source) for extensions in the free version, a serious constraint if we were to use MeVisLab as the basis for providing multiple complex imaging applications to researchers. In addition, non-native extensions were not currently supported.

2.5 medInria

medInria [4] was an imaging platform that focused on algorithms related to image registration, diffusion image processing and tractography. Combined with extensive support for different types of DICOM image formats and its cross-platform capabilities, medInria had excellent annotation capabilities and was extensively used by clinical researchers by virtue of its easy-to-use interface. Nevertheless, because of its inability of assimilating extensions, it became a narrowly focused tool.

2.6 ITK-SNAP

ITK-SNAP [12] was one of the most widely used tools for segmentation and has been used to solve annotation problems in multiple organ systems. It was cross-platform and very well-tested. Nonetheless, because of its inability to support extensions and the architectural complexity of the visualization system, it was not a candidate to use as a base for CaPTk's development.

3 CaPTk's Infrastructure

The development of medical imaging tools that can be clinically relevant is driven by a specific medical need around which algorithms are developed, and involves understanding of a researcher's/clinician's workflow. As illustrated in Fig. 1, the planning begins with defining the clinical requirement of the task to be accomplished, followed by acquiring relevant sample datasets. Using these initial requirements and datasets, an algorithm is designed and then its usability is established using a series of refinements, including a robust visualization and interaction tool-chain.

---

[2] www.mevis.fraunhofer.de.

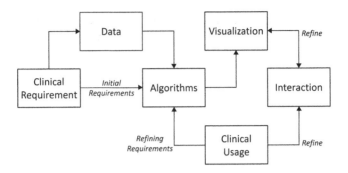

Fig. 1. Data flow diagram for tool development planning. Addition of a GUI tremendously increases the usability of the algorithm. Feedback from clinical usage is used to refine the algorithm and rendering further.

3.1 Functionality

By keeping the data flow practices in mind (Fig. 1), CaPTk has been designed as a general-purpose tool spanning brain, breast, lung, and other cancers. It has a broad, three-level functionality as illustrated in Fig. 2. The first level provides basic image pre-processing tasks such as image input-output (currently supporting NIfTI [13] and DICOM file types), image registration [14], smoothing (denoising) [15], histogram-matching, and novel intensity harmonization for brain magnetic resonance imaging (MRI) scans (namely WhiteStripe [16]), among other algorithms. The second level comprises various general-purpose routines including extensive QIP feature extraction compliant with the Image Biomarker Standardization Initiative (IBSI) [17], feature selection, and a Machine Learning (ML) module. These routines are used within CaPTk for specialized tasks, but are also available to the community as general-purpose analysis steps that sites can use as the basis for pipelines customized for their data. In particular, this level targets extraction of numerous QIP features capturing different aspects of local, regional, and global visual and sub-visual imaging patterns, which, if integrated through the available ML module, can lead to the generation of predictive and diagnostic models. Finally, the third level of CaPTk focuses on specialized applications validated through existing scientific studies keeping reproducibility [18] as the cornerstone of development. These studies include: predictive models of potential tumor recurrence and patient overall survival [19–26], generating population atlases [27], automated extraction of white matter tracts [28], estimation of pseudo-progression for glioblastoma (GBM) [19], evaluating the Epidermal Growth Factor Receptor splice variant III (EGFRvIII) status in GBM [29,30], breast density estimation [31,32], estimation of tumor directionality & volumetric changes in 2 time-points [33], among others.

CaPTk's GUI was designed after multiple interactions with radiologists and other clinical collaborators, targeting single subject analysis primarily for clinical research. Along with the easy-to-use GUI, every specialized application within the CaPTk has an accompanying command-line interface (CLI) executable.

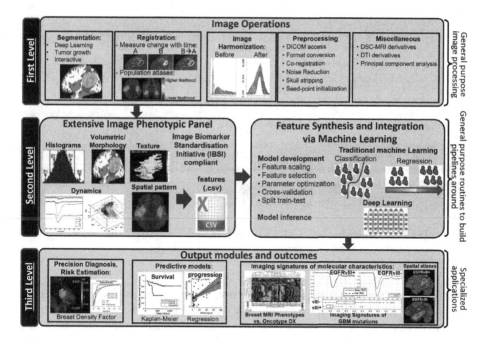

Fig. 2. Diagram showcasing the three-level functionality of CaPTk.

These CLI applications allow the use of CaPTk functions as components of larger, more complicated pipelines and for efficient batch processing of large numbers of images.

3.2 Architecture

The architecture of CaPTk is depicted in Fig. 3. At the lowest level of the architecture (shown as blue in Fig. 3) is the operating system (OS) and its core components, such as system libraries (OpenGL, OpenMP) and hardware drivers. One level above (in green) are the libraries (e.g., ITK, VTK, OpenCV, Qt) that provide the lower level functionality of CaPTk. The next level consists of CaPTk core libraries constituting the algorithmic functionality of CaPTk, including all its three levels previously described and graphically shown in Fig. 2. At the very top (in orange), lie the GUI and the CLI components.

The Insight Toolkit (ITK) [34] has helped take research in the field of medical imaging to new heights by providing a set of common Application Programming Interfaces (APIs) and offered the ability for users to easily read, write, and perform computations with medical imaging datasets. ITK is used in CaPTk for tasks related to medical image registration (via the Greedy algorithm[3]) and segmentation, and ITK functions are components in new image analysis algorithms. The Visualization Toolkit (VTK) [35] has provided powerful

---

[3] sites.google.com/view/greedyreg/about.

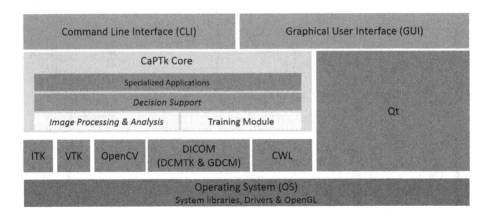

Fig. 3. Detailed diagram of CaPTk's architecture showing the various inter-dependencies between the components. (Color figure online)

functionalities for rendering and interaction of complex datasets. The Open Source Computer Vision Library (OpenCV) [36] has long fostered a healthy community of computer vision researchers and has established an industry-standard library for providing ML algorithms that can be used out-of-the-box, upon which CaPTk is building. For DICOM-related functionalities we leverage the DICOM Toolkit (DCMTK[4]), Grassroots DICOM (GDCM[5]), and have incorporated the DCMQI [37] tool for DICOM-Seg output. Qt[6] is used for GUI related tasks and cross-platform features. We build upon these open source and well-validated tools for CaPTk's development, with major emphasis towards being computationally efficient and fully cross-platform across the major desktop operating systems (i.e., Windows, Linux and macOS).

The architecture of CaPTk provides all source-level applications with efficient access to every imaging action and all common functions, treating imaging data as objects that can be passed between applications. This allows multiple applications access to common tasks (ie., pre-processing) without duplication of code or inefficient I/O. Simultaneously, the CaPTk design model that implements procedures written in C++ as both graphical and command-line applications also enables easy incorporation of stand-alone executables, such as ITK-SNAP[7] and DeepMedic [38], written in any language.

The CaPTk core provides libraries for different specialized applications, such as those developed for brain, breast, and lung cancer. To ensure reproducibility of the algorithms as originally written, each of these specialized tools is designed as a monolithic application, and is independent of the other general-purpose applications.

---

[4] www.dcmtk.org.
[5] gdcm.sourceforge.net.
[6] www.qt.io.
[7] www.itksnap.org.

3.3   Extending CaPTk

CaPTk can be easily extended with customized applications while leveraging all the graphical and algorithmic functionalities present in the application core. There are two ways for a developer to integrate their application into CaPTk[8]:

3.3.1   Application-Level Integration

This is the easiest way to call a custom application from CaPTk's GUI. This can be done for an application written in any language as long as a self-contained executable can be created. Essentially, these applications are called in the same way as they would be executed interactively from the command line. Their dependencies can be provided through the OS platform, distributed as part of the CaPTk distribution (preferable), or can be downloaded from an external site during installation. Once the executable is created, integration in CaPTk requires only minor modification to the build script (CMake) and user interface code. An example of such an integration is the LIBRA application [31,32].

3.3.2   Source-Level Integration

This is the deeper level of application integration and is available if the application is written in C++, possibly incorporating ITK, VTK, and/or OpenCV. To ensure that a source-level integration happens for an application, it needs to be written as a CMake project, i.e., a **CMakeLists.txt** is essential to define the requirements, project structure, generated libraries and/or executable(s), and the corresponding install targets for CaPTk's installation framework. The location of the entire project needs to be added to CaPTk's build path via the appropriate CMake script files. The developer also has the option to add a customized interface for the application. Examples of this kind of integration are the Feature Extraction, Preprocessing and Utilities applications of CaPTk.

3.4   Utilizing CaPTk in Custom Pipelines

The Common Workflow Language (CWL) [39] is an open standard to enable easy description of an executable for analysis workflows and tools in a way that ensures portability and scalability across a wide range of software and hardware environments, ranging from different operating systems to personal machines and High Performance Computing (HPC) environments. CWL has been conceived to meet the needs of high-throughput data-intensive scientific areas, such as Bioinformatics, Medical Imaging, Astronomy, High Energy Physics, and ML.

CaPTk leverages CWL as a means of relaying compact, human- and machine-readable descriptions of applications within the software suite. These descriptions, either in memory or as text files, are used to validate the optional and required inputs to an application. Each native application is specified in the CWL grammar in it's abstract form that includes all possible parameters, and

---

[8] cbica.github.io/CaPTk/tr_integration.html.

CWL is further used to pass the actual arguments used in applying an analysis routine to a specific set of data. These CWL descriptions can improve validity of results and enhance data provenance for end-users by recording all details about data and options used to generate results from any CWL pipeline in a compact, reproducible, verifiable format.

We have contributed the C++ implementation of the CWL parser that is used within CaPTk's CLI[9] to the CWL development community. This CWL parser is an easy and portable tool for developers to structure and write a CLI application (Fig. 4). Our contribution should i) allow the easy construction of pipelines without requiring any additional scripting, and ii) provide the ability to run both local and cloud applications. The reading and writing of a CWL definition file is embedded within the Command Line Parser itself to reduce the effort of writing pipelines.

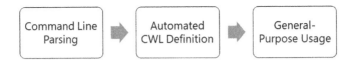

Fig. 4. The principle of CaPTk's CWL CLI parser.

3.5   Online Accessibility

CBICA's Image Processing Portal (IPP[10]) can be used by anyone to run resource-intensive applications of CaPTk on CBICA's HPC servers at no charge. CaPTk can be used to initialize the inputs that are required for all applications that are semi-automated, such as GLISTR [7] and GLISTRboost [8]. In addition, applications that may be impractical to run on a personal machine due to their long-running nature can be run remotely on IPP.

3.6   Code Maintenance

To ensure that good practices in open science principles are followed during the CaPTk development, the source code and issue tracker for CaPTk are maintained publicly on GitHub[11]. The entire history of the source code can be found on the repository. A stable version of CaPTk is compiled and released twice every year, and it includes the source code, as well as self-contained installers for multiple platforms.

CaPTk also follows the software development best practices [40] and employs Continuous Integration and Continuous Deployment (CI/CD) via GitHub and Azure DevOps[12] as mechanisms to encourage a rapid development cycle while

---

[9] github.com/CBICA/CmdParser.

[10] ipp.cbica.upenn.edu.

[11] github.com/CBICA/CaPTk, github.com/CBICA/CaPTk/issues.

[12] dev.azure.com/CBICA/CaPTk.

incorporating work from multiple developers. Any code contributions, either from the CaPTk development team or external users, are merged into the master only after successful completion of CI/CD checks and a code review[13]. Each push into the master GitHub code repository for CaPTk automatically produces a new binary installer for each supported OS that gives users immediate access to the bleeding edge development of CaPTk.

4   Example of General Purpose Applications

4.1   Quantitative Image Phenomic (QIP) Feature Panel

Building around the functionalities provided by ITK [34] and MITK [2], we have made the Feature Panel available in CaPTk as generic as possible, while at the same time maintaining safeguards to ensure clinical validity of the extracted features. For example, all computations are performed in the physical space of the acquired scan instead of the image space. Options to control the Quantization Extent (i.e., whether it should happen for the entire image or only in the annotated region of interest), the Quantization Type (e.g., Fixed Bin Number, Fixed Bin Size or Equal width [17]), resampling rate, and interpolator type, are all defined in the physical space and accessible to the user to alter as needed for a particular study. The user also has the capability to perform lattice-based feature computations [41, 42] for both 2D and 3D images.

There are two broad types of features getting extracted by CaPTk (detailed mathematical formulations can be found at: cbica.github.io/CaPTk/tr_FeatureExtraction.html):

1. First Order Features:
   (a) Intensity-based: Minimum, maximum, mean, standard deviation, variance, skewness and kurtosis.
   (b) Histogram-based: Bin frequency & probability, intensity values at the 5th and 95th quantiles, statistics on a per-bin level.
   (c) Volume-based: Number of pixels (for 2D images) or voxels (for 3D images) and their respective area or volume.
   (d) Morphology-based: Various measures of the region of interest such as Elongation, Perimeter, Roundness, Eccentricity, Ellipse Diameter in 2D and 3D, Equivalent Spherical Radius.
2. Second Order Features:
   (a) Texture Features: Grey Level Co-occurrence Matrix, Grey Level Run-Length Matrix, Grey Level Size-Zone Matrix, Neighborhood Grey-Tone Difference Matrix are part of this section [17].
   (b) Advanced Features: Includes but is not limited to Local Binary Patterns, frequency domain features, e.g., Gabor and power spectrum [41].

---

[13] en.wikipedia.org/wiki/Code_review.

All features are in conformance with the Image Biomarker Standardisation Initiative (IBSI) [17], unless otherwise indicated within the documentation of CaPTk. Most implementations are taken from ITK [34] or MITK [2] with additions made to ensure conformance when the physical coordinate space rather than the image space, which is one of the major sources of variation in between CaPTk and comparative packages, such as PyRadiomics [43]. Other sources of variation include the type of binning/quantization used and the extent on which it is applied. The user also has the option of performing batch processing with multiple subjects[14].

4.2 ML Training Module

We have based our machine learning module on OpenCV's ML back-end[15] and exposed the utility of Support Vector Machines (SVM) using Linear and Radial Basis Function kernels to the user, offering a grid search optimization for c and g. Additional kernels and algorithms will also be offered later. In addition, there is the capability to perform k-fold cross validation and to split the training and testing sets by percentages (Fig. 5).

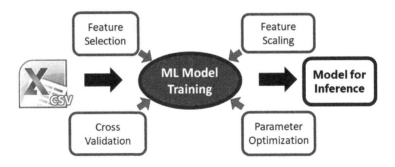

Fig. 5. The principle of CaPTk's machine learning module.

5 Future Directions

We developed CaPTk to address a gap at the time in the existing available tools and to make our preexisting advanced computational studies available to clinicians and imaging researchers in a tool that did not require a computational expert. The growing number of users of CaPTk and the reach of specialized applications within the suite validates our initial decision to produce a software package with interface features designed to make the specialized algorithms accessible in a way that we could not have accomplished with toolkits available when CaPTk was initiated. However, feedback from this wider set of researchers

---

[14] cbica.github.io/CaPTk/tr_FeatureExtraction.html.
[15] docs.opencv.org/master/dd/ded/group__ml.html.

and clinicians has exposed some limitations in the current version of CaPTk that we intend to address in its next major revision, i.e., CaPTk v.2.0.

Importantly, the variability and inconsistency among imaging headers has turned out to be a serious challenge in medical imaging. CaPTk is currently able to handle NIfTI file format and has limited support for handling DICOM files (which is currently based on DCMTK[16] and GDCM[17]) and does not support network retrieval from DICOM-compliant Picture Archiving and Communication Systems (PACS[18]). Over the course of its evaluation by clinicians, it has become evident that seamless integration with PACS is a necessary element for CaPTk's future.

Furthermore, a graphical interface for cohort selection and a batch processing interface, combined with the use of CWL to provide data provenence for each step in a pipeline, would also be essential to the construction of replicable research studies. In order to achieve this goal, integration with a data management, archival, and distribution system such as XNAT [10] is essential.

Finally, integration with state-of-the-art tools that rely on server-side communication such as the Deep Learning based NVIDIA-Clara annotation engine[19], cannot be easily accomplished with CaPTk in its current state.

We have recently surveyed the state of other image processing applications and toolkits and conducted extensive outreach to prospective collaborators as we begin to design the architecture for the next generation of the **Ca**ncer imaging **P**henomics **T**oolkit. While we continue to support and develop the current implementation of CaPTk, plans for the 2.0 branch include support (via CWL) for web- and cloud-based services and leveraging the current and updated functionality of MITK [2] and/or Slicer Core in the future.

6 Conclusions

We developed the CaPTk to provide a library for advanced computational analytics based on existing published multi-disciplinary studies, to meet the need at the time in the field of medical image processing applications for a suite that could be easy extended with components using any programming language. We consider medical image analysis to be much more than just a collection of algorithms; to ensure success of a tool, having the capability to visualize and interact with the original and processed image datasets is equally important. Building upon the powerful segmentation and registration framework provided by ITK and the rendering and interaction capabilities (particularly for 3D datasets) provided by VTK, we have built a light-weight application, CaPTk, that can visualize and interact with different datasets and also provide a solid annotation pipeline, thereby making it a powerful tool for use by clinical researchers. By

---

[16] www.dcmtk.org.

[17] gdcm.sourceforge.net.

[18] en.wikipedia.org/wiki/Picture_archiving_and_communication_system.

[19] news.developer.nvidia.com/nvidia-clara-train-annotation-will-be-integrated-into-mitk.

enabling easy integration of computation algorithms regardless of their programming language, we have further ensured that computational researchers have a quick path for potential translation to the clinic.

Acknowledgments. CaPTk is primarily funded by the Informatics Technology for Cancer Research (ITCR)[20] program of the National Cancer Institute (NCI) of the NIH, under award number U24CA189523, as well as partly supported by the NIH under award numbers NINDS:R01NS042645, NCATS:UL1TR001878, and by the Institute for Translational Medicine and Therapeutics (ITMAT) of the University of Pennsylvania. The content of this publication is solely the responsibility of the authors and does not represent the official views of the NIH, or the ITMAT of the UPenn.

References

1. Kikinis, R., Pieper, S.D., Vosburgh, K.G.: 3D slicer: a platform for subject-specific image analysis, visualization, and clinical support. In: Jolesz, F.A. (ed.) Intraoperative Imaging and Image-Guided Therapy, pp. 277–289. Springer, New York (2014). https://doi.org/10.1007/978-1-4614-7657-3_19

2. Wolf, I., et al.: The medical imaging interaction toolkit. Med. Image Anal. **9**(6), 594–604 (2005)

3. Link, F., Kuhagen, S., Boskamp, T., Rexilius, J., Dachwitz, S., Peitgen, H.: A flexible research and development platform for medical image processing and visualization. In: Proceeding Radiology Society of North America (RSNA), Chicago (2004)

4. Toussaint, N., Souplet, J.-C., Fillard, P.: MedINRIA: medical image navigation and research tool by INRIA (2007)

5. Davatzikos, C., et al.: Cancer imaging phenomics toolkit: quantitative imaging analytics for precision diagnostics and predictive modeling of clinical outcome. J. Med. Imaging **5**(1), 011018 (2018)

6. Rathore, S., et al.: Brain cancer imaging phenomics toolkit (brain-CaPTk): an interactive platform for quantitative analysis of glioblastoma. In: Crimi, A., Bakas, S., Kuijf, H., Menze, B., Reyes, M. (eds.) BrainLes 2017. LNCS, vol. 10670, pp. 133–145. Springer, Cham (2018). https://doi.org/10.1007/978-3-319-75238-9_12

7. Gooya, A., et al.: GLISTR: glioma image segmentation and registration. IEEE Trans. Med. Imaging **31**(10), 1941–1954 (2012)

8. Bakas, S., et al.: GLISTRboost: combining multimodal MRI segmentation, registration, and biophysical tumor growth modeling with gradient boosting machines for glioma segmentation. In: Crimi, A., Menze, B., Maier, O., Reyes, M., Handels, H. (eds.) BrainLes 2015. LNCS, vol. 9556, pp. 144–155. Springer, Cham (2016). https://doi.org/10.1007/978-3-319-30858-6_13

9. Zeng, K., et al.: Segmentation of gliomas in pre-operative and post-operative multimodal magnetic resonance imaging volumes based on a hybrid generative-discriminative framework. In: Crimi, A., Menze, B., Maier, O., Reyes, M., Winzeck, S., Handels, H. (eds.) BrainLes 2016. LNCS, vol. 10154, pp. 184–194. Springer, Cham (2016). https://doi.org/10.1007/978-3-319-55524-9_18

10. Marcus, D.S., Olsen, T.R., Ramaratnam, M., Buckner, R.L.: The extensible neuroimaging archive toolkit. Neuroinformatics **5**(1), 11–33 (2007). https://doi.org/10.1385/NI:5:1:11

---

[20] itcr.cancer.gov.

11. McAuliffe, M.J., et al.: Medical image processing, analysis and visualization in clinical research. In: Proceedings 14th IEEE Symposium on Computer-Based Medical Systems, CBMS 2001, Bethesda, MD, USA, pp. 381–386 (2001). https://ieeexplore.ieee.org/document/941749

12. Yushkevich, P.A., et al.: User-guided 3D active contour segmentation of anatomical structures: significantly improved efficiency and reliability. Neuroimage **31**(3), 1116–1128 (2006)

13. Cox, R., et al.: A (sort of) new image data format standard: NIfTI-1: we 150. Neuroimage **22** (2004). https://www.scienceopen.com/document?vid=6873e18e-a308-4d49-b4aa-8b7f291c613c

14. Yushkevich, P.A., Pluta, J., Wang, H., Wisse, L.E., Das, S., Wolk, D.: Fast automatic segmentation of hippocampal subfields and medial temporal lobe subregions in 3 tesla and 7 tesla T2-weighted MRI. Alzheimer's Dement. J. Alzheimer's Assoc. **12**(7), P126–P127 (2016)

15. Smith, S.M., Brady, J.M.: Susanâ"a new approach to low level image processing. Int. J. Comput. Vis. **23**(1), 45–78 (1997)

16. Shinohara, R.T., et al.: Statistical normalization techniques for magnetic resonance imaging. NeuroImage: Clin. **6**, 9–19 (2014)

17. Zwanenburg, A., Leger, S., Vallières, M., Löck, S.: Image biomarker standardisation initiative. arXiv preprint arXiv:1612.07003 (2016)

18. Wilkinson, M.D., et al.: The FAIR guiding principles for scientific data management and stewardship. Sci. Data **3**, 160018 (2016)

19. Akbari, H., Bakas, S., Martinez-Lage, M., et al.: Quantitative radiomics and machine learning to distinguish true progression from pseudoprogression in patients with GBM. In: 56th Annual Meeting of the American Society for Neuroradiology, Vancouver, BC, Canada (2018)

20. Akbari, H., et al.: Imaging surrogates of infiltration obtained via multiparametric imaging pattern analysis predict subsequent location of recurrence of glioblastoma. Neurosurgery **78**(4), 572–580 (2016)

21. Macyszyn, L., et al.: Imaging patterns predict patient survival and molecular subtype in glioblastoma via machine learning techniques. Neuro-oncology **18**(3), 417–425 (2015)

22. Akbari, H., et al.: Pattern analysis of dynamic susceptibility contrast-enhanced MR imaging demonstrates peritumoral tissue heterogeneity. Radiology **273**(2), 502–510 (2014)

23. Rathore, S., et al.: Radiomic signature of infiltration in peritumoral edema predicts subsequent recurrence in glioblastoma: implications for personalized radiotherapy planning. J. Med. Imaging **5**(2), 021219 (2018)

24. Akbari, H., et al.: Survival prediction in glioblastoma patients using multiparametric MRI biomarkers and machine learning methods. In: ASNR, Chicago, IL (2015)

25. Rathore, S., Bakas, S., Akbari, H., Shukla, G., Rozycki, M., Davatzikos, C.: Deriving stable multi-parametric MRI radiomic signatures in the presence of inter-scanner variations: survival prediction of glioblastoma via imaging pattern analysis and machine learning techniques. In: Medical Imaging 2018: Computer-Aided Diagnosis, vol. 10575, p. 1057509. International Society for Optics and Photonics (2018)

26. Li, H., Galperin-Aizenberg, M., Pryma, D., Simone II, C.B., Fan, Y.: Unsupervised machine learning of radiomic features for predicting treatment response and overall survival of early stage non-small cell lung cancer patients treated with stereotactic body radiation therapy. Radiother. Oncol. **129**(2), 218–226 (2018)

27. Bilello, M., et al.: Population-based MRI atlases of spatial distribution are specific to patient and tumor characteristics in glioblastoma. NeuroImage: Clin. **12**, 34–40 (2016)
28. Tunç, B., et al.: Individualized map of white matter pathways: connectivity-based paradigm for neurosurgical planning. Neurosurgery **79**(4), 568–577 (2015)
29. Bakas, S., et al.: In vivo detection of EGFRvIII in glioblastoma via perfusion magnetic resonance imaging signature consistent with deep peritumoral infiltration: the φ-index. Clin. Cancer Res. **23**(16), 4724–4734 (2017)
30. Akbari, H., et al.: In vivo evaluation of EGFRvIII mutation in primary glioblastoma patients via complex multiparametric MRI signature. Neuro-oncology **20**(8), 1068–1079 (2018)
31. Keller, B.M., et al.: Estimation of breast percent density in raw and processed full field digital mammography images via adaptive fuzzy c-means clustering and support vector machine segmentation. Med. Phys. **39**(8), 4903–4917 (2012)
32. Keller, B.M., Kontos, D.: Preliminary evaluation of the publicly available laboratory for breast radiodensity assessment (LIBRA) software tool. Breast Cancer Res. **17**, 117 (2015). https://doi.org/10.1186/s13058-015-0626-8
33. Schweitzer, M., et al.: SCDT-37. Modulation of convection enhanced delivery (CED) distribution using focused ultrasound (FUS). Neuro-Oncology **19**(Suppl 6), vi272 (2017)
34. Yoo, T.S., et al.: Engineering and algorithm design for an image processing API: a technical report on ITK-the insight toolkit. Stud. Health Technol. Inform. **85**, 586–592 (2002)
35. Schroeder, W.J., Lorensen, B., Martin, K.: The visualization toolkit: an object-oriented approach to 3D graphics. Kitware (2004)
36. Bradski, G.: The OpenCV Library. Dr. Dobb's J. Softw. Tools (2000). https://github.com/opencv/opencv/wiki/CiteOpenCV
37. Herz, C., et al.: DCMQI: an open source library for standardized communication of quantitative image analysis results using DICOM. Cancer Res. **77**(21), e87–e90 (2017)
38. Kamnitsas, K., et al.: Efficient multi-scale 3D CNN with fully connected CRF for accurate brain lesion segmentation. Med. Image Anal. **36**, 61–78 (2017)
39. Amstutz, P., et al.: Common workflow language, v1.0 (2016)
40. Knuth, D.E.: Computer programming as an art. Commun. ACM **17**(12), 667–673 (1974)
41. Gastounioti, A., et al.: Breast parenchymal patterns in processed versus raw digital mammograms: a large population study toward assessing differences in quantitative measures across image representations. Med. Phys. **43**(11), 5862–5877 (2016)
42. Zheng, Y., et al.: Parenchymal texture analysis in digital mammography: a fully automated pipeline for breast cancer risk assessment. Med. Phys. **42**(7), 4149–4160 (2015)
43. Van Griethuysen, J.J., et al.: Computational radiomics system to decode the radiographic phenotype. Cancer Res. **77**(21), e104–e107 (2017)

Author Index

Printed in the United States
By Bookmasters